International Conference on Electronics, Communications and Control Engineering (ICECC 2018)

Journal of Physics: Conference Series Volume 1026

Male, Maldives
6 – 8 March 2018

ISBN: 978-1-5108-6405-4
ISSN: 1742-6588

Printed from e-media with permission by:

Curran Associates, Inc.
57 Morehouse Lane
Red Hook, NY 12571

Some format issues inherent in the e-media version may also appear in this print version.

Printed by Curran Associates, Inc. (2018)

For permission requests, please contact the Institute of Physics
at the address below.

Institute of Physics
Dirac House, Temple Back
Bristol BS1 6BE UK

Phone: 44 1 17 929 7481
Fax: 44 1 17 920 0979

techtracking@iop.org

Additional copies of this publication are available from:

Curran Associates, Inc.
57 Morehouse Lane
Red Hook, NY 12571 USA
Phone: 845-758-0400
Fax: 845-758-2633
Email: curran@proceedings.com
Web: www.proceedings.com

Table of contents

Volume 1026

2018 International Conference on Electronics, Communications and Control Engineering (ICECC)

6–8 March 2018, Avid College, Maldives

Accepted papers received: 8 May 2018 Published online: 4 June 2018

Preface

2018 International Conference on Electronics, Communications and Control Engineering (ICECC)

Conference Committee

Peer review statement

Papers

Data Analysis and Security

A Compression Instruction Set Design based on RISC-V for Network Packet Forwarding

Z Cao, Q Lv, Y Wang, M Wen, N Wu and C Zhang.....1

Multi-period data envelopment analysis models and resource allocation: A case study

J Jablonsky, L Marek and P Berka.....12

Research on the method of flexible data bus technology of the monitoring of the hidden engineering audit

Rong Chen and Luqian Yu.....18

Coverage Test Technology Based on ONESPIN Verification Platform

Yawen Yang, Shan Zhou and Lu Kong.....27

Efficient block cipher mode for NVM

A Khodjanov, F Rustamov and J Yun.....33

Electronic Information Technology and Design

Grasp control of a prosthetic hand through peripheral neural signals

E Noce, C Gentile, F Cordella, A L Ciancio, V Piemonte and L Zollo.....38

Dynamic Mechanical Analysis of a Power Electronic Gate Driver Board

Thomas Schriefer and Maximilian Hofmann.....44

A survey on multipliers, adders and adiabatic logic styles suitable for power reduction

S Jagadeesh Babu and R Sivakumar.....51

Design of a noninvasive diabetes detector based on acetone recognition

Y C Jiang, M J Sun and R J Xiong.....56

Design and fabrication of a non-invasive, wireless system for monitoring needle insertion during epidural puncture

E Schena, C Massaroni, D Formica, A Mattei, V Piemonte, P Saccomandi, R Setola and M Carassiti.....64

Compact Model of Short-channel Cylindrical ballistic Gate-All-Around MOSFET Including the Source-to-drain Tunneling

H Cheng, Z Zhang, Z Yang and Z Liu.....70

Design of an Inductor-less Direct AC-AC Converter Realizing 1/4x and 4x Conversion

K Eguchi, F Asadi, K Kuwahara and I Oota.....77

Electronic and Communication Engineering

The analysis of IEEE 802.11 DCF protocol based on LEO satellite Wi-Fi network

Wen He, Xiaofeng Tao and Defeng Ren.....85

The Design of Broadband Multi-target Folding Jammer Based on a Periodic Non-uniform LFM Local Oscillator

Qi Zhang, Jun Zhu, Bin Tang and Huan Yang.....93

Design of Automatic Storage System Monitoring Software Based on OPC Communication

Zhang Peng, Zhang Teng and Zhou Guangcheng.....102

Design and implementation of GPS baseband correlator

Qingming Yi, Min Shi, Zhiji Lin and Chong Luo.....108

Intelligent Predictive Food Traceability Cyber Physical System in Agriculture Food Supply Chain

Rui-Yang Chen.....117

Control Theory and Control Engineering

Optimizing and Modeling for Plastic Injection Molding Process using Taguchi Method

T Kiatcharoenpol and T Vichiraprasert.....124

A High-efficiency FPGA-based Accelerator for Convolutional Neural Networks using Winograd Algorithm

Y Huang, J Shen, Z Wang, M Wen and C Zhang.....129

Contributions from the Drone Delivery System in Thailand to Environmental Pollution

J Koiwanit.....141

Novel Vulnerability Metrics for Interdependent System based on System Controllability

L Faramondi, G Oliva and V Piemonte.....146

A Sitting Posture Surveillance System Based on Kinect

Si He.....156

A Chinese Chessboard Calibration Method in Chess-Playing Robot by Machine Vision Sensing

Jian Jiao, Changcun Huang, Hongliang Liu and Guo Zhang.....163

Application Study of Fuzzy ARTMAP Neural Network in Robot Servo System

Guoxuan Chi.....174

Design of cricket control system based on visual inspection

Liqian Wang.....180

Electronic Circuits and Power Engineering

Thinking About the Development of Offshore Wind Farm in Northeast China

Xia Lin, Peng Wang and Yonghai Yu.....185

Study on the Application of Electronic Equipment in Voice Disease Prevention

Yingying Zhang.....190

Parameter Reconstruction of a Circuit System from Partial Eigeninformation

Wenting Wan.....195

The Identification Method of Transformer Excitation Inrush Current Based on the Decay Characteristics of Aperiodic Component

Jianqin Feng, Pei Zhang and Zhifei Chen.....203

Research on Inrush Current Mechanisms of Single Phase Transformers

Jianqin Feng, Pei Zhang and Zhifei Chen.....213

IOP Conf. Series: Journal of Physics: Conf. Series **1026** (2018) 011001 doi:10.1088/1742-6596/1026/1/011001

PREFACE

The 2018 International Conference on Electronics, Communications and Control Engineering, ICECC 2018, was held on the campus of Avid College in Male, Maldives during March 6–8, 2018. The Maldives are an archipelago of 1,192 coral islands grouped into 26 coral atolls (200 inhabited islands, plus 80 islands with tourist resorts) in the Indian Ocean, which was a delightful place for the conference.

At ICECC 2018, 50 submissions were received and 30 papers were selected for presentation. Participants from 12 countries made the conference truly international in scope. It was devoted to serve as a platform for scientists, researchers, engineers and developers from a wide range of electronics, communications and control engineering areas to exchange ideas and applications. The conference was co-sponsored by International Association of Computer Science and Information Technology and Avid College.

There were 3 keynote speeches and 1 plenary speech covering the different areas of the conference: Prof. Chin E. Lin (National Cheng Kung University, Taiwan) talked on Airborne Robotics in Drone Applications, Prof. Houssain Kettani (Florida Polytechnic University, USA) on Advances in High Performance Computing and Improvements on Monte Carlo Simulation Techniques, Prof. R. Sivakumar (R.M.K. Engineering College, India) on Medical Image Fusion using Stationary Wavelet Transform and Prof. Ming June Tsai (National Cheng Kung University, Taiwan) on A Novel Definition of the ZMP via Screw Theory.

I would like to express my deep appreciation to conference local chairs, Abdulla Musthaq Rashad, Avid College, Maldives and Babu Rajagopal, Ministry of Education, Maldives for their support. They also dispatched more volunteers to host the conference. I would like to express my deep appreciation to Abdulla Rasheed, Avid College for arranging local press coverage and food arrangements.

Given the rapidity with which science is advancing, we expect that the future conferences will be as stimulating as this most recent one was, as indicated by the contributions presented in this proceedings volume.

All in all, the ICECC2018 in Maldives was very successful. The keynote speeches and the plenary speech bridged the gap between the different fields of Electronics, Communication and Control, making it possible for researchers in a given area to gain insight into new areas. Also, included among the speakers were several young scientists, namely, postdocs and students, who brought new perspectives to their fields.

Conference Chair
Prof. R. Sivakumar, R.M.K. Engineering College, India
March 20, 2018

IOP Conf. Series: Journal of Physics: Conf. Series **1026** (2018) 011002 doi:10.1088/1742-6596/1026/1/011002

Conference Committee

Advisory Chair

Chin E. Lin, National Cheng Kung University, Taiwan

Conference Chairs

Houssain Kettani, Florida Polytechnic University, USA

R. Sivakumar, R.M.K. Engineering College, India

Program Committee Chairs

Ming June Tsai, National Cheng Kung University, Taiwan

Iickho Song, Korea Advanced Institute of Science and Technology, Korea

Local Chairs

Babu Rajagopal, Ministry of Education, Maldives

Abdulla Musthaq Rashad, Avid College, Maldives

Organizing Committees

Abdulla Rasheed, Avid College, Maldives

Narersh Kumar, Universiti Malaysia Kelantan, Malaysia

Guest Editors

Houssain Kettani, Florida Polytechnic University, USA

R. Sivakumar, R.M.K. Engineering College, India

Iickho Song, Korea Advanced Institute of Science and Technology, Korea

Technical Committees

Khairul Azhar Mat Daud, University Malaysia Kelantan, Malaysia

Xiuhai Cui, Harbin Institute of Technology, China

Mohana Sundaram Muthuvalu, Universiti Teknologi Petronas, Malaysia

Mehul Mahrishi, Rajasthan Technical University, India

Hassan Soleimani, Universiti Teknologi Petronas, Malaysia

Vijay Janyani, Malaviya National Institute of Technology, India

Ratna Wardani, Yogyakarta State University, Indonesia

Chunhua Liu, City University of Hong Kong, Hong Kong

Alpesh Vala, Charotar University of Science and Technology, India

Rushan Ziatdinov, Keimyung University, South Korea

Mohammed M. Bait-Suwailam, Sultan Qaboos University, Oman

Chih-Hung Wang, National Chiayi University, Taiwan

Chiew-Foong Kwong, University of Nottingham, Ningbo, China

Mohammed Sowket Ali, Bangladesh Army University of Science and technology, Bangladesh

Shahrul Azmi bin Mohd Yusof, Universiti Utara Malaysia, Malaysia

Manish Bhardwaj, SRM University, India

Tossapol Kiatcharoenpol, KMITL, Thailand

Mauro Capocelli, University Campus Biomedico of Rome, Italy
Vincenzo Piemonte, University Campus Biomedico of Rome, Italy
Jarotwan Koiwanit, King Mongkut's Institute of Technology Ladkrabang, Thailand
Yuchen Jiang, Northwestern Polytechnical University, China
C Gurudas Nayak, Manipal Institute of Technology, India
Gabriele Oliva, Università Campus Bio-Medico di Roma, Italy
Radzi Ambar, Universiti Tun Hussein Onn Malaysia, Malaysia
Emiliano Schena, University Campus Biomedico of Rome, Italy
Rohaya Binti Latip, Universiti Putra Malaysia, Malaysia
Francesca Cordella, Università Campus Bio-Medico di Roma, Italy
Surekha Kamath, MIT, Manipal, India
Joobeom Yun, Sejong University, South Korea
Madya Dr Muhammad Mahadi Bin Abdul Jamil, Universiti Tun Hussein Onn Malaysia, Malaysia
V.I George, Manipal Institute of Technology, India
Loredana Zollo, Università Campus Bio-Medico di Roma, Italy
Hamid Ali Abed Al-asadi, Basra University, Iraq

IOP Conf. Series: Journal of Physics: Conf. Series **1026** (2018) 011003 doi:10.1088/1742-6596/1026/1/011003

Peer review statement

All papers published in this volume of *Journal of Physics: Conference Series* have been peer reviewed through processes administered by the proceedings Editors. Reviews were conducted by expert referees to the professional and scientific standards expected of a proceedings journal published by IOP Publishing.

IOP Conf. Series: Journal of Physics: Conf. Series **1026** (2018) 012001 doi:10.1088/1742-6596/1026/1/012001

A Compression Instruction Set Design based on RISC-V for Network Packet Forwarding

Z Cao[1], Q Lv[1], Y Wang[1], M Wen[1], N Wu[2], C Zhang[1]

[1] College of Computer, National University of Defense Technology, Changsha 410073, China
[2] NetForward Micro-Electronic Corporation, Chengdu 610031, China

Abstract. The multi-core processor of RISC for network packet forwarding has been limited by the on-chip storage space. As more and more cores are implemented in one chip, the storage resources allocated by each core on the chip become less and less, as well as the conflict of visiting RAM has become more prominent. Therefore, the use of more compact instruction size will reduce the number of visits, and get a higher instruction cache hit rate, thereby improving the performance of the application and energy efficiency. Based on the open source instruction set RISC-V, this paper proposes a compressed ISA for the network packet forwarding. It is proved by experiment that the new instruction set which customized by this method has higher compression efficiency and better system performance for the network packet forwarding applications.

1. Introduction

With the continuous development of network technology, the network speed of the current backbone generally can achieve to 40Gbps or more. At the same time, there is a very high demand for the speed on network packet forwarding of network equipment due to the rapid growth of network speed [1]. The current network equipment mainly uses ASIC and network processor. However, since the high cost and long design period by using ASIC, it has been used in the network equipment less and less. On the contrary, most of the network equipment manufacturers choose the network processor, because of its advantages of low cost and high flexibility. In order to improve the processing ability of network equipment, network manufacturers have to use the multi-core architecture [2,3,4], such as Alcatel-Lucent FP series [5], Marvell's Xelerated series [6], EZchip NP Series [7], and so on. Until now, the number of cores can be reached to dozens, or even hundreds.

By implementing the network processor of the multi-core architecture, the processing capacity for packet forwarding has been improved. Meanwhile, it also brought some problems, for example, higher cost, less storage space for each core, higher energy dissipation, and other issues.

- **Cost problem:** Firstly, even if we put aside the complex and bad historical design problem of the existing mature business ISA (Instruction Set Architecture), we also need to consider the high usage fees and strict use restrictions. Secondly, some open source instruction sets, such as OpenRISC, will be slowly developed and lack of maintenance.
- **On-chip storage space problem:** the number of PEs (Processing Element, PE) in one chip increases rapidly, but the manufacturing process improves slowly, therefore, each PE gets less storage space, such as the sizes of instruction cache and data cache will be drastically reduced.

At the same time, the cache hit rate reduces, visit conflict increases and the system performance reduces as well.

- **Energy dissipation problem:** For most specific processor cores, launching the instruction stream and cache misses are two major reasons for energy dissipation [8]. The paper [9] pointed out that in an ordinary five-stage pipeline, the access and instruction cache misses may cause 40% of the energy dissipation, however main memory accesses and processor stalls will consume more energy than cache misses.

To solve the problems above, the design method of the ASIP (Application Specific Instruction set Processor) may be a good choice: customizing the hardware with the existing instruction set, such as FPGA or other programmable hardware, help separate the design from hardware implementation. With customization, it is easier to optimize the design of hardware and to improve the processing performance in network packet forwarding. In order to implement the customization, this paper proposes the following solutions:

- The open source instruction set RISC-V has obvious advantages: Firstly, the RISC-V instruction set allows anyone to use and modify it for free; Secondly, unlike the instruction set which has long history, there is no historical and compatible issues for RISC-V; Thirdly, modular design also makes it easy to add instructions and modify; Last but not the least, RISC-V is designed for the modern computing devices, such as cloud computing, data centers, network equipment and embedded devices, light quantitative and modular design bring us a simpler and controllable hardware implementation. Besides, the RISC-V contains extension sets of I (Integer), C (Compression), M (integer Multiple / Division), A (Atomic), F (single-float precision) and D (double-float precision). The I set is a complete collection, and the others are the extension sets. The C set is a superset of the I set to increase instruction stream density.
- Compressing instruction set can increase the instruction stream density, and thus it can reduce the number of fetching instructions, improve the instruction cache hit rate, and even the number of RAM access. This equates to increase the size of instruction cache, improve processing performance and reduce the energy dissipation.

Based on the RISC-V instruction set, this paper attempts to find out the frequency of each instruction by analyzing the applications in network packet forwarding domain, and then replace the instructions which have the highest frequency with some short word instructions, so the instruction density is increased, and thus to solve the problems which are caused by the multi-core architecture in network processor.

The contributions of this paper are summarized as follows:

- Test and analyze the applications in the network packet forwarding, and find out the common characteristics, and then provide the basis for customizing the instruction set in this field;
- Custom the RISC-V instruction set by cutting, adding and changing some instructions, so that high density instruction steam can be provided;
- Modify the compiler so as to support running and testing the customized instruction sets.
- The rest of this paper is organized as follows: section 2 describes the work of the compression instruction set. The third section describes how to design the compression instruction set for the network packet forwarding field. The fourth section describes the experiment. And the fifth section is the summary and future work.

2. Background

This section is going to introduce compressed ISA related work, such as the comparison between CISC (Complex Instruction Set Computer) and RISC (Reduced Instruction Set Computer), and then two existing compression methods and their typical cases.

For the initial computer architecture, smaller storage space and expensive storage devices severely limit the size of the application instruction stream, so the CISC was originally designed to use the variable-length instructions: according to the frequency of various instructions occurred in different applications, the higher the frequency, the shorter the code is used. At the same time, according to its operands or the different ways of addressing mode to shorten its encoding length, such as DEC VAX [10] and Intel X86 instruction set.

For the instruction sets that follow the RISC principle [11], they have always been to choose performance rather than code size. An unavoidable problem with loose instruction encoding is that the density of the code is smaller. Compared to the CISC instruction set, the program that uses the RISC instruction set will require more instructions to complete the same function. With the extensive use in embedded systems with the RISC instruction set, as well as the development of multi-core architecture, the demand for improving the code density is becoming increasingly important. To meet the demands, some RISC vendors have extended their instruction sets to support a subset of their base ISA's functionality by using short instruction words, typically 16 bits. In general, there exist two compression methods: Statistical-based Compression and Dictionary-based Compression [12,13].

2.1. Statistical-based Compression

Statistical-based compression uses the frequency of singleton characters to choose the size of the codewords that will replace them. Frequent characters are encoded using shorter codewords so that the overall length of the compressed text is minimized. Huffman encoding of text is a well-known example. These short instructions have a straightforward mapping to one or more instructions in the base ISA. MIPS16 [14] and ARM Thumb [15] are notable examples of this approach.

Thumb is actually a separate instruction set added to the standard ARM RISC instruction set, which consists of approximately 36 16-bit instructions, including basic additions, subtractions, loops, and jump instructions. The MIPS16e instruction set includes a set of 16-bit standard MIPS instruction as well. Thumb and standard ARM instruction sets, as well as the MIPS16e and standard MIPS instruction sets can not be mixed. It must be explicitly switched between the compression instruction and the standard instruction via a mode switch instruction. Switching not only increases the complexity of the hardware, generates additional switching code, but also reduces the system performance by switching them. By using the Thumb and MIPS16e instructions to replace the standard ARM and MIPS 32-bit instructions, you can reduce the size of most program code by about 20% to 30%.

ARM has targeted improvements and proposed Thumb-2 compression instruction set [16]. Unlike Thumb, Thumb-2 is a complete set of compression instructions that can completely replace the ARM standard instruction set, running 16-bit and 32-bit mixed code without switching. But this also increases the implementation of the hardware. In general, Thumb-2 provides less code compression efficiency, but less performance loss.

2.2. Dictionary-based Compression

Dictionary-based compression selects entire phrases of common characters and replaces them with a single codeword. The codeword is used as an index into the dictionary entry which contains the original characters. Compression is achieved because the codewords use fewer bits than the characters they replace. For example, replace frequent instructions or instruction sequences with an index into a dictionary, which stores the decompressed instructions. Though these schemes are effective at reducing code size, the dictionary lookup adds latency and offsets the energy reduction from compression.

IBM's CodePack [17] system uses the most complex code compression techniques. Unlike Thumb and MIPS16e, the CodePack system completely compresses the code as if it were running WinZip in PowerPC. CodePack parses and compresses the entire program, and the generated user code must be decompressed and executed in a compressed version. Despite its complexity, CodePack offers 20% to 30% of space savings, similar to other compression technologies. However, CodePack has some other effects. Since each compressed program has its own compact key, CodePack is essentially both a compression system and an encryption system. This also means that the compressed PowerPC program is not binary code compatible. In addition, CodePack generates two key values for each program because the upper and lower 16 bits of the instruction are compressed separately.

The variable-length instruction set has been used to improve code density for a long time, but since the original RISC ISA does not have enough opcode space for these unplanned compression instructions, they have been treated as a new ISA Development. RISC-V is designed to support compression instructions from the outside, leaving enough opcode space for RVC (RISC-V Compressed), so that it can be added as a simple extension along the basic ISA (along with other standard extensions). RISC-V compression instruction set is based on the statistical-based compression method. The idea of RVC is to reduce code size for embedded applications, improve performance and energy efficiency for all applications.

3. Customized Compression Instruction Set Design

3.1. Minimizing the Instruction Set

For the specific domain of network packet forwarding, the usual operation is to deal with the packet header, and the processing of each arrival packet is basically the same. In order to compress the scale of the application code in network packet forwarding, the first thing is to verify the distribution of different instructions in this field.

Unlike desktop applications and multimedia applications, there is no a standard unified benchmarking program to evaluate the performance of the network. Some of the popular test sets, such as Commbench [18], NpBench [19], NetBench [20], and EEMBC's network 2.0 [21], have different considerations and designs, and the evaluation conditions are not completely the same. Considering the specific field of network message forwarding in this paper, it refers to several above mentioned typical benchmarks, and extract some typical cases of network message forwarding as a test assembly in this experiment. The test assembly run independently on the target machine in the event of detachment from the operating system. This paper chooses four common applications in network packet forwarding field, including DRR, FRAG, RTR, and CRC. Details are shown in Table 1.

Table 1. The applications of the test assembly.

Application	Content
DRR	QoS queue scheduling algorithm, by controlling the use of different types of packets to link bandwidth, so that different data streams get different levels of service
FRAG	IP packet fragmentation technology
RTR	Routing Table finding algorithm in three—layer Forwarding
CRC	The cyclic redundancy check algorithm， calculating the checksum in the link layer

Compile these programs with a complete RISC-V instruction set (G set, which contains the IMAFD sets), and then obtain the distribution of various instruction types. The distribution of the various types of instructions is shown in Table 2.

Table 2. Instruction Set Distribution.

Applications	I Set	F/D Sets	M Set	A Set
DRR	99.61%	0.18%	0.21%	0%
FRAG	99.61%	0.17%	0.22%	0%
RTR	98.85%	0.97%	0.18%	0%
CRC	99.62%	0.17%	0.21%	0%

From the analysis results in Table 2, it shows that the application for network packet forwarding uses very few RISC-V extended instructions. Therefore, when designing ASIP for network packet forwarding, it is possible to consider replacing hard floating point element with soft floating point element, and so on, so as to achieve the purpose of minimizing the instruction set, and reducing the complexity of the hardware structure, as well as reducing development costs and improving the system performance.

3.2. Customize the compression Instruction Set

In RISC-V, RVC (RISC-V Compressed set) is a variable-length instruction set extension, and it is a superset of the RISC-V ISA, encoding the most frequent instructions in half the size of a RISCV instruction; the remaining functionality is still accessible with full-length instructions. RVC programs are 25% smaller than RISC-V programs, fetch 25% fewer instruction bits than RISC-V programs, and incur fewer instruction cache misses [9].The 32-bit RVC instructions are shown in table 3.

Table 3. 32-bit RVC Instructions.

Inst[15:13] Inst[1:0]	000	001	010	011	100	101	110	111
00	ADDI4SPN	FLD	LW	FLW	Reserve	FSD	SW	FSW
01	ADDI	JAL	LI	LUI/ADDI16SP	MISC-ALU	J	BEQZ	BNEZ
10	SLLI	FLDSP	LWSP	FLWSP	J[AL]R/MV/ADD	FSDSP	SWSP	FSWSP
11				>16 bits (Standard Instructions)				

According to the design in Section 3.1, even if the RISC-V extended instruction set has been removed and only the I collection has been retained, the I collection also contains a large number of instructions. Since the application for network packet forwarding has a relatively simple application of the desktop system Function, the use of the frequency of the instructions must be obviously different. In this section, it re-customs RVC based on the I instruction set, so that it has a higher compression ratio. After compiling the four applications (DRR, CRC, RTR and FRAG) in the test set according to the RV32I standard, the frequency distribution of each instruction and the number of the registers on this basis can be obtained. The results are shown in Figure 1 and Figure 2.

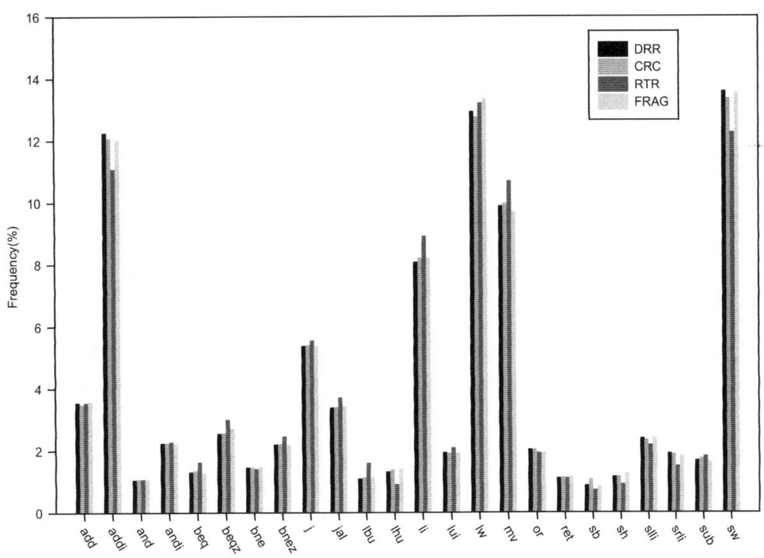

Figure 1. The top 24 instructions in the highest frequency

IOP Conf. Series: Journal of Physics: Conf. Series **1026** (2018) 012001 doi:10.1088/1742-6596/1026/1/012001

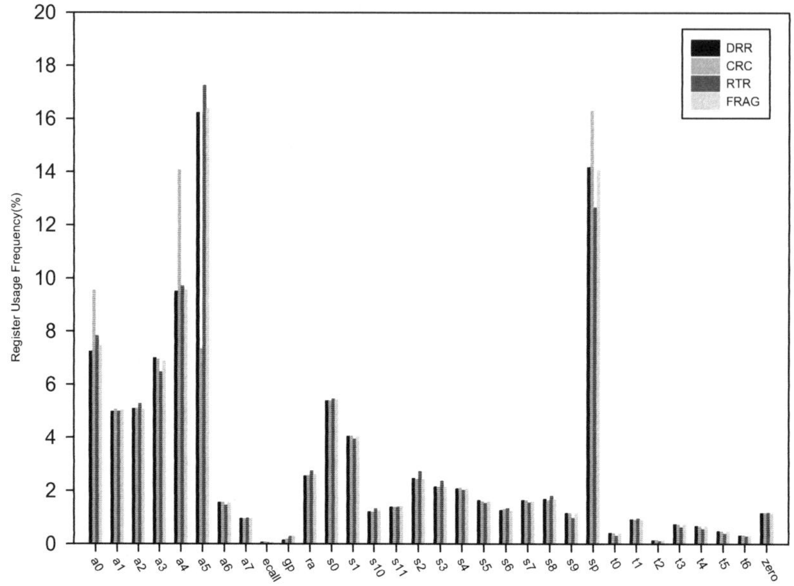

Figure 2. Register usage ratio

It can be seen from Figure 1 that the high frequency standard instructions used for network packet forwarding applications do not match the instructions defined by the RVC instruction set. In other words, if the application for network packets is forwarded only in accordance with the instructions defined by RVC Compression, the compression ratio would not be high. Due to the coding format, RVC instruction code number is less than RISC-V standard instruction set itself. How to effectively select the standard instruction to compress is worthy of studying. For one thing, adding compression instructions which are not defined in RVC but frequently used would bring a better performance in the specific application situation. For another, the compression instructions in original RVC which don't provide good compression effect can be deleted or modified so that the final ISA could be more effective.

The following is re-compiling the RV32I based on the results of the distribution, and the design criteria are based on the following principles and facts:

- The repeatability of the registers in the instruction. This repetition is particularly noticeable in I-type instructions and R-type instructions for 3 operands.
- The locality of the register. Compression instructions because the instruction length is limited, so the register can also be limited, usually only 8. As can be seen in Figure 2, the use of registers is localized.
- Most of the immediate or address offsets are small.
- The use of special registers in registers, such as zero registers, ABI link registers, or ABI stack registers

According to the instruction usage frequency in Figure 1 and the register usage ratio shown in Figure 2, the list of RVC instructions to be customized is shown in Table 4.

Table 4. Customized RVC.

inst[15:13] inst[1:0]	000	001	010	011	100	101	110	111
00	ADDI4SPN	***LBU***	LW	***SH***	Reserve	***SB***	SW	***ANDI***
01	ADDI	JAL	LI	LUI/ADDI16SP	MISC-ALU	J	BEQZ	BNEZ
10	SLLI	***LBUSP***	LWSP	***SHSP***	J[AL]R/MV/ADD	***SBSP***	SWSP	***Reserve***

In Table 4, the bold texts represent the new compression instructions, while the underlined texts are instructions for the original instructions to be modified, and the rest are the original RV32C instructions.

3.3. Compiler Modification

Since the ISA has been modified, the compiler need to be migrated correspondingly to translate the high-level programs to target machine codes, and then the CPU can implement the functions. The process of machine code generation is shown in figure 3.

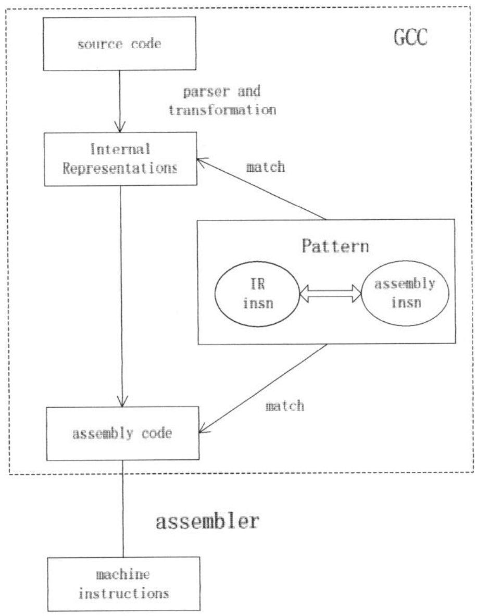

Figure 3. The process of machine code generation

The GCC compiler translates the source code into an intermediate representation, making it easy for the compiler to optimize. After optimization, the compiler generates the corresponding assembly code based on the assembly instructions in the target platform machine description file and the GCC intermediate representation template. And then use the assembler to convert the assembly code into machine code.

With ISA modified, riscv.md, the machine description file, needs to be changed as well. A typical method is adding the compression instruction template, which allows the compiler to generate compression instructions. On the other hand, the assembler should be changed so that it can recognize the compression instructions and generate the corresponding machine code.

4. Experiment and Result

In order to obtain the real compression effect, this paper designs a complete hardware and software experimental environment. It acquires the relevant data, and evaluate the performance after applying the new compression instruction set through simulating the real network environment for message forwarding.

4.1. Hardware Experimental Environment

This paper adopts the high-performance hardware FPGA board and the single core supporting RVC, RVC-N and RVI. The specific configuration is as follows:

- Xilinx official board VC709: There is a high performance XC7VX690T Virtex-7 FPGA chip onboard, as well as SFP+ and UART interface. SFP+ converter, which achieves the conversion of the optical port and the electric port, is also used in the experiment.

- The CPU core adopts the RISC-V SOC system written by Sergey Khabarov [22], and it is modified according to actual requirements. Since the original design is 64-bit architecture and does not support RVC, it is necessary to change it to 32-bit architecture, and provide support for RVIC or RVIC-N. The CPU is designed to support unaligned instruction storage.

4.2. Software Experimental Environment
The software experiment environment includes test assembly, PC test program, compiler, and so on.
- Network Test Assemblies: This paper chooses four common applications in network packet forwarding field, including DRR, FRAG, RTR, and CRC.
- Test software for sending and receiving: This experiment adopts network test software, which is provided by NetForward Micro-Electronic Corporation, to carry out sending and receiving test for the target and to simulate the normal network communication data.
- Compiler: For comparison, the RVIC-based test programs and RVIC-N-based test programs need be compiled. Therefore, compilers of different functions need to be configured.

4.3. Experimental Design and Results
This paper analyzes the differences between RVC-N and RVC instructions in the application of network message forwarding, from the aspects of static code compression rate, dynamic code compression rate (energy consumption), and performance and instruction failure rate.

4.3.1. Static Code Compression Ratio. Static code size impacts instruction memory costs and start-up overhead, and it correlates with instruction cache miss rates [9]. One of the goals in defining the RVC-N is thus to reduce static code size in network domain. In [9], the standard RVC and RISC-V, MIPS, ARM, MIPS16, Thumb, Thumb-2 and X86 were compared based on SPEC CPU2006 benchmark suit. The results show that static compression ratio of RVC ranks second only after Thumb and Thumb-2. In this paper, only the standard RVC and RVC-N are compared based on the network test set to determine that whether RVC-N in the field of network packet forwarding has a better effect.

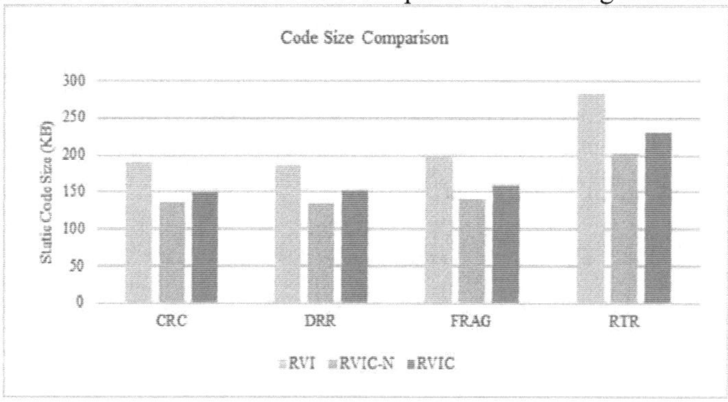

Figure 4. Comparison of static code compression ratio

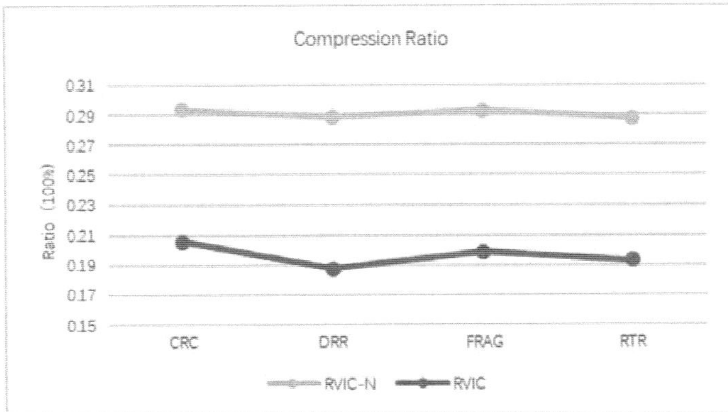

Figure 5. Static compression ratio

As shown in Figure 4, the code size of four test programs has been significantly reduced through the use of RVIC-N or RVIC code compression. The comparison results in Figure 5 also show that RVIC-N has a higher compression efficiency than RVIC.

4.3.2. Dynamic Code Compression and Energy Consumption. In an idealized processor model, dynamic code size equals to the amount of data fetched from instruction memory. Reducing dynamic code size can thus reduce instruction fetch energy, as fewer total bits are retrieved from instruction memory. There is a standard task in this experiment, and count the amounts of fetching instructions by using different compression instruction sets when the task is finished. By comparing the amounts, the result illustrate which compression instruction set has better performance. The comparison result is shown in Figure 6.

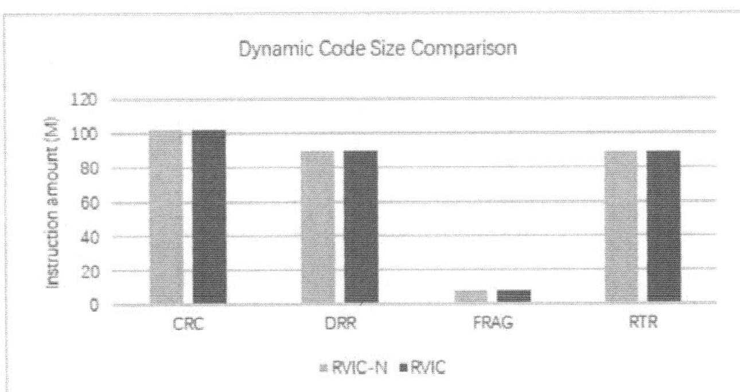

Figure 6. Dynamic Code Size Comparison

From figure 6, there is no big error between the dynamic code sizes, which is out of the expectation. By analyzing the design, it may be caused by the storage mode: the 32-bit instruction is stored on any 16-bit boundary.

4.3.3. Instruction Cache Missing Rate and Performance. Compression instructions usually will result in a smaller instruction cache failure rate. Because the instruction length is reduced, a Cache Line can save more instructions, or it means that more instructions can be contained in the same size instruction cache. Compression instructions can improve the hit rate. Meanwhile, it can improve system performance through reducing the number of CPU pauses for waiting for instructions. In order to find the role and performance of RVC in instruction cache failure rate, the instruction caches of different sizes is set and the mapping method is direct mapping mode. The instruction missing ratios of standard RVC and RVC-N are tested individually in the dynamic operation process.

Table 5. Missing ratios with different cache sizes

Cache	CRC	DRR	FRAG	RTR

Size(Bytes)	RVIC-N	RVIC	RVIC-N	RVIC	RVIC-N	RVIC	RVIC-N	RVIC
256	0.22%	5.27%	2.32%	4.59%	3.94%	3.76%	4.33%	9.45%
512	0.21%	1.25%	1.32%	4.58%	2.52%	0.50%	3.81%	5.47%
1024	0.19%	0.23%	0.30%	4.57%	1.99%	0.22%	1.53%	5.06%
2048	0.13%	0.18%	0.09%	0.05%	0.17%	0.17%	1.03%	2.30%
4096	0.08%	0.11%	0.06%	0.03%	0.11%	0.08%	0.56%	1.50%
8192	0.05%	0.04%	0.02%	0.01%	0.04%	0.06%	0.19%	0.92%
16384	0.02%	0.02%	0.01%	0.00%	0.03%	0.01%	0.03%	0.77%

5. Conclusions

This paper proposes an approach to custom RISC-V compression instruction set, and presents a custom instruction set for network packet forwarding. By analyzing the applications for network packet forwarding, the frequently used RISC-V instructions can be found out. Based on the results, unnecessary RISC-V extended instruction sets are cut, so as to reduce the complexity of hardware design and improve the system performance. Besides, the frequently used instructions are added to the compression instruction set, and the low frequently used instructions are removed from the original compression instruction set. From the experimental results, the result shows that the modified RISC-V compression instruction has higher compression efficiency in the field of network packet forwarding. What's more, it also can improve the instruction cache hit rate and system performance, and reduce the energy dissipation. This approach which customized the compression instruction set for network packet forwarding can also be used for other fields.

The current work of this paper is still relatively shallow. It just plays a fundamental role for the compression instruction set customization in the network packet forwarding field. In order to further enhance the effectiveness and expand its practicality, the future work is as follows: First, this article only uses four basic network applications as benchmarks to evaluate the distributions of instructions, which is lack of analysis of more comprehensive network applications. The following work is carrying out subsequent analysis of more network application testing procedures. Second, in the test of instruction cache hit rate, considering more different cache size of the configuration options and selecting a different mapping mode to implement the instruction cache will provide the basis for selection of parameters for the multi-core architecture. Last, the CPU core used in this paper is non-commercial mature product, and the ROM and RAM it used are implemented by on-chip resources, which differs from the real SOC systems. The next step is considering the introduction of off-chip ROM and RAM, such as FLASH and DDR chips.

6. Acknowledgments

The authors gratefully acknowledge supports from National Key Research and Development program under No.2016YFB1000401 and No.2016YFB1000403; National Nature Science Foundation of China under NSFC No.61502509 and 61402504. This work was also supported by NetForward Micro-Electronic Corporation.

7. References

[1] Zhang X, Sun Z and Zhang M 2006 An Efficient Packet Scheduler for Modern Network Processors: Guarantee Load Balanceing and Packet Ordering IEEE 2006 Workshop on High Performance Switching and Routing Ponzan Poland June 7-9

[2] Peyravian M, Calvignac J, Sabhikhi R 2003 IBM—PowerNP Network Processor *http://www.ibm.com*

[3] Intel IXP1200 Network Processor Family - Hardware Reference Manual 2001 *http://www.intel.com*

[4] Cisco ASR 1000 Series Aggregation Services Routers *http://www.cisco.com*

[5] FP3: Breakthrough 400G network processor *http://www.alcatel-lucent.com/products/fp3*

[6] Marvell Introduces New Xelerated Network Processors and Traffic Management Solutions for the Mobile Internet *http://www.marvell.com/*
[7] EZchip NP-5 Product Brief *http://www.ezchip.com*
[8] Katevenis M, Sherburne R, Patterson D, and Sequin C 1984 The RISC II Micro-Architecture *Journal of VLSI & Computer Systems* vol1, p138–152
[9] Waterman A 2011 Improving Energy Efficiency and Reducing Code Size with RISC-V Compressed *Technical Report* No. UCB/EECS-2011-63
[10] Strecker W 1978 VAX-11/780—A Virtual Address Extension to the DEC PDP-11 Family *In AFIPS Spring Conference*
[11] Patterson D, Sequin C 1981 RISC I: A Reduced Instruction Set VLSI Computer *In Proceedings of the 8th Annual International Symposium on Computer Architecture* (ISCA '81 IEEE Computer Society Press), pp443-457
[12] Lefurgy C, Bird P, Chen I, and Mudge T 1997 Improving Code Density Using Compression Techniques. *In Proceedings of the 30th Annual International Symposium on Microarchitecture* p 194–203
[13] Li T, Wu J, Lam S K, Srikanthan T and Lu X 2010 Selecting Profitable Custom Instructions for Reconfigurable Processors *J. Syst. Architect. Special Issue on Hw/Sw Co-Design* **56** p 340-351
[14] MIPS 2010 Technologies Inc MIPS64 Architecture for Programmers Volume IV-a: The MIPS16e Application-Specific Extension to the MIPS64 Architecture *http://www.mips.com/products/architectures/mips64/*
[15] ARM Ltd. 2008 ARMv6-M Architecture Reference Manual *http://infocenter.arm.com/help/topic/com.arm.doc.ddi0419c/*
[16] Phelan R 2003 Improving ARM Code Density and Performance *ARM Limited*
[17] Lefurgy C, Piccininni E and Mudge T 1999 Evaluation of A High Performance Code Compression Method *International Symposium on Microarchitecture* (Micro-32 Proceedings IEEE Xplore) p 93-102
[18] Wolf T and Franklin M 2000 CommBench - A Telecommunications Benchmark for Network Processors *IEEE International Symposium on PERFORMANCE Analysis of Systems and Software* (IEEE), pp154-162
[19] Lee B K and John L K 2003 NpBench: A Benchmark Suite for Control Plane and Data Plane Applications for Network Processors *International Conference on Computer Design* Proceedings. (IEEE), pp226-233
[20] Mangione-Smith B, Memik G and Mangione-Smith W H NetBench: A Benchmarking Suite for Network *Proc. Int'l Conf. Computer Aided Design*, pp39-43
[21] Dinhtrong T T, Bieman J M 2005 The FreeBSD Project: A Replication Case Study of Open Source Development *IEEE Transactions on Software Engineering* vol.31 (6), pp481-494
[22] *https://github.com/sergeykhbr/riscv_vhdl*

IOP Conf. Series: Journal of Physics: Conf. Series **1026** (2018) 012002 doi:10.1088/1742-6596/1026/1/012002

Multi-period data envelopment analysis models and resource allocation: A case study

J Jablonsky[1], L Marek[1], P Berka[1]

[1] University of Economics, Prague, W. Churchill Sq. 4, 13067 Praha 3, Czech Republic

Abstract. Data envelopment analysis is a general tool for efficiency and performance measurement of the set of decision making units based on solving linear optimization problems. The paper deals with using this group of models for resource allocation analysis among Czech economic faculties within 7 years period from 2008 until 2014. The resources in the period t are divided among faculties according to their teaching and research performance in the period $(t - 1)$. The teaching performance is measured by the number of students and number of graduated in the given period. The research performance is defined according to the national rules by the number of so called RIV points. The overall performance level in the given period is estimated using data envelopment analysis model with constant returns to scale assumption. The input in the period t is partly based on the efficiency or super-efficiency scores from the preceding period. Super-efficiency score given by Andersen and Petersen model is used for discrimination among efficient units. The results of the allocation in the last period are compared to the ones calculated to the standard methodology. Numerical experiments are realized using own MS Excel add-ins and LINGO modelling and optimization system.

1. Introduction

Allocation of a limited amount of resources among the set of decision making units (DMUs) is an important task that finds its applications in various areas of human activities. One of the typical problems of this nature is allocation of funds among higher education institutions which is a problem that significantly influences all activities of the units under evaluation. The common practice in the Czech Republic consists in dividing of available resources among universities and then among faculties according to their performance in the past one or several periods. The overall performance of the universities (faculties) is given by their partial performances in various activities carried out. Among the main activities of higher educational institutions belong teaching, research, international relations, co-operation with partners, etc. Of course teaching and research are the most important ones in this context. That is why evaluation of teaching and research performance of the set of educational institutions (let us denote them further units or faculties) is a very important problem and the methodology used for this evaluation must reflect and respect many requirements.

The main aim of this paper is to contribute to the discussion in this field and try applying not so widely used techniques. The common practice in the Czech Republic uses basic multiple criteria decision making approaches, i.e. assigns expert weights to the defined criteria that describe various aspects of overall performance and then applies simple additive weighting. The given performance score is a basis for a future allocation of funds. In the last several tens of years a more sophisticated methodological approaches for efficiency and performance evaluation have been proposed and verified by various researchers. Data Envelopment Analysis (DEA) models introduced by Charnes et al. [1] belong without any doubts among the most theoretically developed and applied techniques. Applications of DEA models in higher education are numerous. Let us mention at least three of them – Beasley [2], Thanassoulis et al. [3] and Jablonsky [4].

The paper is organized as follows. Section 2 defines shortly basic DEA models and introduces multi-period DEA models which is a class of models that evaluates the performance of the set of units within several consecutive periods. An iterative approach that combines results of the performance evaluation in the period t and the amount of funds allocated to the unit in the period t (based on the evaluation in the period $(t-1)$) as the main input variable in the period $(t+1)$ is introduced in Section 3. The proposed approach is illustrated on the real data set of 19 Czech economic faculties in multiple periods from 2008 until 2014. The given results are summarized in the final section of the paper.

2. Multi-period DEA models

Traditional DEA models analyse relative technical efficiency of the set of n DMUs that are described by m inputs and r outputs in one period. The efficiency score θ_q of the DMU_q is defined as the weighted sum of outputs divided by the weighted sum of inputs as follows:

$$\theta_q = \sum_{k=1}^{r} u_k y_{kq} \Big/ \sum_{i=1}^{m} v_i x_{iq} \tag{1}$$

where u_k, $k = 1, 2, \ldots, r$ is the positive weight of the k-th output, v_i, $i = 1, 2, \ldots, m$ is the positive weight of the i-th input, and x_{ij}, $i = 1, 2, \ldots, m$, $j = 1, 2, \ldots, n$ and y_{kj}, $k = 1, 2, \ldots, r$, $j = 1, 2, \ldots, n$ are non-negative values for the DMU_j of the i-th input and the k-th output respectively. Traditional DEA models maximize the efficiency score (1) under the assumption that the efficiency scores of all other DMUs do not exceed 1 (100%). The presented problem is not linear in the objective function but it can be converted into a linear optimization problem and then solved easily. The transformation consists in maximization of the nominator or minimization of the denominator in expression (1). In some cases, it is more convenient working with dual problem to the linearized version of the model described above. This model, often referenced as input oriented envelopment CCR (Charnes, Cooper and Rhodes) model is formulated as follows:

$$\begin{aligned}
&\text{minimize} && \theta_q^{CCR} && \\
&\text{subject to} && \sum_{j=1}^{n} x_{ij}\lambda_j + s_i^- = \theta_q^{CCR} x_{iq}, && i = 1,2,\ldots,m, \\
& && \sum_{j=1}^{n} y_{kj}\lambda_j - s_k^+ = y_{kq}, && k = 1,2,\ldots,r, \\
& && \lambda_j \geq 0, && j = 1,2,\ldots,n.
\end{aligned} \tag{2}$$

The optimal objective function value of model (2) is lower than 1 for inefficient units and equals 1 for the units weakly or fully efficient. In order to rank efficient units many models based on various principles have been proposed. An important group of these models are so called super-efficiency models. The first model of this nature was introduced in Andersen and Petersen [5]. Its input oriented formulation is very close to model (2). In Andersen and Petersen (AP) model the weight of the DMU_q λ_q is set to zero. It causes that the DMU_q is removed from the set of units and the efficient frontier changes its shape after this removal. Super-efficiency score θ_q^{AP}, that is greater than 1 for units CCR fully efficient, measures the distance of the unit DMU_q from the new efficient frontier. Its value expresses how many times the inputs may increase (in input oriented version of the model) in order the evaluated unit remains efficient.

Traditional DEA models (1) and (2) evaluate the efficiency or performance of the set of units in one time period only according to the values of input and output variables but it is clear that the performance in the period t can depend not only on the inputs in this period but on the inputs in one or several preceding periods. That is why various attempts how to solve this problem have been proposed in the past. This group of models is called multi-period DEA models. The first model of this class was introduced by Park and Park [6]. Unfortunately, their model (further as PP model) is not a real multi-period model because it returns just the best efficiency score of the unit under evaluation within all periods considered. It is the model oriented on the best period. It is clear that according to this property the PP model cannot be used as a tool for future allocation of resources. Its mathematical formulation can be found e.g. in Park and Park [6] or Jablonsky [4]. Similarly to the PP model it is possible formulate models that are oriented on the worse period of the unit under evaluation or the model that returns average efficiency score over all periods. Their formulation is given in Jablonsky [4]. In addition, this paper contains an original SBM multi-period model that measures efficiency during all periods using undesirable slacks of inputs and outputs. Nevertheless, this model does not take into account interconnections among particular periods. In order to do it, the appropriate model is not linear and its solution is questionable. The DEA model that uses efficiency score given in the period (t–1) as the only input variable in the period t is as follows:

$$
\begin{aligned}
\text{minimize} \quad & \sum_{q=1}^{n}\sum_{t=2}^{T}\theta_q^t \\
\text{subject to} \quad & \sum_{j=1}^{n}\theta_j^{t-1}\lambda_{jq}^t \le \theta_q^t\theta_q^{t-1}, && t=2,3,\ldots,T, i=1,2,\ldots,m, q=1,2,\ldots,n, \qquad (3)\\
& \sum_{j=1}^{n} y_{kj}^t\lambda_{jq}^t \ge y_{kq}^t, && t=1,2,\ldots,T, k=1,2,\ldots,r, q=1,2,\ldots,n,\\
& \theta_q^1 = 1, && q=1,2,\ldots,n,\\
& \lambda_{jq}^t \ge 0, && t=1,2,\ldots,T, j=1,2,\ldots,n. \ q=1,2,\ldots,n,
\end{aligned}
$$

where the DMUs are described by the same set of outputs and one input in T consecutive time periods $t = 1, 2, \ldots, T$, and assume that y^t_{kj}, $k = 1, 2, \ldots, r$, $j = 1, 2, \ldots, n$ are the values of the k-th output in the t-th period of the DMU_j. The only input in the period t is the efficiency score of the units in the period (t–1) and its values for the first period are set to 1 for all units. θ_q^t is the aggregative efficiency score of DMU_q in the period t. It is clear that model (3) is not linear in the first set of constraints (moreover, the number of variables and constraints can be very high) and that is why it cannot be solved easily. To overcome this problem we propose in the next section of the paper an iterative approach for analysis of multi-period systems.

3. Resource allocation – a case of Czech economic faculties

In this section we will test the hypothesis that the multi-period DEA analysis can generate results usable for resource allocation of the set of homogenous faculties. For this purpose the set of 19 Czech public economic faculties was investigated. Each of the faculties is described by two inputs - number of academic staff and labour costs. Three most important performance characteristics as outputs are taken into account – the number of students, the number of graduated, and the number of RIV points generated by the faculty in particular years (RIV points is an aggregated characteristic that measures the quality of publication outputs of individuals and/or educational units in the Czech Republic). The data set is available since 2007 until 2013. Table 1 contains efficiency scores θ^{CCR} and super-efficiency scores θ^{AP} for all faculties and periods. The best (highest) values are bolded.

Table 1. Efficiency (super-efficiency) scores in all periods.

Faculty	2007	2008	2009	2010	2011	2012	2013	Avg
FSV UK	1.4722	**1.7058**	1.1984	1.6965	1.1766	1.5307	1.3887	1.4527
EkF JČU	0.6730	0.6825	**0.9610**	0.6937	0.7666	0.6818	0.6175	0.7252
FSE UJEP	0.7713	0.6978	0.7977	0.7678	0.9260	0.8206	**0.9849**	0.8237
ESF MU	0.6921	0.6466	0.9762	1.0216	**1.0444**	1.0315	0.9765	0.9127
OPF SU	0.6206	0.6824	0.8831	**0.9331**	0.9072	0.8495	0.8372	0.8162
FE ZČU	0.9104	0.8333	0.8441	0.9530	**1.0022**	0.9381	0.8750	0.9080
HF TUL	0.4978	0.5363	0.6122	0.6490	0.5923	**0.8037**	0.6573	0.6212
FES UP	0.8354	0.6271	**0.8623**	0.8270	0.7358	0.6568	0.7726	0.7596
FP VUT	**1.0290**	0.9581	0.8996	0.8446	1.0065	1.0142	0.9721	0.9606
EkF VŠB	**0.9919**	0.9457	0.8883	0.8779	0.9000	0.9682	0.8525	0.9178
FME Zlín	1.3748	1.4282	1.3081	1.3163	**1.6378**	1.0406	1.4069	1.3590
FFU VŠE	**1.1239**	1.0924	0.8858	0.8394	0.8629	0.8370	0.7877	0.9184
FMV VŠE	0.7464	0.6835	0.6572	0.6340	0.7071	**0.7236**	0.6744	0.6895
FPH VŠE	0.9634	**0.9881**	0.8743	0.8047	0.8138	0.7390	0.6931	0.8395
FIS VŠE	0.7533	0.7983	0.8435	0.8182	0.8443	**0.9640**	0.8495	0.8387
NH VŠE	**1.5045**	1.4318	1.0453	0.9920	0.9135	1.0390	0.9321	1.1226
FM VŠE	0.7225	0.7049	0.8124	0.8441	0.8253	**0.9756**	0.9482	0.8333
PEF ČZU	0.9038	1.1285	1.3193	1.3149	1.3638	**1.4432**	1.3190	1.2561
PEF MZLU	0.8037	0.8461	**0.9368**	0.7982	0.8450	0.9348	0.8331	0.8568

4. Resource allocation – a case of Czech economic faculties

The results presented in Table 1 show that only two faculties are efficient in all periods (FSV UK and FME Zlín). Application of the PP multi-period model (orientation on the best period) leads to the following ranking: FSV UK (1.7058), FME Zlín (1.6378), NH VŠE (1.5045), …, FMV VŠE (0.7236). The multi-period model oriented on the worse period (see Jablonský [4]) generates the same ranking in first three places but the worse faculty is changed in this case, i.e. FSV UK (1.1766), FME Zlín (1.0406), NH VŠE (1.0390), …, HF TUL (0.4978). The same ranking (first three and last place) is given by average efficiencies over all periods (the last column of Table 1). The efficiency (super-efficiency) scores in the period *t* express the level of relative performance of the faculties according to the given model and variables (inputs/outputs) taken into account but they just hardly can be used as indicators for allocation of resources in the future periods because the results do not consider interconnections among the periods. In order to try overcoming this drawback we propose the following simple iterative procedure:

- The set of outputs contains the same three variables as in the previous experiments, i.e. the number of students, the number of graduated and the number of RIV points.
- The set of inputs includes two variables – the number of employees and the labor costs but this second variable is given in a different way comparing to the previous model. In 2007 the values of this variable correspond to the reality. In the other periods it is the result of a simple formula. The amount of 50 % of the labor costs in the period *t* is calculated according to the labor costs' share in the previous period and the remaining 50 % according to the efficiency scores' share in this period.

- The efficiency scores are calculated using model (2) and, if necessary, super-efficiency scores using the AP model derived from model (2). Either two mentioned inputs are considered or just labor costs derived using the presented way are taken into account. The results of allocation of resources among faculties according to the given approach are presented in Table 2. The first 8 columns of the table contain differences in allocation of funds given by the model with two inputs and original allocation by government methodology. The positive/negative values indicate that the allocation by the model is higher/lower than the reality. The last column of Table 3 contain the same information but given by the model with one input variable (labor costs) only. The results show that there are faculties overfunded in all periods (e.g. PEF ČZU and FMV VŠE). In the contrary some of the faculties are underfunded in all or almost all periods (e.g. FME Zlín, NH VŠE, and others).

Table 2. Differences in real funding and funding by the model.

Faculty	2008	2009	2010	2011	2012	2013	2014	2014a
FSV UK	5997	17544	10269	20844	-7166	-6503	-17364	-16567
EkF JČU	20431	14085	15480	10396	7587	1016	-1228	1642
FSE UJEP	3691	4762	4864	8954	8491	13237	15948	10085
ESF MU	-9655	-12901	-7325	-1177	2112	996	5945	3112
OPF SU	-8758	-9311	-4260	-8153	-6975	-5020	1934	6253
FE ZČU	7804	12856	15178	13311	13740	16804	16230	9831
HF TUL	-4088	-4645	-1970	-960	-199	5278	1094	1860
FES UP	-643	-2569	3251	5887	-4	6462	6626	7111
FP VUT	6321	8784	10445	7175	6786	6426	5138	5013
EkF VŠB	-17022	-17904	-19730	-15926	-12632	-14866	-15413	-10018
FME Zlín	18709	20030	19368	15750	23395	18230	22479	17873
FFU VŠE	6616	8052	3143	-327	83	-487	-212	-118
FMV VŠE	-15821	-18543	-24991	-30903	-26917	-30064	-34109	-29677
FPH VŠE	-852	-1517	-2978	-6524	-11828	-14165	-8428	-4523
FIS VŠE	-1322	113	-1668	-4084	-1161	-1731	-5411	-2075
NH VŠE	22668	25877	21707	20693	17412	17566	18870	11498
FM VŠE	7779	13082	16868	17407	20597	20076	19965	10512
PEF ČZU	-37916	-55918	-55652	-49018	-34498	-28167	-27135	-22003
PEF MZLU	-3937	-1876	-2001	-3346	1177	-5088	-4926	194

5. Conclusions

Evaluation of efficiency and performance of higher educational institutions is of a high importance and the results of this evaluation can serve as starting points for allocation of limited resources among these institutions. The commonly used practice in the Czech Republic is based on weighting of several criteria among which the most important is the number of students and research performance of the institution in the last five years measured by RIV points. The paper compares this methodology with DEA approach – a traditional CCR input oriented model and its super-efficiency modification with five (or four) variables is applied. The given results show significant differences in funding of the faculties under evaluation according to the official methodology and the DEA models. Of course the proposed models do not cover all criteria considered by the official methodology but the differences in funding can just partly be explained by this drawback. A future research in this field can be focused on extension of the number of variables in the model and application of other DEA models. SBM models, e.g. Tone's efficiency and super-efficiency models (see Tone [7]), could be an interesting alternative to the traditional DEA models.

Acknowledgements
The research is supported by the Czech Science Foundation, project no. 16-01821S.

References
[1] Charnes A Cooper WW and Rhodes E 1978 Measuring the efficiency of decision making units. *Eur. J. Oper. Res.* **2** 429–44.
[2] Beasley JE 1995 Determining Teaching and Research Efficiencies. *J. Oper. Res. Soc.* **46** 441–52.
[3] Thanassoulis E Kortelainen M Johnes G and Johnes J 2011 Costs and efficiency of higher education institutions in England: a DEA Analysis. *J. Oper. Res. Soc.* **62** 1282–97.
[4] Jablonsky J 2016 Efficiency analysis in multi-period systems: an application to performance evaluation in Czech higher education. *Cent. Eur. J. Oper. Res.* **24** 283–96.
[5] Andersen P and Petersen NC 1993 A procedure for ranking efficient units in data envelopment analysis. *Manage. Sci.* **39** 1261–64.
[6] Park KS and Park K 2009 Measurement of multiperiod aggregative efficiency. *Eur. J. Oper. Res.* **193** 567–80.
[7] Tone K 2002 A slacks-based measure of super-efficiency in data envelopment analysis. *Eur. J. Oper. Res.* **202** 32-41.

Research on the method of flexible data bus technology of the monitoring of the hidden engineering audit

Rong Chen[1], Luqian Yu[2]

1.Nanjing Audit University, Audit S&T Research institution, Nanjing, China
2.Nanjing Audit University, School of Government Audit, Nanjing, China

E-mail: Chenr2001_125@163.com; ylq180256@163.com

Abstract. Hidden engineering is the difficulty of the precision tracking audit of public engineering. Single sensor detection means that it is difficult to reveal the internal state of the project systematically. Multi-sensor detection equipment is the new method of hidden engineering audit to track the internal state of the hidden engineering in all directions, all-weather and all day. The disadvantages of all kinds of non-destructive testing system in integration are "system fragmentation, difficult collaborative operation, poor user experience effect, higher building operation cost". They became major bottleneck in the development of non-destructive network detection. A new data integration bus framework of distributed object is proposed. Various detector level, detection device level, detection system level with configuration integration build a sound, light, electricity, heat, magnetic multi-waveband non-destructive endoscopic fusion technology framework, synthesize the information of the fragments that multi-sensor detection obtained, and achieve to visualize the internal state of hidden engineering.

1. Introduction

Public engineering audit has started to use ultrasonic, electromagnetism, eddy current, laser, ground penetrating radar(GPR), video surveillance, Beidou navigation, and other non-destructive testing means to visualize the internal data of hidden engineering; single non-destructive detection sensors can also detect the internal status data of some hidden engineering, however, due to the fragmentation of the detected data, these methods cannot perform systematic tracking of the internal state of hidden engineering. IO application system have become part of people's daily life given the development of new technology, especially intelligent sensor technology, electronic communication technology, data fusion technology and big data technology. This study proposes a framework for flexible data bus technology based on the non-contact engineering audit and integrated multiple heterogeneous sensors and solves the problem of detecting the internal state of public engineering through network. This study forms a complete, unified, and continuous internal status imaging with integrated data fragmentation.

2. Research status of composite nondestructive testing

Composite nondestructive detection fusion algorithm began in the early 90s, the Sun team [1] studied the registration problem of eddy current and ultrasonic image, this research is from the signal (image) of the physical meaning of the conversion of graphics and image registration are usually different, the process of trying to achieve through a method called Q transform, which requires the establishment of mathematics two methods of energy transfer model. The Heideklang team [2] proposed a pixel level fusion algorithm combining thermal imaging, magnetic flux leakage and eddy current elements, and evaluated 29 kinds of signal processing algorithms. The experimental results show that the false alarm rate of the detection results is reduced from 1.65% to 0.28%. The P Cotič team [3] is aiming at the sensitivity of concrete penetration depth, resolution and material performance, an unsupervised clustering method of GPR and infrared image fusion with two source elements is proposed. The experimental results show that the clustering method can effectively improve the defect level detection ability. The team Thirunavukkarasu [4] compared Laplacian Pyramid, wavelet fusion, statistical reasoning, Bayesian fusion and principal component analysis. The experimental results show that SNR based on Bayesian fusion improves 33.75dB and 3.22 information entropy. The team Heideklang [5] proposed a high-level algorithm for fusion of eddy current, GMR and thermal imaging. The experimental results show that this method maintains the same sensitivity level, and the readiness exceeds the single sensor detection characteristic. Based on the porosity, water saturation, modulus of elasticity, compressive strength, chloride content and carbonation, the team Marie-Aude [6] put forward probability and fuzzy mathematics for data fusion, and achieved good results. In order to improve processing speed, the client / server mode, browser / server mode, single computer computing mode. The literature [7] puts forward the processing of parallel data fusion.

3. Visualization and detection concept of audit network in hidden engineering

3.1. Application mode of audit visualization of network exploration

Audits of large buildings, bridges, highways, railways, and other infrastructure projects are characterized by a large number of projects and complicated engineering. To facilitate the efficiency of hidden engineering audit and reduce the workload of the auditor, self-propelling robots; fixed or artificial loading platforms, such as unmanned aerial vehicles, robots, PTZ preset, and artificial platforms can be used with the integration of non-destructive multi-source sensors, external monitoring sensors; and data detection in engineering activities to detect the stream of real-time data. These data will be transmitted to the data fusion center for parallel and distributed processing. This process will form a unified and continuous hidden engineering with internal two- and three-dimensional visualization of data through a variety of communication tools, such as cable, wireless technology, and satellite.

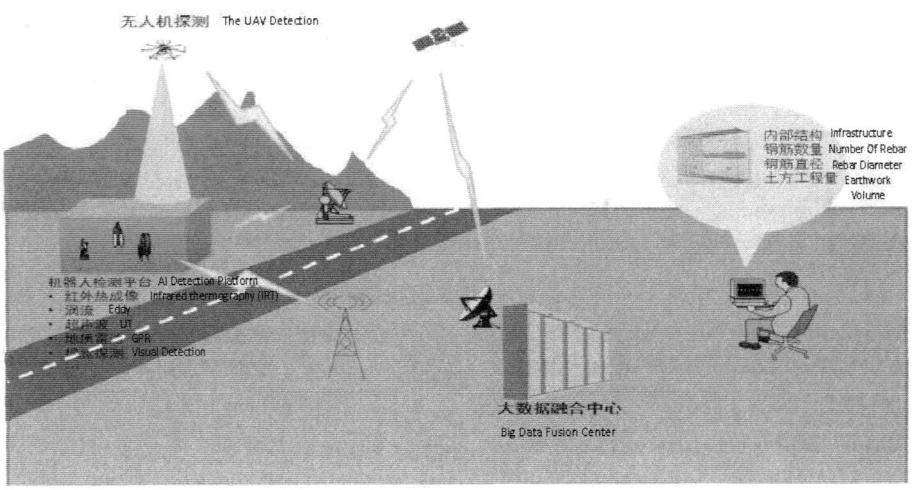

Figure 1. Visualization system of a hidden engineering network

As Figure 1 shows, the visualization system of network detection audit is composed of a sensor detection subsystem, data fusion subsystem, wide area communication subsystem, and the 3D analysis and display subsystem. The sensor detection subsystem covers all kinds of non-destructive testing equipment (including multiple sensitive detection modules for integrated equipment), video surveillance equipment, and 3D scanning equipment. The data fusion subsystem covers data calculation based on supported frame, data fusion processing services, and cluster computing system. The wide area communication subsystem is composed of satellite communication, wireless communication system, and cable communication system. The three-dimensional analysis and display subsystem is composed of data analysis and display subsystem.

3.2. Integration Model of Network Monitoring Technology

The traditional multi-sensor network detection system is based on the construction of the platform as center. Each subsystem in the building considers the function and construction goal of the system instead of the entire system. Thus, this system does not support sharing of sensors, controllers, and other devices. This system is characterized by repeated construction, lower level of information sharing, poor user experience, and other issues.

The multi-sensor detection model in large-scale public engineering audit highlights the top layer design. Each element of the system, namely sensors, controllers, and alarm, are nodes in the network. Resource interconnection, information sharing, collaborative operation, flexible organization and arrangement of sensors, controllers, and other devices in the environment of entire network form a distributed engineering audit system.

4. Integration method of flexible data bus

4.1. Basic concepts

A public engineering audit system is composed of a processing application service for multiple, distributed heterogeneous sensor. This system requires data integration bus technology with flexible wide area to shield the difference between subsystems/equipment location, communication protocol, programming language, hardware equipment and control mode. While the business used in the development and integration of the system, this technology does not be required to focus the equipment, protocol and control mode of the data. This system only needs to define control logic to realize the development of interactive interface network and the distributed and integrated network audit monitoring system.

This study presents an open data bus technology, namely, ODIBus and Open Data Integration Bus, which is a set of distributed, high real-time, micro kernel bus equipment with intelligent data integration. This technology includes an embedded software bus, management debugging tools, and configuration hardware platform with corresponding standard specification composition. This technology consists of distributed real-time software bus and an intelligent configuration hardware platform. Distributed real-time software bus uses real-time distributed object communication (ODC) to realize sensor and equipment level, system level of plugging and playing, flexible reorganization, on-demand grouping functions, server version (ODIBus/S), and embedded edition (ODIBus/E). The configuration hardware platform uses the analog front end (AFE) of sensor technology to realize the flexible assembly of the sensor/controller based on user requirements.

The traditional engineering system network is based on BACnet, Lonworks, Profinet, and other field bus technology. Application integration is performed by retaining the internal structure of the system. ODIBus integrates heterogeneous buses and software. The establishment of an integrated communication networking environment is based on ethernet, which includes industrial and wireless ethernet, unified infrastructure. The application system is realized by interconnecting the built-in micro kernel ODIBus/E software and the upper machine ODIBus/S equipment to determine the integration principle. This principle is based on the configuration of hardware development of the sensor and the controller to achieve an intelligent system and IP. The principle is shown in Figure 2.

IOP Conf. Series: Journal of Physics: Conf. Series **1026** (2018) 012003 doi:10.1088/1742-6596/1026/1/012003

Compared with the traditional integration technology, the ODIBus bus can realize "location transparency, transparent protocol, transparent language, transparent devices, transparent control mode,"and reduce the difficulty of system development and integration. The key points of the ODIBus are real-time optimization and low costs of large concurrent transmission, hot plug, low power consumption, reconfiguration, and the availability of a framework for a lightweight, intelligent hardware.

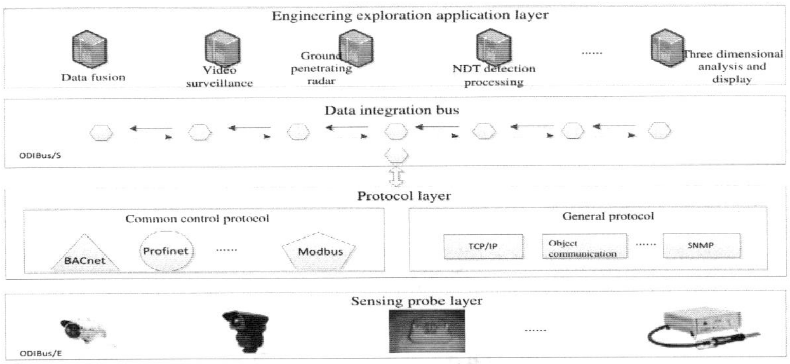

Figure 2 Schematic diagram of ODIBus application mode

4.2. Distributed, Real-time Software Bus

4.2.1. ODC mechanism

ODC is the middleware that integrates Common Object Request Broker Architecture (CORBA) objects and data objects (DDS). These objects require CORBA mechanism, whereas complex data objects are distributed using DDS mechanism.

1. CORBA objects: CORBA is a common service located between the operating system and application software. The foundation and the core of the ORB architecture and its center are connected to the client and the data server.
2. Data distribution service (DDS) distribution technology: DDS aims to develop a new generation of distributed real-time specification of communication middleware technology.
3. Communication mechanism of ODC: ODC is the integration of CORBA and DDS objects. These objects use ORB mechanism for logging requests. DDS is then used to distribute data.The mechanism is shown as Figure 3.
   - Data objects and ORB objects are defined as IDL files that support each other.
   - The theme files compiled by the writer, the reader class file, and the IDL of ORB generate a client stub and service framework with generating the corresponding part of the code that can be used by each compiler, which is determined by writing their IDL platform.
   - Data objects and the basic environment of ORB are initialized. Publishers complete the registration server, registration data type, and registration theme. Subscribers then complete the registration server and subscription topic before sending data.
   - Application after integration may utilize the API function of DDS to publish and subscribe the data.

IOP Conf. Series: Journal of Physics: Conf. Series **1026** (2018) 012003 doi:10.1088/1742-6596/1026/1/012003

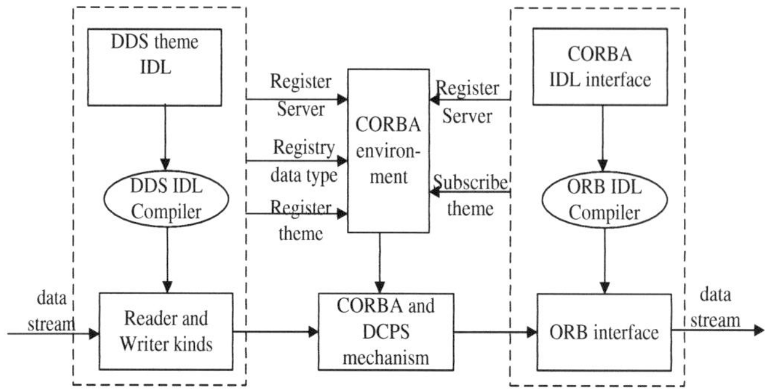

Figure 3. Integrated CORBA and DDS communication mode

4.2.2. Distributed real-time software bus server

Distributed real-time software bus is part of the ODIBus software. A software bus with real-time distributed, asynchronous object communication mechanism provides an object bus, a communication protocol conversion module, driver library, enterprise application integration module, real-time database, ODIBus language, ODIBus kit, ODIBus virtual machine (OVM), a compiler, and software debugging tools for graph development. ODIBus uses a graphic development tool for real-time online debugging. A simple dragging component module is used to achieve a variety of functions to connect data flow and control logic. The application development of users can then be completed and feedback results can be obtained in real time.

ODIBus uses Java facing object technology in the form of ODIBus tool kit released from the control logic module as Figure 4.

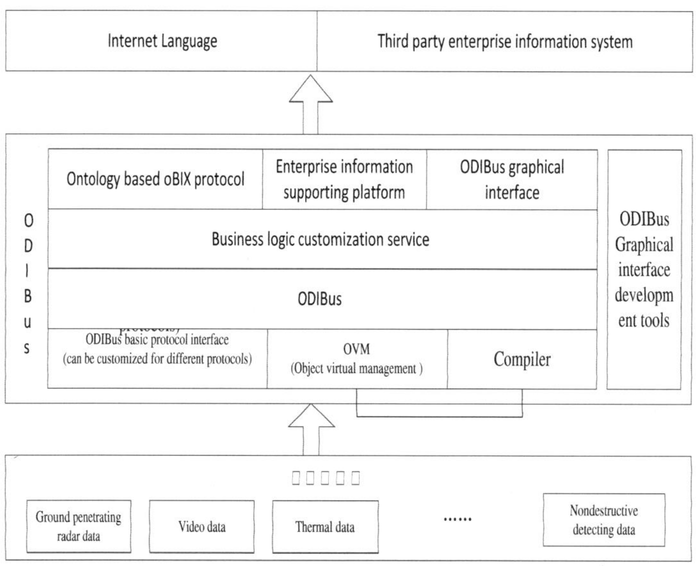

Figure 4. Bus software schematic of ODIBus software

4.2.3. Distributed real-time bus (embedded version)

ODIBus/E includes perform plug and play, communication processing, data integration agent, positioning and processing, control information processing, equipment registration, and other basic software and sensor information processing and applications. The low power consumption of intelligent configuration hardware platform includes platforms with the plug and play function. This functionality supports the sensing and software module, which in turn supports controller. ODIBus/E is designed by layered architecture, which includes embedded software and application layers.

Embedded basic software layer includes the plug and play (PNP) module, communication processing module, distributed object module, and positioning module. The application layer includes detection information processing, control information processing, and the fusion module. The OBDIBus/E meets data object distribution and object component request function through object-oriented middleware and OBDIBus/S platform. Figure 5 shows the embedded ODIBus software bus.

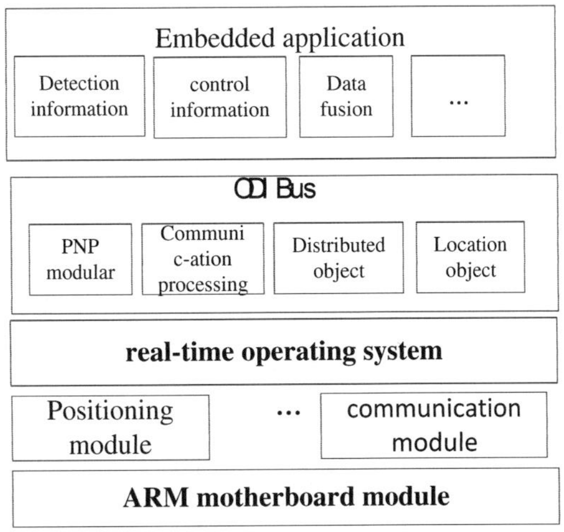

Figure 5. Embedded ODIBus software bus

4.2.4. Hardware framework of intelligent configuration for embedded network

The configuration hardware platform of intelligent network is part of the hardware of ODIBus. This platform can achieve the intelligent controller/sensor flexible configuration of the network. The configuration hardware platform of intelligent network is composed of a low-cost ARM motherboard, ethernet module, Wi-Fi communication module, ZigBee module, and front-end analog circuit selection and the hardware diagram of ODIBus is shown in the Figure 6.

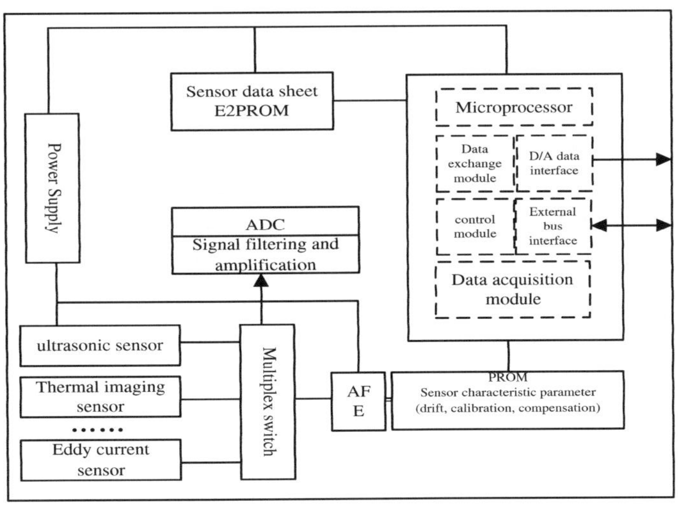

Figure 6. Hardware diagram of ODIBus

5. Separation of business architecture and the software development model of technical architecture

IOP Conf. Series: Journal of Physics: Conf. Series **1026** (2018) 012003 doi:10.1088/1742-6596/1026/1/012003

The traditional engineering control system integrator must grasp each device or programming interface of the subsystem, the system operation mode, control mode, and the position of the equipment in the design of business architecture. Thus, this process consumes a significant amount of energy.

ODIBus is in an integrated and transparent engineering audit system. Thus, business developers do not need to consider the underlying architecture. Programming interface can be directly used, which is provided by ODIBus, to realize distributed flexible integration and improve the rapid deployment capability of the system.

Business developers of project audit will encapsulate the business object as data objects and release it into the tool kit components, which will be run on an intelligent hardware platform. The entire network system will then obtain equipment information and will be accessed in the system. System developers use graphic debugging tools to facilitate rapid grouping and system generation. Figure 7 shows the system developers and system integrators.

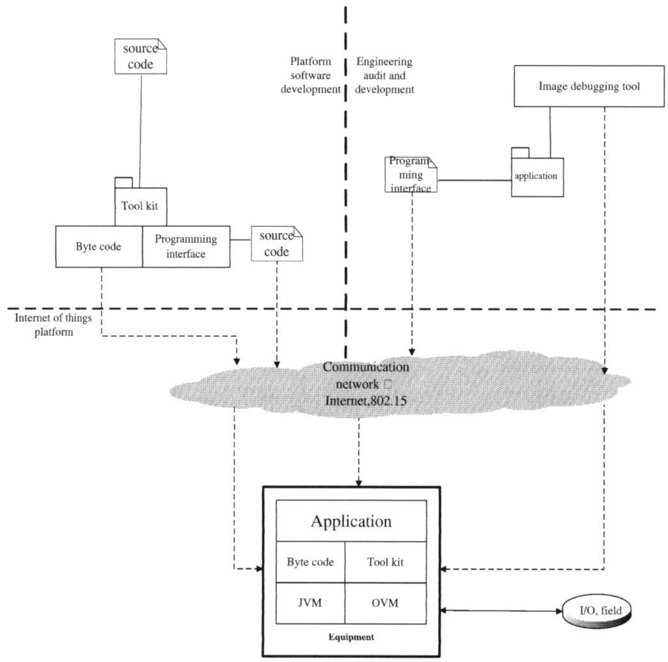

Figure 7 System developers and system integrators

6. System Integration Application Based on Flexible Data Bus Integration Technology

6.1. Integrated application system of flexible data bus

ODIBus can realize the adaptation function of heterogeneous devices. These devices have three forms, namely Internet of things, gateway agent embedded with intelligent IP controller, and sensor equipment. ODIBus bus develops software functions, such as a variety of protocol conversion engine configuration. ODIbus optimizes real-time transmission control and micro memory agent software based on all kinds of communication hardware. Figure 8 shows the typical application model for ODIBus integration.

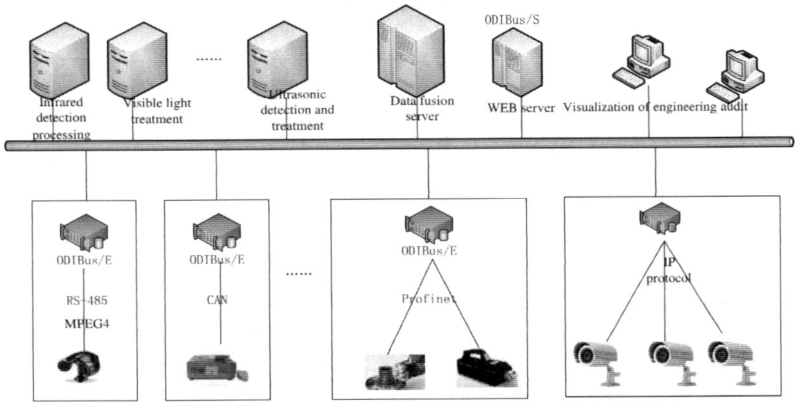

Figure 8. Visualization of the overall solution for audit project based on ODIBus

6.2. *Integrated application of performance data for flexible data bus*

Traditional real-time application runs on a real time line (e.g., BACNet, Profinet, and Modbus) to ensure the real-time performance of communication.

The main service mechanism cannot easily support real-time applications, especially complicated real-time applications. OBDIBus receives request for granularity in the resource control method and provides the running environment for request execution, which is the basic control unit in the implementation of control. Figure 9 shows the performance test for data distribution. The ordinate denotes delay (MS), whereas the horizontal coordinate denotes the number of concurrent.

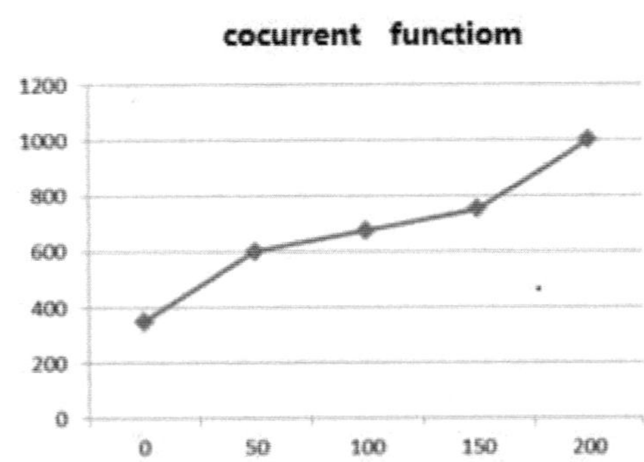

Figure 9. OBDIBus performance and typical application interface

7. Conclusion

According to the internal state information to expose the hidden engineering demand, using NDT sensor, video detection, 3D scanning and other heterogeneous sensors, combined with big data online, real-time, high speed data fusion algorithm can realize data online, remote visualization of audit of public project, due to the heterogeneous multi sensor and a large amount of data, common data fusion architecture is difficult to deal with complex data stream, therefore it is needed to consider the calculation model of flow data, data fusion research framework based on cloud computing, which can solve the problem of large amount of data.

This thesis is a systematic integration of the subject, which mainly focuses on the integration of complex and heterogeneous sensor networks. The detection of the internal state of concealed works requires complex multi-sensor detectors for network detection. Equipment/systems for sensor detection often contain their own hardware and software and processing system. We propose a real-time flexible integration bus method to integrate multi-sensor application system in public works based on the real-time communication mechanism of objects (including data object). This method is based on the demand for multilevel configuration integration of heterogeneous detection systems.

Reference

[1] K Sun,S Udpa,L Udpa,T Xue,W Lord. Registration Issues in the Fusion of Eddy Current and Ultrasound NDE Data Using Q-Transforms. *Springer US*, **1996, pp. 813 ~ 820**

[2] R Heideklang,P Shokouhi. Multi-sensor image fusion at signal level for improved near-surface crack detection. *Ndt & E International*, **2015, vol71, pp.16-22**

[3] P Cotič, Z Jagličić, E Niederleithinger, M Stoppel, V Bosiljkov. Image Fusion for Improved Detection of Near-Surface Defects in NDT-CE Using Unsupervised Clustering Methods. *Journal of Nondestructive Evaluation*,**2014, vol33(3), pp.384-397**

[4] S Thirunavukkarasu, BPC Rao, AK Soni, SS Ahmed. Comparative performance of image fusion methodologies ineddy current testing. *Research Journal of Applied Sciences Engineering & Technology*, **2012, vol24(24), pp.16-22**

[5] R Heideklang, P Shokouhi. Quantitative Multi-modal NDT Data Analysis. *Aip Conference*, **2014, vol1581(1), pp.1928-1932**

[6] Ploixa M, Garniera V, Breysseb D, et al. NDE data fusion to improve the evaluation of concrete structures[J]. *NDT&E International*, **2011, 44(5): 442-448**

[7] Quadri S A, Sidek O. Development of heterogeneous multi-sensor data fusion system to improve evaluation of concrete structures[J]. *International Journal of Image& Data Fusion*, **2014, 5(2): 97-108**

IOP Conf. Series: Journal of Physics: Conf. Series **1026** (2018) 012004 doi:10.1088/1742-6596/1026/1/012004

Coverage Test Technology Based on ONESPIN Verification Platform

YANG Yawen[1], ZHOU Shan[1], KONG Lu[1]

[1] Technology and Engineering Center for Space Utilization Chinese Academy of Sciences, Beijing, China

Abstract. Coverage test technology is a common software testing technology, which is the basic requirement of software testing. The coverage analysis can quantify the completeness of the test vector. This paper introduces and compares two kinds of mainstream coverage analysis techniques, code coverage and functional coverage. On the basis of existing theories, the working principle of ONESPIN verification platform for VHDL/Verilog/System Verilog language coverage is studied in this paper and the specific testing procedures with a practical example are presented. At last, the covered code, uncovered code and the software defect in the test results are analyzed.

1. Introduction

Along with the rapid growth of the scale and complexity of the integrated circuit design, the complexity of the verification work and the verification time increase. In the process of simulation verification, it is not workable to simulate all the input combinations, and how to balance the simulation verification time and the completeness of the test is a practical problem. Compared with the traditional random verification, currently constraint random test vector can improve the quality and quantity of verification, but it can also result in some areas in the design that are never detected, while others are always repeated. So how do you evaluate the test vectors? When does the validation work end? These questions need to be answered by coverage analysis indicators. Coverage is a necessary indicator to evaluate the quality of verification. Coverage can quantify the completeness of the simulation incentive set. Coverage can diagnose which part of the design has not been fully verified, which can help the generation of subsequent test vectors.

Currently, there are two main methods based on coverage analysis: one is the code coverage analysis, and the other one is the functional coverage analysis. In the following, the first section introduces the two main coverage analysis methods and common coverage collection tools; the second section introduces the working principle and workflow of the ONESPIN verification platform; the third section gives the example to introduce the specific test steps of the ONESPIN verification platform and analyzes the results; at last, the paper makes summary and outlook.

2. Coverage test technology

2.1. Code coverage

Code coverage testing technology originated in software testing is a kind of common white-box testing technology and its analysis object is mainly code. In the test process, the simulation tools, such as

Mentor's Questa and Synopsys's VCS, capture code coverage in the dynamic simulation process by inserting special PLI (Programming Language Interface) tasks in the HDL code to get the code coverage metrics by calculating the proportion of the executed source code, and then the simulation tools provide graphical and report files feedback on statement coverage, branch coverage, condition coverage, expression coverage, toggle coverage, finite state machine coverage [1].

In general, after the tester testing all functional points in the process of functional testing, it's hard to make sure all the source code has been ran. Code coverage testing techniques will be very effective for business scenario. Code coverage can detect the redundant state, the security protection state, the wrong state, and the undetected state of the code, detect the fault and incomplete conditions of the test vectors and increase the maintainability of the code.

As the design scale increases, the code coverage analysis becomes less and less efficient. It can only detect whether the code is fully executed, but cannot detect whether the software's function is correct, so another coverage analysis method - functional coverage is introduced.

2.2. Functional coverage

Functional coverage focuses on design of architecture, is used to check design's features, and primarily aims at the overall design's function rather than the HDL code. Functional coverage can effectively describe the boundary logic and the combination of data and control signals, and indicate which functions have been tested. Functional coverage has become an important criterion for measuring verification work. The key of functional coverage analysis is that verification engineers manually define the verification tasks according to the DUT (Design Under Test). The definition of verification tasks determines the coverage space, and the completeness of verification tasks determines the comprehensiveness of functional coverage analysis. After defining the verification tasks, the verification engineers typically use the verification methodology to describe the verification tasks [2][3].

Currently, simulation software and verification tool can automatically identify professional hardware logic functional verification language, such as SVA (System Verilog Assertion) and PSL (Property Specification Language) and perform corresponding calculations, but as the design scale is getting bigger, the functional points are getting more, the test scenarios are getting more complex, the signal numbers as well as the possible combinations of the signals are explosive growth [4][5]. The key problems of verification are obvious as following:

- It takes a lot of manpower and time to build and debug the verification environment.
- It takes a lot of manpower and time to design test case set, and it is difficult to ensure the completeness and correctness of the test case set.
- It is difficult to assess the adequacy of the functional verification coverage, that is, the logical functional domain represented by RTL code is greater than/equal to/less than the functional test case set.

The verification platform of the Germany ONESPIN company combines the formalized verification engine with the ABV based functional verification methodology, which can effectively solve the above problems

3. ONESPIN verification platform

3.1. General concepts and verification principle

Germany ONESPIN verification platform provides formal functional verification for RTL modules. The verification platform provides Windows/Linux versions, supports VHDL/Verilog/System Verilog language and supports SVA/PSL/ITL/TIDAL/OVL verification language library [6]. The architecture of ONESPIN verification platform is as "Figure. 1", which has the following characteristics:

- The most technically advanced formal platform available
- High performance, capacity, optimized formal model
- High proof depth, convergence, broad range algorithms
- Automate engine application for best results and heuristic director

IOP Conf. Series: Journal of Physics: Conf. Series **1026** (2018) 012004 doi:10.1088/1742-6596/1026/1/012004

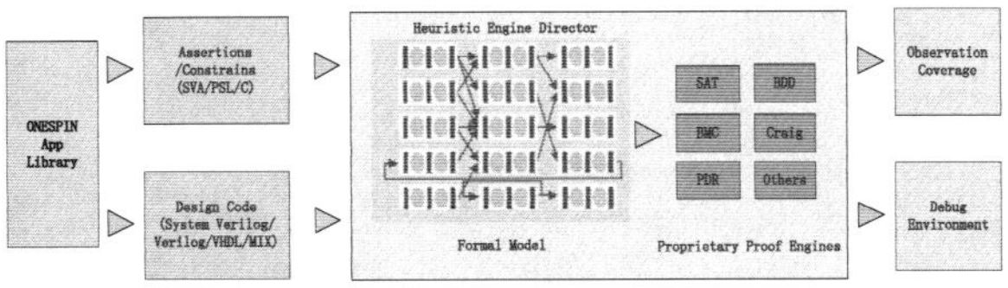

Figure 1. The architecture of ONESPIN

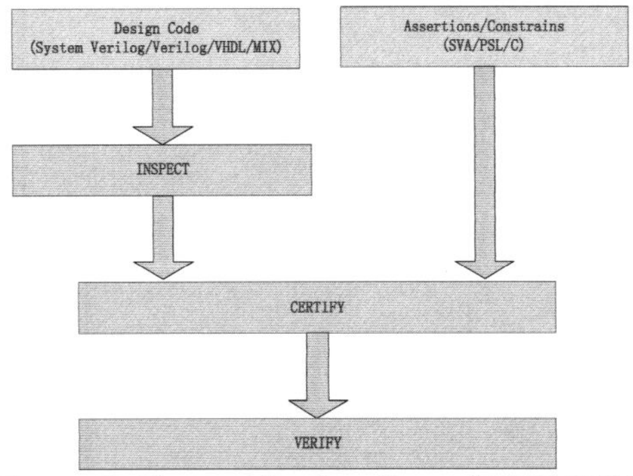

Figure 2. The verification flow of ONESPIN

3.2. Verification flow

The verification platform includes three main tools: INSPECT, CERTIFY and VERIFY. ONESPIN verification can be performed in accordance with the process as "Figure. 2".

1) Use INSPECT to find the low-level defects in the RTL code, and make a preliminary analysis of the dead code and the uncovered code.

2) Use professional hardware logic functional verification language such as SVA or PSL to accurately describe the functional and timing requirements, and then use CERTIFY to check the completeness and correctness of the functional requirement suite.

3) Use VERIFY to perform functional verification, check the consistency of the RTL code and the functional requirement, and generate detailed coverage information.Using VERIFY, it is possible to answer the following key questions of verification:

- How good are my assertions – which parts of the RTL code are actually covered and passed? Which parts are covered but failed? Which parts are not covered and should be targeted by additional assertions?
- How good are my constraints – is there some unintended over-constraining?
- How much of my verification plan has been completed?

All the coverage data can be exported to standard UCIS format data, and for all the problems and the errors found, VERIFY can automatically generate the counterexample, and display in a waveform diagram which linkage with code and show the process of the problems.

4. Application example

This section applies this platform to verify specific instance, analyze and statistic the results.

4.1. The design overview

The selected design is a voting system. The key signals in the design are explained in "Table 1".

Table 1. The Key signals in the design.

	Signals	I/O	Explanation
1.	i_clk	input	clock
2.	i_rst_n	input	reset, low is active
3.	x[2:0]	input	multi-channel base signal
4.	x_q[2:0]	input	multi-channel quality signal
5.	bp[2:0]	input	multi-channel secondary signal
6.	bv[2:0]	internal	internal signal, judge which channel is selected
7.	om [1:0]	output	om [1]: the quality bit output of the voting om [0]: the logic rusult of the voting

IOP Conf. Series: Journal of Physics: Conf. Series **1026** (2018) 012004 doi:10.1088/1742-6596/1026/1/012004

4.2. Specific operation steps

Step I: The platform reads and compiles the RTL code, and then run tool - INSPECT to detect defects. The results are as “Figure. 3” and “Figure. 4”. As shown in the results, there are no low-level defects and dead code in the design.

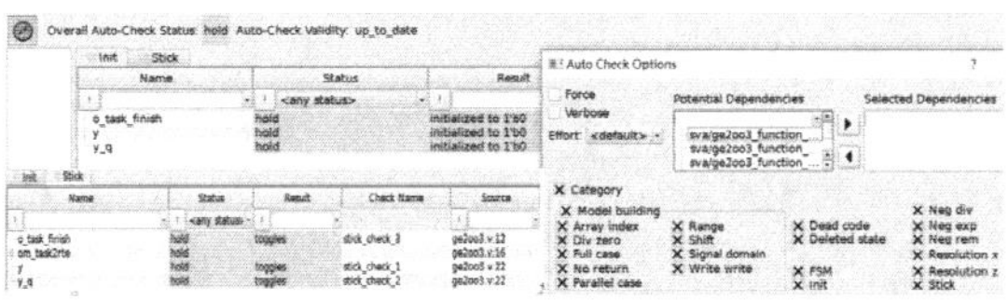

Figure 3. INSPECT result for low-lever defects

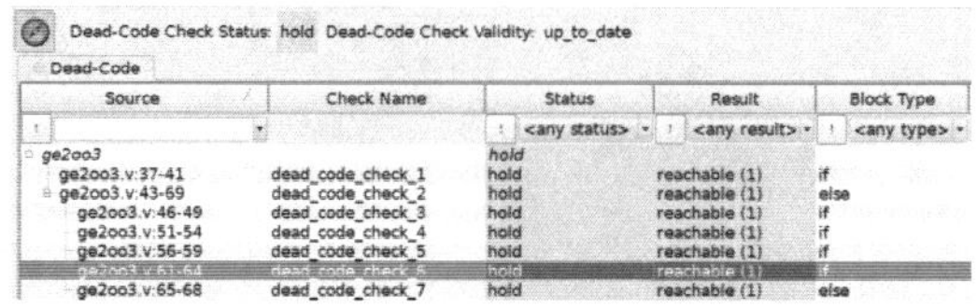

Figure 4. INSPECT result for dead code

Step II: Analyze functional requirement and use SVA to describe the requirement manually. The functional requirement and the corresponding SVA description are as shown in “Table 2” [7].

Table 2. Functional requirement and corresponding SVA Description

<table>
<tr><th></th><th>Functional requirement</th><th>SVA</th></tr>
<tr><td>1.</td><td>Channel voting: if bp[i] is enable and x_q[i] is unable, then bv[i] is active.</td><td><pre>// **
 property BV_CALCULATION_LOGIC;
    disable iff (!i_rst_n)
    bv==(~im_rte2task_in_x_q&im_rte2task_in_bp);
  endproperty : BV_CALCULATION_LOGIC
  BV_CALCULATION_LOGIC_1 : assert property(@ (posedge i_clk) BV_CALCULATION_LOGIC);
// **</pre></td></tr>
<tr><td>2.</td><td>Logic voting: if all bits of bv are enable and at least two bits of x are enalbe, then the value of om is 2’b01.</td><td><pre>// **
 property GE2oo3_ALL_valid_TISP_0;
   @ (posedge i_clk) disable iff (!i_rst_n)
  ($countones(bv)==3 && $countones(im_rte2task_in_x)>=2)|=> om_task2rte==2'b01;
 endproperty : GE2oo3_ALL_valid_TISP_0
 GE2oo3_ALL_valid_TISP_0_0 : assert property (GE2oo3_ALL_valid_TISP_0);
// **</pre></td></tr>
<tr><td>3.</td><td>Logic voting: if bv and x don’t meet the conditions in 2 above, then the value of om is 2’b00.</td><td><pre>// **
 property GE2oo3_ALL_valid_TISP_1;
   @ (posedge i_clk) disable iff (!i_rst_n)
  ($countones(bv)==3 && $countones(im_rte2task_in_x)<2)|=> om_task2rte==2'b00;
 endproperty : GE2oo3_ALL_valid_TISP_1
 GE2oo3_ALL_valid_TISP_1_0 : assert property (GE2oo3_ALL_valid_TISP_1);
// **</pre></td></tr>
<tr><td>4.</td><td>Logic voting: if at least two bits of bv are enable and at least one bit of x is enalbe, then the value of om is 2’b01.</td><td><pre>// **
 property GE2oo3_2_valid_TISP_0;
   @ (posedge i_clk) disable iff (!i_rst_n)
  ($countones(bv)==2 && $countones(im_rte2task_in_x)>=1 &&
  im_rte2task_in_x[f_whichtwoisselected(bv[0],bv[1],bv[2])]!=1)|=> om_task2rte==2'b01;
 endproperty : GE2oo3_2_valid_TISP_0
 GE2oo3_2_valid_TISP_0_0 : assert property (GE2oo3_2_valid_TISP_0);
// **</pre></td></tr>
<tr><td>5.</td><td>Logic voting: if bv and x don’t meet the conditions in 4 above, then the value of om is 2’b00.</td><td><pre>// **
 property GE2oo3_2_valid_TISP_1;
   @ (posedge i_clk) disable iff (!i_rst_n)
  ((bv[0]+bv[1]+bv[2])==2 && $countones(im_rte2task_in_x)<1)|=> om_task2rte==2'b00;
 endproperty : GE2oo3_2_valid_TISP_1
 GE2oo3_2_valid_TISP_1_0 : assert property (GE2oo3_2_valid_TISP_1);
// **</pre></td></tr>
<tr><td>6.</td><td>Logic voting: if at least two bits of bv are unable, then the value of om is 2’b11.</td><td><pre>// **
 property GE2oo3_1or0_valid_TISP_0;
   @ (posedge i_clk) disable iff (!i_rst_n)
   $countones(bv)<=1|=> om_task2rte==2'b11;
 endproperty : GE2oo3_1or0_valid_TISP_0
 GE2oo3_1or0_valid_TISP_0_0 : assert property (GE2oo3_1or0_valid_TISP_0);
// **</pre></td></tr>
</table>

Step III: Run tool - CERTIFY to check the SVA, the result is as “Figure. 5”. As shown in the result, the SVA requirement suite is not complete. Though the waveform generated by CERTIFY, the requirement description of reset behavior and status output signal o_task_finish is not found in the SVA requirement suite. The supplementary functional requirement and the corresponding SVA description are as shown in “Table 3”.

Run tool - CERTIFY to check the SVA again, all seven type of check are successful, that is to say the SVA requirement suite is complete and correct. Take a type check result for example which is shown as “Figure. 6”.

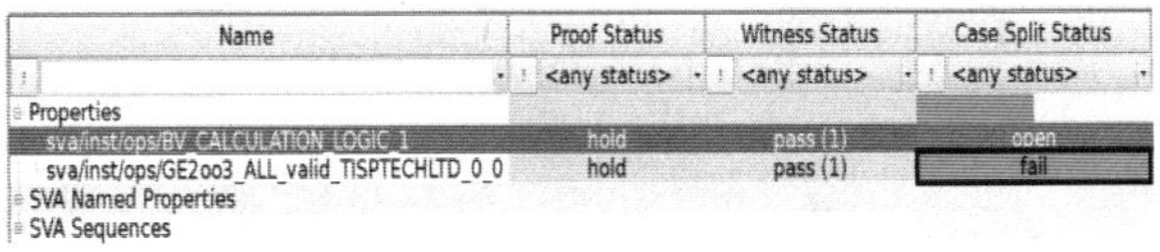

Figure 5. CERTIFY result

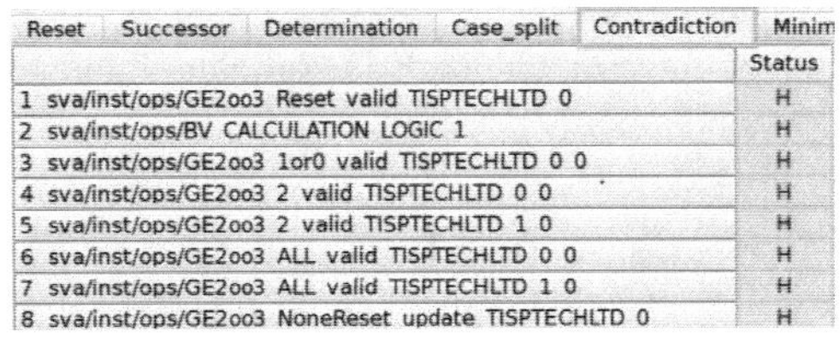

Figure 6. CERTIFY result for contradiction

Table 3. The supplementary functional requirement and SVA.

<table>
<tr><th></th><th>Functional requirement</th><th>SVA</th></tr>
<tr><td>7.</td><td>Reset: if i_rst_n is active, then the value of om is 2'b00 and the value of o_task_finish is 1'b0.</td><td>// **
property GE2oo3_Reset_valid_TISPTECHLTD;
@ (posedge i_clk)
|i_rst_n |= om_task2rte==2'b00 && o_task_finish<=1'b0;
endproperty : GE2oo3_Reset_valid_TISPTECHLTD
GE2oo3_Reset_valid_TISPTECHLTD_0: assert property (GE2oo3_Reset_valid_TISPTECHLTD);
// **</td></tr>
<tr><td>8.</td><td>status output signal: if i_rst_n is not active, the state of o_task_finish is same as the state of signal i_task_start.</td><td>// **
property GE2oo3_NoneReset_update_TISPTECHLTD;
@ (posedge i_clk) disable iff (!i_rst_n)
i_task_start |=> o_task_finish==$past(i_task_start);
endproperty : GE2oo3_NoneReset_update_TISPTECHLTD
GE2oo3_NoneReset_update_TISPTECHLTD_0: assert property (GE2oo3_NoneReset_update_TISPTECHLTD);
// **</td></tr>
</table>

Step IV: Use tool - VERIFY to perform functional verification and generate coverage information. The verification results, the coverage information and the counterexample are shown as "Figure. 7", "Figure. 8"and "Figure. 9". In the 1st round verification, there is an inconsistency between the code and the SVA. 21.05% statements and 14.29% branches in the code has not been reached.

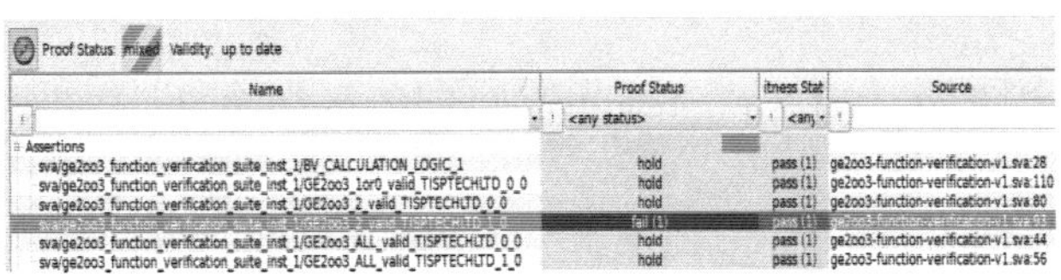

Figure 7. The 1st verification result

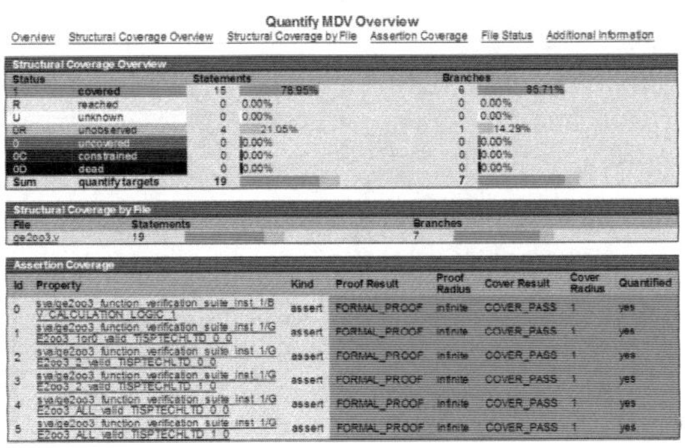

Figure 8. The 1st coverage information

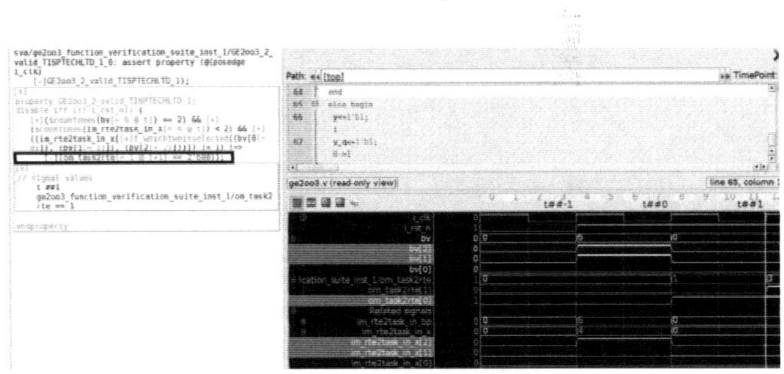

Figure 9. Counterexample waveform

The designers analyze and modify the code based on the waveforms and reports provided by VERIFY, and then submit the code to the verification engineers.

Step V: Use tool - VERIFY to perform regression verification. The final verification result and the coverage information are shown as "Figure. 10" and "Figure. 11". The results mean the code is consistent with the SVA, the code and the function domains are covered.

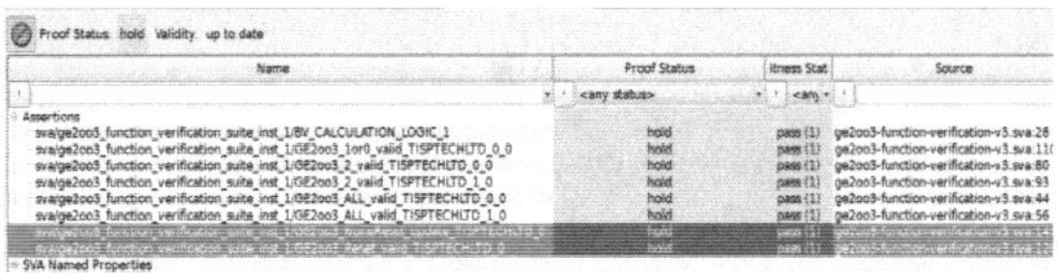

Figure 10. The regression verification result

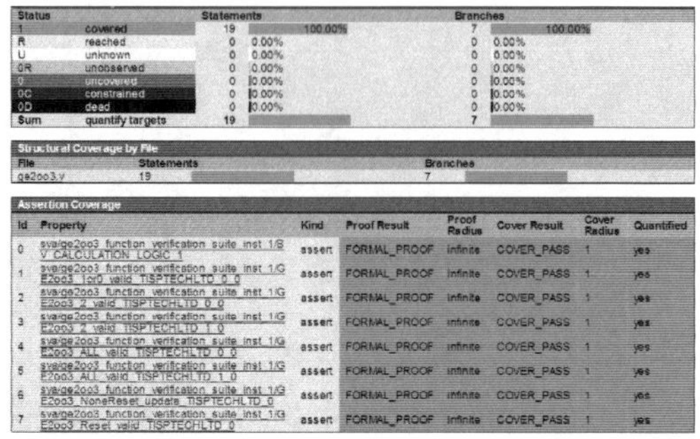

Figure 11. The regression coverage information

5. Conclusion

In this paper, an automated coverage method is proposed based on ONESPIN. Compare to traditional method, verification engineers can use ONESPIN verification platform in the coverage test process without manually construct test vectors, without manually analyze functional assertion's quality, without manually build verification environment. The whole verification process is highly automated and highly visible, which greatly reduces the verification time and effectively solves the common problems in the coverage test.

6. References

[1] Mentor Graphics, Questa SIM Reference Manual, 2011.

[2] Martinez A, Barron R, and Molina H 2014, "Automated functional coverage directed for complex system," VLSI-SOC.

[3] Harry Foster, "Trends in functional verification: a 2014 industry study," DAC, 2015.

[4] Hassan Sohofi, Zainalabedin Navabi, "Assertion-Based Verification for System-Level Designs," ISQED, 2014.

[5] Rott J S, Marriott P, and Graham M 2015, "Navigating the Functional Coverage Black Hole: Be More Effective At Functional Coverage Modeling," In *Design and Verification Conference(DVCon)*.

[6] OneSpin, User Manual: OneSpin 360™ Version 2016_12.

[7] OneSpin, OneSpin SystemVerilog Assertions Quick Reference Card.

Efficient block cipher mode for NVM

A Khodjanov[1], F Rustamov[1] and J Yun[1]

[1] Department of Computer and Information Security, Sejong University, 209 Neungdong-ro, Gawngjin-gu, Seoul 05006, South Korea

E-mail: jbyun@sejong.ac.kr (Corresponding author)

Abstract. The usage of non-volatile memory (NVM) storage devices is rapidly increasing in consumer and enterprise systems, providing high performance and low energy consumption compared to hard disk drives. Lately it is observed that NVM storage devices are widely used in cloud computing as well. However, existing disk encryption methods are designed for hard disk devices and an efficient block cipher mode of operation for NVM storage devices has not yet been studied in detail. In this work we have analyzed and evaluated performance of CBC, CTR and XTS-AES block cipher modes for whole disk encryption solutions and found that CTR and XTS-AES modes outperform CBC mode under the same scenarios.

1. Introduction

Storage technology is evolving very rapidly. The next generation of storage technology Non-Volatile Memory (NVM) has attracted more and more attention from both industrial and academic researchers [1]. NVM storage technology is just like HDDs, but more importantly has DRAM-like performance and byte-addressability. In NVM storage devices, the smallest write unit is one page (such as 4KB) and write operations can only be performed out-of-place, since data blocks have to be erased before new data can be written. Random writes especially can cause internal fragmentation of NVM storage devices and thus lead to a higher frequency of expensive erase operations. Besides performance degradation, the lifetime of NVM storage devices can also be dramatically reduced by random writes.

Security of data in storage devices is one of the main issues in computer security. Securing storage devices using encryption is a common method to deal with this problem. There are several disk encryption methods that are available, which include software-based encryption, controller based encryption and internal disk encryption. These methods provide solutions for encrypting data in motion and at rest. However, existing disk encryption methods are designed for hard disk devices, and a separate study on block cipher performance for NVM storage devices is needed.

Among many encryption/decryption standards, Advanced Encryption Standard (AES) is the most widely used symmetric cryptographic algorithm, which has been chosen as a secure encryption algorithm by the National Institute of Science and Technology (NIST) [2]. Existing encryption technologies like BitLocker [3] are using AES-CBC block cipher mode and cannot be applied to the NVM type of storage devices considering that it will impose security and performance degradation. Taking this into account, we have conducted an experimental study to measure the performance of different block cipher modes and tried to find the most efficient one for NVM.

IOP Conf. Series: Journal of Physics: Conf. Series **1026** (2018) 012005 doi:10.1088/1742-6596/1026/1/012005

2. Block ciphers for NVM

2.1. Block cipher

In cryptography, a **block cipher** is an algorithm that converts fixed-length groups of N bits, called *blocks*, to a cipher block of N bits with an unvarying transformation that is specified by a symmetric key [4]. Block ciphers operate as important elementary components in the design of many cryptographic protocols, and are widely used to implement encryption of bulk data. There are five modes of operation for AES. Each of these modes has different characteristics. The five modes are ECB (Electronic Code Book), CBC (Cipher Block Chaining), CFB (Cipher Feedback), OFB (Output Feedback), and CTR (Counter).

In this paper, we will cover CBC mode since it is well known and mostly used in disk encryption and proposed CTR mode. This mode is amenable to parallelization due to the fact that you can split the message into chunks, each chunk having a range of counter values associated with it, and encrypt (or decrypt) each chunk independently. In addition to this, we have proposed XTS-AES mode because this mode is the most common if you are encoding random accessible data (like a hard disk or RAM).

2.1.1. Cipher Block Chaining (CBC) mode

The Cipher Block Chaining (CBC) is a block cipher mode of operation that is known to be the most commonly used mode in disk encryption nowadays. This mode is a secrecy mode whose encryption procedure highlights the joining ("chaining") of the plaintext obstructs with past ciphertext squares [4]. Disk encryption in CBC mode is shown in Fig. 1, it may encrypt each sector with a different initialization vector (IV), which is an important constraint. In CBC mode, encryption is not conducted in parallel. This is a major reason for performance degradation compared with other modes.

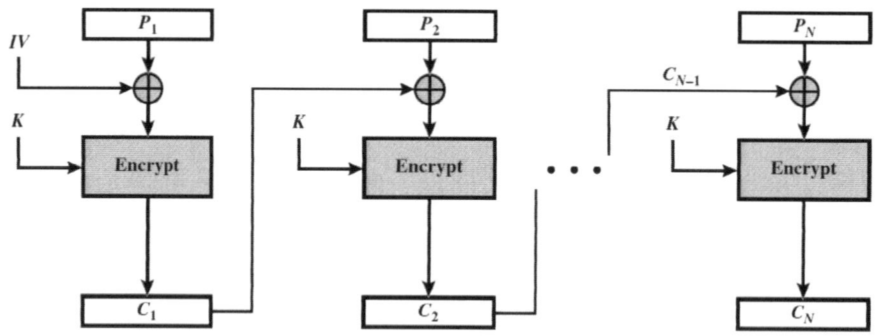

Figure 1. CBC mode (Encryption) [4].

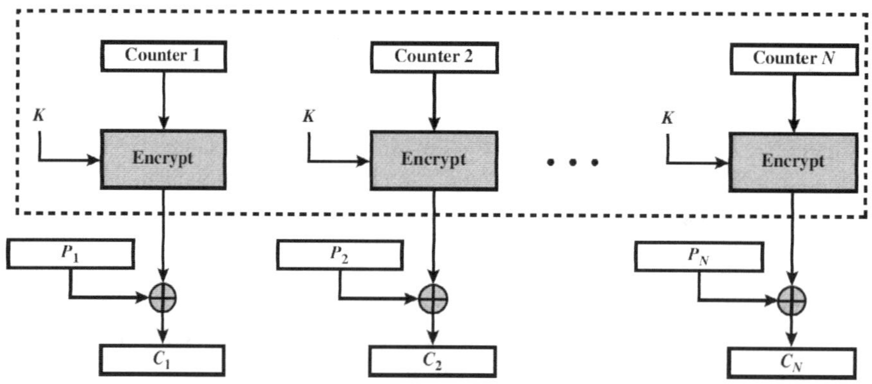

Fig. 2. CTR mode (Encryption) [4].

2.1.2. Counter (CTR) mode
Fig. 2 shows the encryption in CTR mode. Counter mode turns a block cipher into a stream cipher. It generates the next keystream block by encrypting successive values of a "counter". The counter can be any function that produces a sequence which is guaranteed not to repeat for a long time, although an actual increment-by-one counter is the simplest and most popular.

2.1.3. XTS-AES mode
The XTS-AES mode is based on the concept of a tweakable block cipher, introduced in [5][6]. Fig. 3 depicts the encryption of XTS-AES mode. XTS-AES mode assures that the same plaintext data unit will encrypt to two different ciphertext data units for two different data unit positions. This is because of the tweak of tweakable block cipher, so XTS-AES mode has better grade for security [7][8].

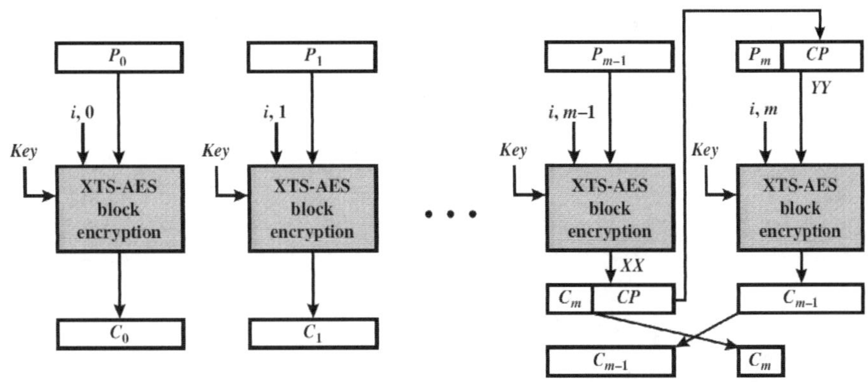

Fig. 3. XTS-AES mode (Encryption) [4].

2.2. Efficient block cipher module for NVM
Based on our study analysis it was found that CTR mode has more advantages of all other chaining modes approved by NIST. Unlike other modes of encryption and decryption, CTR mode can be done in parallel on multiple blocks of plain or cipher data. The encryption/decryption throughput is only limited by the amount of parallelism that is achieved and this mode can be efficiently implemented on the hardware level. CTR mode makes it possible to use very low latency encryption since it allows users to pre-calculate the key streams used in XOR functions. With CTR mode the *i*th block of plain or cipher data can be processed in random-access fashion, which is a very common access pattern for disk read/writes.

3. Performance evaluation
To assess different block cipher modes we have setup a Linux 4.4 machine. The synthetic benchmark platform consists of an Intel i5-4460 CPU and an 8GB DDR3 memory. Since commercialy available NVM is limited at the time of conducting our experiments, we used DRAM as a storage device. A well-known benchmarking tool Flexible IO (FIO) [9] was used in our performance measurements.

3.1. Benchmark results

Experimental results show that CTR mode and XTS-AES mode well outperform CBC mode in most of the cases we were interested in. Fig. 4 illustrates the results of sequential read and write operations. It shows that there is not much difference in performance for sequential read but in sequential write in CTR mode, it dramatically outperforms CBC mode for blocks larger than 4KB.

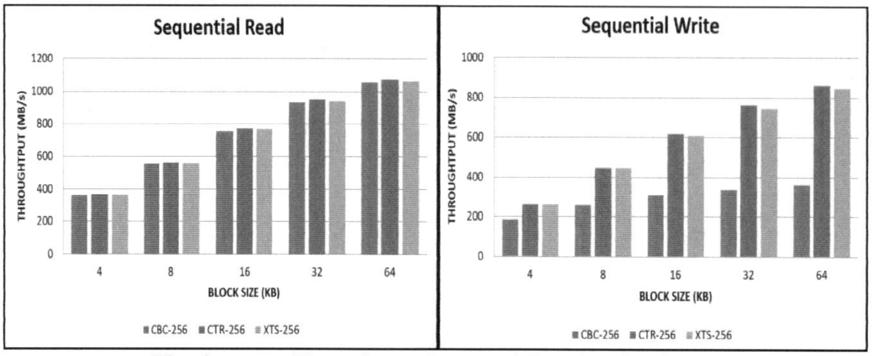

Fig. 4. Throughput of sequential read and write.

The second set of experiments compare throughput rates for random read and write. Fig. 5 illustrates that there is not much difference in random read performance as well, but again random write is far ahead in CTR mode and XTS-AES mode.

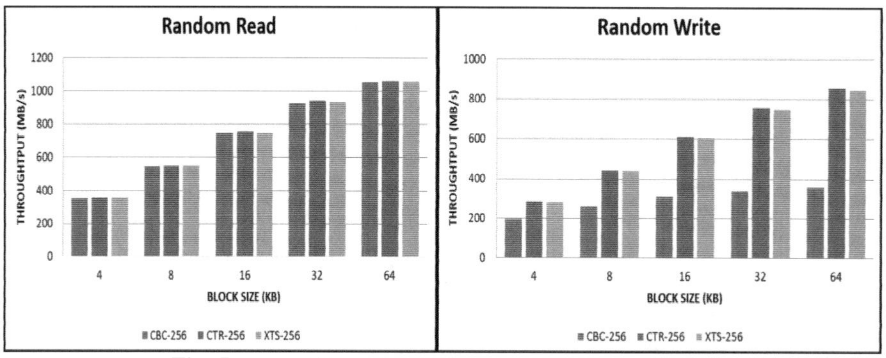

Fig. 5. Throughput of random read and write.

Our final set of experimental analysis was aimed to prove that CTR brings less latency on the encryption and CPU utilization when parallel encryption and decryption was enabled. Fig. 6 demonstrates much lower latency rates for CTR mode than in CBC mode. XTS-AES mode has 7% more overhead than CTR mode on average. The CPU high utilization in CTR mode illustrates that during synthetic benchmark all cores were used for encryption operations and therefore boost the throughput rate.

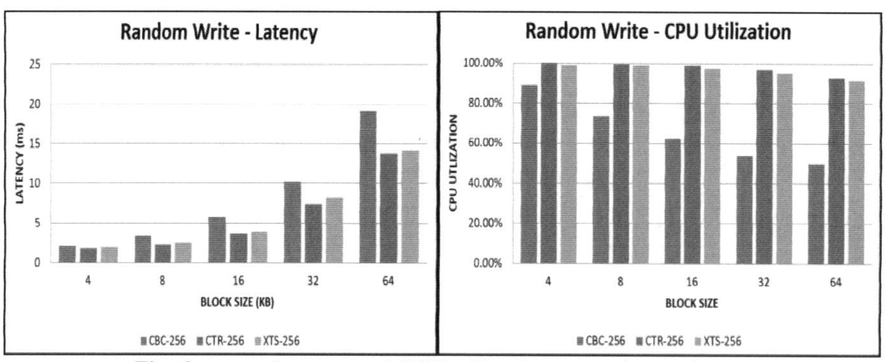

Fig. 6. Latency and CPU utilization on random write.

3.2. Security strength

CTR mode works faster than XTS-AES mode but CTR mode is malleable – an attacker can flip plaintext bits by simply flipping the corresponding ciphertext bits. XTS-AES is less malleable and it does the XOR operation twice on the record, once before encryption and once after the encryption, therefore it provides high-grade security. Although XTS-AES mode has 7% more overhead than CTR mode, we recommend to use XTS-AES mode because of its security. With technology advances in NVM development this speed drawback is very negligible. In addition, in the absence of authentication or access control, XTS-AES mode provides more protection than other approved confidentiality-only modes against unauthorized manipulation of the encrypted data [10].

4. Conclusions

In this paper, we have evaluated the performance of AES encryption algorithm with different block cipher modes to find the most efficient mode for NVM storage encryption. It is clear that CTR mode has performance improvement with lower latency. The study also illustrates CTR mode outperforms CBC mode due to the support of parallel encryption and decryption operations. Results indicate that there is only a negligible difference between CTR and XTS-AES modes. They reached almost the same percentage and number. Therefore, XTS-AES is the most suitable block cipher mode for NVM storage given the efficiency and protection.

5. Acknowledgments

This work was supported by the National Research Foundation of Korea (NRF) grant funded by the Korea government (MSIP) (No.2015R1C1A1A02036511).

6. References

[1] Chen C, *et al.* 2016 Fine-grained metadata journaling on NVM In *2016 IEEE 32nd Symposium on Mass Storage Systems and Technologies*, pp. 1-12 (DOI: 10.1109/MSST.2016.7897077).
[2] NIST: *Announcing the ADVANCED ENCRYPTION STANDARD (AES)*, Nov. 2001, http://nvlpubs.nist.gov/nistpubs/FIPS/NIST.FIPS.197.pdf
[3] Ferguson N 2006 *AES-CBC + Elephant diffuser A Disk Encryption Algorithm for Windows Vista*, Microsoft Corporation https://css.csail.mit.edu/ 6.858/2010/readings/bitlocker.pdf
[4] Stallings W 2010 *Cryptography and Network Security Principles and Practice* (Prentice Hall. Boston, 5th Edition, 192.
[5] Liskov M, *et al.* 2011 Tweakable Block Ciphers *Journal of Cryptology* 24(3), pp. 588-613 (DOI: 10.1007/s00145-010-9073-y).
[6] Minematsu K, et al. 2009 Generation and Extension of XEX Mode *IEICE Trans. Fundamentals of Electronics, Communications and Computer Sciences* **E92-A** 2 (DOI: 10.1587/transfun.E92.A.517).
[7] IEEE Computer Society: "IEEE Standard for Cryptographic Protection of Data on Block-Oriented Storage Devices," IEEE Std 1619™-2007 (2008), pp. 1-32, (DOI: 10.1109/IEEESTD.2008.4493450).
[8] Dworkin M 2010 Recommendation for Block Cipher Modes of Operation: The XTS-AES Mode for Confidentiality on Storage Devices *NIST Special Publication* 800-38E (DOI:10.6028/NIST.SP.800-38E).
[9] Flexible I/O Tester, https://github.com/axboe/fio
[10] Rogaway P 2011 Evaluation of Some Blockcipher Modes of Operation, Technical report, Cryptography Research and Evaluation Committees (CRYPTREC) for the Government of Japan.

IOP Conf. Series: Journal of Physics: Conf. Series **1026** (2018) 012006 doi:10.1088/1742-6596/1026/1/012006

Grasp control of a prosthetic hand through peripheral neural signals

E Noce[1], C Gentile[1], F Cordella[1], A L Ciancio[1], V Piemonte[2], L Zollo[1]

[1] Unit of Biomedical Robotics and Biomicrosystems, Università Campus Bio-Medico di Roma, Via Álvaro del Portillo 21, 00128 Rome, Italy
[2] Unit of Chemical-physics Fundamentals in Chemical Engineering, Università Campus Bio-Medico of Roma, Via Álvaro del Portillo 21, 00128 Rome, Italy

Abstract. The use of neural electrodes to stimulate the Peripheral Nervous System (PNS) of upper limb amputees is giving promising results in restoring tactile feedback. The same interfaces could be used to record the motor activity originated from the brain and transferred to the muscles. In this paper, the possibility to control a prosthetic hand by means of neural signals acquired through tf-LIFE4 electrodes implanted in a human subject was investigated. A Support Vector Machine (SVM) algorithm was adopted to classify two common demanded grasps. The obtained classes were converted into reference positions for a position-and-slippage control strategy that guarantees to perform stable grasps with a prosthetic hand avoiding slippage events. The achieved results showed an accuracy of the classifier higher than 90% and a success rate of the control strategy equal to 100%.

1. Introduction

In the development of prosthetic hands, to take into account the prosthesis user needs is of paramount importance. The analysis of user priorities [1] revealed that two of the most desired features are an intuitive and reliable interface and to perform stable grasps avoiding slippage events.

The most commonly adopted interfaces between the user and the prosthetic device are the electromyographic (EMG) sensors. In fact, commercially available prosthetic hands operate under EMG signals without providing any type of sensory feedback to the user [1], [2]. Invasive neural interfaces allow to restore the bidirectional communication with the central nervous system and to re-establish a physiological control of prosthetic hand thanks to the high level of selectivity. The activity of the nerve fibers can be detected from the fascicle originally innervating the missing limb and used to control an artificial hand [3][4]. The main problems of using the neural signal (ENG) to control a prosthetic hand are related to the low amplitude of the signal, the crosstalk with the muscular activity and the electrical interference in such low amplitude source [5]. All these problems reduce the quality of the ENG signal and make it difficult to extract user intentions from it.

In the literature, force control approaches were proposed to ensure stability during grasp. For instance, in [6], a PI force control was used with a velocity inner loop, in [7] a force control based on a neural network was developed to compensate the non-linearity of hardware components. The main limitations of these approaches are that they cannot hinder the possible object slippage. It was proved that the introduction of slippage information improves the grasp stability [8], [9], [10].

This paper aims at i) investigating the use of ENG signals to control a prosthetic hand by adopting the same pattern recognition techniques applied to EMG signals; ii) proposing and testing on a real prosthetic hand, i.e. the IH2 Azzurra, a position-and-slippage control approach able to move the hand in the desired configuration managing slippage events. The motion commands (i.e. classes related to power and pinch grasps) obtained from the pattern recognition algorithm are converted into reference positions to be given in input to the control strategy.

The paper is structured as follows: in section 2, the methods adopted for collecting and processing data and the control strategy are described; in section 3, the results are shown and discussed. Finally, in section 4 conclusion and future work are reported.

2. Methods and materials

2.1. Data recording and ENG classification

ENG data were acquired from tf-LIFE4 intraneural electrodes inserted longitudinally in the median and ulnar nerve of a male amputee during a human experimentation that took place in Rome in 2009 [3]. The study was approved by the local Ethics Committee and the competent Ministry. Data were recorded by means of the Grass QP511 Quad AC, amplified by 10000 and filtered between 100 Hz and 10000 Hz. The sampling frequency was set to 48000 Hz. Data acquired during the execution of power and pinch grasps were used in the following analysis.

In order to compute the envelope of the neural recordings, the data were filtered with a 4-th order Butterworth filter between 700 Hz and 5000 Hz [3]. A spike detection algorithm was adopted to identify the action potentials inside the recordings. To enhance the presence of spikes inside the recording, the energy E is computed as

$$E[i] = \frac{1}{W} \sum_{i-\frac{W}{2}}^{i+\frac{W}{2}} (x_i - x_m)^2$$

where W is the length in samples of the window in which the energy is computed and xm is the signal mean value [11]. In this work, the window W is set to 120 samples. Secondly, an amplitude threshold was used to detect the spikes. It was computed by means of the standard deviation of the entire signal, calculated from the Mean Absolute Deviation [12], multiplied by a constant factor set to 9 [13]. The W and the constant factor values are experimentally retrieved to reduce false positives and false negatives thus improving the algorithm performance. Once the amplitudes A and the time bins T of the spikes were saved, they were used to compute the envelope of the ENG according to the formula

$$eENG[i] = \frac{\alpha A_i + \beta A_{i+1}}{2} e^{-\frac{T_{i+1}}{T_i}}$$

where α and β are weight factors between 0 and 1. In [13] it was shown that, imposing $\alpha = 0.5$ and $\beta = 0.4$, it is possible to highly correlate the EMG and ENG envelope concurrently recorded. The ENG was classified by means of the Support Vector Machine (SVM) algorithm implemented in Matlab using the libsvm 3.22 library, available online. The Root Mean Square (RMS) features were extracted from a windowed portion of the recorded signals. The obtained vector of features was segmented in two sets, the Training (TRs = 75%) and the Cross-Validation set (CVs = 25%). The SVM was optimized varying the C and γ parameters from 2^0 to 2^{10} and from 0.1 to 3 respectively [14]. Every time the training is performed by means of the TRs with a defined pair (C, γ), the generated model is tested via the CVs. The pair (C, γ) that guarantees the highest accuracy is applied to the testing data.

2.2. Control law

A position-and-slippage control strategy was implemented. The control scheme adopted for each finger of the prosthetic hand is shown in Figure 1. The control aims at using the class from the classifier for moving the hand fingers to reach the desired position (a priori known) [15] by online compensating for possible object slippage. The information about the object slippage are obtained by applying the slippage detection algorithm proposed in [10] on the voltage information obtained by Force Sensing Resistor (FSR) sensors positioned on the thumb and index fingertips. The control approach has been adapted for underactuated fingers. The voltage measured by the FSR sensors was converted in force values after an appropriate sensor calibration. These force signals were constantly monitored during the trials in order to avoid breaking the grasped object.

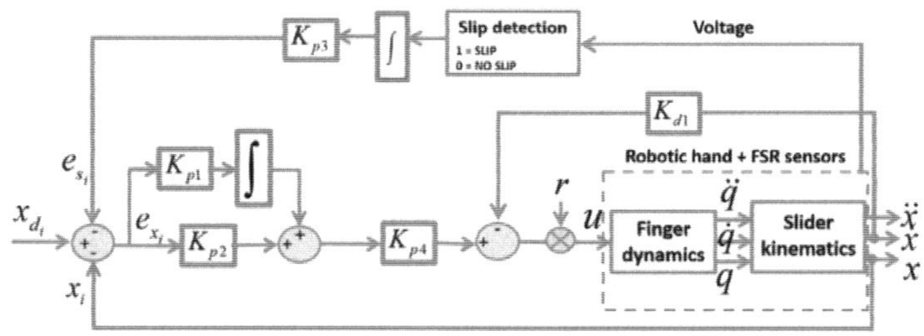

Figure 1. Block scheme of the control strategy applied to one underactuated finger.

In Figure 1, x_{d_i} represents the reference position to be reached by the i-th hand finger, e_{x_i} is the position error for the i-th finger slider obtained from a Proportional-Integrative (PI) position control

$$e_{x_i} = x_{d_i} + e_{s_i} - x_i \ ,$$

x_i is the actual position derived from the hand position sensors, K_{p1} and K_{p2} are the controller gains, K_{p3} is a constant that regulates the weight a slip-dependent additional contribution e_{s_i} has in the control. This additional contribution has the aim of managing slippage events [8], and is expressed as

$$e_{s_i} = K_{p1} \int_0^{t_f} \alpha \, dt.$$

α is the binary signal equal to 0 if slip does not occur and equal to 1 when slip occurs and t_f is the final integration time. The integration of this signal will guarantee a smooth increment of the applied grasping force in presence of slippage, as demonstrated in [8]. The obtained position error e_{x_i} should be led to zero by a Proportional-Derivative (PD) control in the slider space described by the equation

$$u = r\left[K_{p4} e_{x_i} - K_{d1} \dot{x}\right],$$

where r is the vector of the pulley radii, $K_{p4}$is the proportional gain, K_{d1} is the derivative gain and x, $\dot{x}$ and $\ddot{x}$ are the slider position, velocity and acceleration, respectively [16].

2.3. Setup and experimental protocol

The proposed architecture was tested on the IH2 Azzurra anthropomorphic robotic hand [17] (Figure 2a). The 5-finger robotic hand has 11 Degrees of Freedom (DoFs), 5 of which are active. A slider located in the hand palm is responsible for the Flexion/Extension (F/E) of each underactuated finger, whereas a universal joint is used for linking the thumb Adduction/Abduction (A/A) motor, positioned in the palm, to the TM joint. The finger position is measured by means of an incremental encoder.

FSR sensors were glued on the thumb and on the index fingertips that were covered with silicon material simulating a cosmetic glove (Figure 2a). In particular, the silicon material was molten on a 3D printed model of the fingers. The obtained silicon cups were placed onto each fingertip to cover and protect the surface of the FSRs. The slippage events were detected in real-time by applying the slippage detection algorithm proposed in [10] to the voltage signal acquired by the FSR sensors positioned on the thumb and index fingertips.

The testing sets are represented by 10 acquisitions (named T1 to T10) in which the power grasp and pinch grasp were randomly performed. To increase the performance of the classifier, a voting system was applied. Instead of classifying one sample per time, the predicted class is obtained by counting the highest number of the predicted class in a span of 5 predictions. Even if it slows the classification, it is a way to improve dramatically the performance of the classifier.

The obtained classes are converted in reference position to be given in input to the prosthesis control. The control strategy was experimentally tested on the IH2 Azzurra prosthetic hand. The following objects were grasped by the hand with a pinch and a power grasp, respectively: a ball of 35mm diameter and a glass of 75 mm diameter. The objects to be grasped were located in front of the hand, which was positioned on a support. Once grasped, 4 slippage events were induced to the grasped object by applying an impulsive force to the objects. Ten repetitions of the same grasping action were performed. The starting A/A angle of the thumb was imposed on the basis of the human being behaviour [18]. In particular, for the pinch grasp, it was set to 1.656 rad and for the power grasp it was set to 1.845 rad. The finger controller was implemented in C programming language under Windows 10. The communication between the hand and the PC was performed by means of an USB port. The gains of the position control were imposed equal to $K_{p1} = 1$, $K_{p2} = 4$, $K_{p3} = 80$, $K_{p4} = 0.1$, $K_{d1} = 0.004$.

3. Experimental results

The training process of the classifier was able to identify a pair (C,γ) that maximizes the accuracy of the CVs. In particular, the pair chosen and later used for the Testing set is (2^{10}, 0.47). Once the model for the SVM was generated with these values, the classification algorithm was applied to the Testing sets. The accuracy ranged from 86.67% (T2) to 92.67% (T5). The total accuracy of the classifier across the tests is 90.13% ± 2.24%.

In Figure 2 and Figure 3, the results obtained on two representative grasping trials (corresponding to the two classified grasping configurations, i.e. pinch and power grasps) are shown.

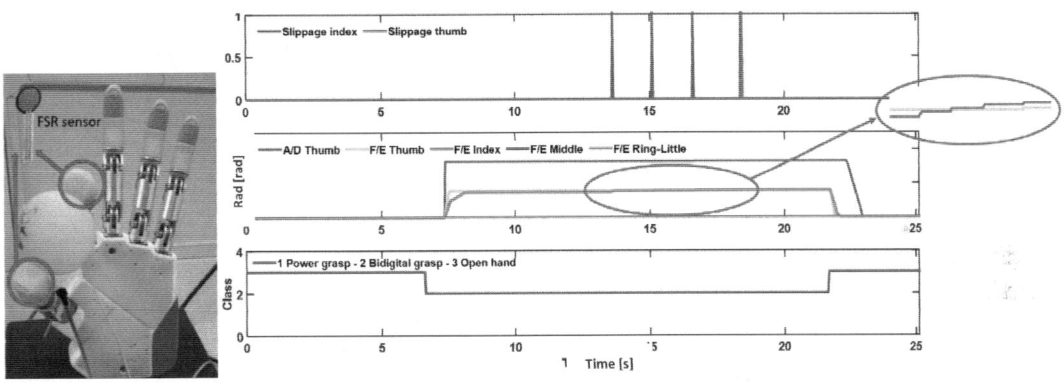

Figure 2. (a) Grasping configuration. The position where the FSR sensors are located are encircled in red. (b) Detected slippage events (the slippage event detected by the sensor positioned on the index finger is shown in blue, whereas the one detected by the thumb is in red), hand joint angles (the different colours correspond to the different fingers), grasping class (i.e. output of the classification algorithm).

In particular, the grasping configurations are shown in Figure 2a and in Figure 3a. In Figure 2b and in Figure 3b, the detected slippage events, the hand joint angles and the grasping class are reported for the pinch and the power grasp, respectively. It is important to outline that grasping forces are acquired from the sensors placed on both the fingers, i.e. thumb and index. From the reported plots it is possible to note that, when the object slippage is detected by the sensor embedded in the index and/or in the thumb fingertips (outlined in blue and in red in the top of Figure 2a and of Figure 3a, respectively), a variation of the F/E angle of the index Metacarpophalangeal (MCP) joint and of the thumb Trapeziometacarpal (TM) joint occurs, as evident in the zoomed part of the signal in Figure 3b and in Figure 4b. The 20 grasping trials were performed with a success rate (i.e. ratio between the number of compensated slippage events and the number of trials) of 100%. Therefore, it is possible to conclude that the proposed control approach guarantees a stable grasp also in presence of unforeseen events.

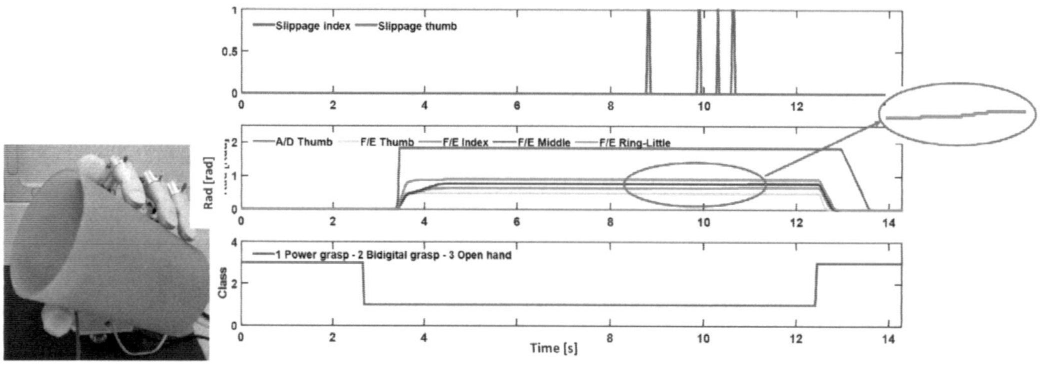

Figure 3. (a) Grasping configuration. (b) Detected slippage events (the slippage event detected by the sensor positioned on the index finger is shown in blue, whereas the one detected by the thumb is in red), hand joint angles (the different colours correspond to the different fingers), grasping class (i.e. output of the classification algorithm).

4. Conclusion

In this paper the feasibility to use neural signals to control a prosthetic hand was demonstrated. The neural signals were processed to obtain their envelopes in order to reduce the variability of the raw signals. The proposed approach allows the ENG envelope to be treated as the EMG one, a signal that is more commonly used in pattern recognition. The classified gestures (i.e. power and pinch grasps) were given in input to a position-and-slippage control architecture applied on a prosthetic hand. The obtained results showed an accuracy of the classifier above 90% and a success rate of the proposed position-and-slippage control strategy of 100 %. Future work will be devoted to increase the number of detected and classified hand gestures and to include the patient in the control loop by giving him/her the possibility to actively control the grasping force.

5. Acknowledgments

This work was supported partly by the Italian Institute for Labour Accidents (INAIL) with PPR2 (CUP: E58C13000990001), RehabRobo@work (CUP: C82F17000040001), PPR AS 1/3 and PCR 1/2 (CUP: E57B16000160005) projects and partly by the European Project H2020/AIDE (CUP: J42I15000030006).

6. References

[1] Cordella F, et al. 2016 *Front. in Neur.* vol. **10**, no. 209. DOI: 10.3389/fnins.2016.00209
[2] Ciancio A L, et al. 2016 *Front. in Neur.* vol. **10**, no. 116. DOI: 10.3389/fnins.2016.00116
[3] Rossini P M, et al. 2010 *Clin. Neur.* vol. **121**, pp 777-83
[4] Dhillon G S, et al. 2004 *J. of Hand Surg.* vol. **29**, pp 605-15
[5] Micera S, et al. 2010 *IEEE Review in Biom. Eng.* vol. **3**, pp. 48-68
[6] Jo J, Kim S K, et al. 2012 *IEEE Conf. on Aut. Scien. and Eng.*
[7] Pasluosta C F, et al. 2012 *Ass. Techn.* vol. **24**, pp. 196–208
[8] Engeberg E D, et al. 2012 *IEEE/ASME Trans. on Mech.* vol. **18**, pp. 376–385
[9] Sriram G, et al. 2014 *IEEE Conf. on El./Inform. Techn.* pp. 225–230
[10] Cordella F, et al. 2016 *IEEE Int Conf Robot Autom*
[11] Rutishauser U, et al. 2006 *J. of Neur. Meth.* vol. **154**, pp. 204-24
[12] Quiroga R Q, et al. 2004 *Neur. Comp.* vol. **16**, pp. 1661-87
[13] Noce E, et al. 2016 *IEEE Int Conf Biomed Robot and Biomech*
[14] Hsu C W, et al. 2003 *Technical report, National Taiwan University*

[15] Cordella F, et al. 2012 *Int J Interact Des Manuf* 6:113. DOI: 10.1007/s12008-012-0149-9
[16] Zollo L, et al. 2007 *IEEE/ASME Trans. on Mech.* vol. **12**, pp. 418–429
[17] http://www.prensilia.com/index.php?q=en/node/40
[18] Flanagan J R, et al. *Exp. Brain Res.* vol. **95**, pp. 131–143

IOP Conf. Series: Journal of Physics: Conf. Series **1026** (2018) 012007 doi:10.1088/1742-6596/1026/1/012007

Dynamic Mechanical Analysis of a Power Electronic Gate Driver Board

Thomas Schriefer[1] **and Maximilian Hofmann**[2]

[1] Chair of Electron Devices, Friedrich-Alexander University Erlangen-Nuremberg, Cauerstraße 6, 91058 Erlangen, Germany
[2] Fraunhofer Institute for Integrated Systems and Device Technology IISB, Schottkystraße 10, 91058 Erlangen, Germany

E-mail: thomas.schriefer@fau.de

Abstract. Vibration-induced loads stress electronical equipment particularly in automotive applications. If these high cycle oscillations excite the natural frequencies of electrical devices or circuit boards, malfunctions by material fatigue are likely to occur and the functional performance of the electrical system is affected. The assessment of the load impact, which is typically initiated by random noise of frequencies up to 2000 Hz, demands the full description of the system's dynamic-mechanical behaviour. In this paper a multi-dimensional approach of analytical, numerical and experimental methods is presented to evaluate the vibrational resistance of a power electronic circuit board applied in an automotive drive inverter.

1. Introduction

The advancing electrification of the automotive powertrain leads to an increased use of electrical components in harsh environments. Electrical functionalities are shifted to the site of action in order to reduce mechanical interfaces and to increase the degree of integration. As the new environmental conditions are often neglected in the development process, product life expectancy is likely to be incorrectly predicted. An electrical drive with integrated power electronics that is mounted in the unsprung wheel-hub is a typical example. Beside thermal stresses, the mechatronic system is affected by vibrational loads, which are caused by road undulations and rotational engine forces. Figure 1 shows that in unsprung positions electronic systems face significantly higher acceleration forces in comparison to sprung masses. If the mechanical resonances of the electrical components are excited, acceleration fields stress solder joints. Typical failure modes are high-cycle fatigue by material ageing, brittle fracture by exceeding static load limits and delamination between material interfaces. Similarly to electrical devices, electrical connectors are subjected to oscillatory loads, whose micro-movements rupture the refinement of the mating surfaces. As a result of this, electrical resistance is increased and the transmitted electrical signal is likely to be modified, which is particularly critical in safety-related applications.

Due to the high number of physical effects and geometrical details, dynamic-mechanical characterization of electronic systems is very complex. Beside masses and stiffness factors, damping constants must be known, which are different for every setup and cannot be taken from generic databases. In most cases, analyses are made for circuit boards or electronic devices [1-4] to estimate the general vibration resistance of printed circuit boards (PCB) or to explicitly determine solder

IOP Conf. Series: Journal of Physics: Conf. Series **1026** (2018) 012007 doi:10.1088/1742-6596/1026/1/012007

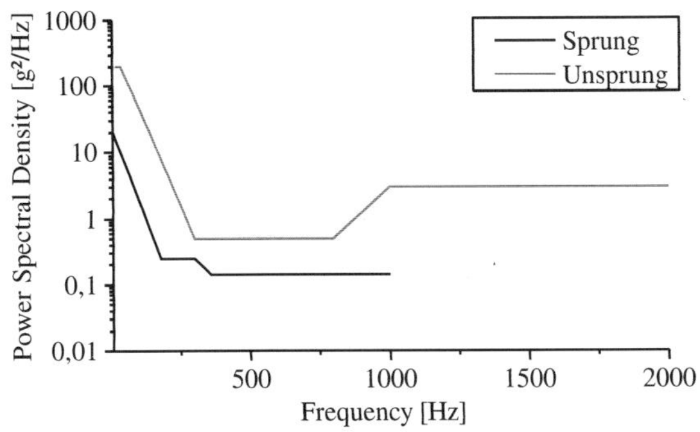

Figure 1. ISO16750-3 load profiles for electronic equipment [2]

Figure 2. Experimental setup comprising shaker and device under test

fatigue by experiments and numerical simulation. Typically, the numerical simulations are developed in the frequency domain [5, 6], which only considers linear material behaviour. [7] concludes that these models provide good results as long as the dynamic input parameters are known to the full extent. The experimental procedure to determine these modal parameters is described in [6], while the possibility to describe electric devices by the theory of oscillations is suggested in [8, 9]. The publications have in common that they do not perform holistic analyses with experimental, numerical and analytical simulations.

In this publication, the gate driver of an automotive drive converter is dynamically characterized. The driver provides the controlling current for the power switches converting direct (DC) to alternating current (AC). Mechanical vibrations stressing the electrical interfaces can lead to system malfunction. The driver is therefore experimentally characterized to extract material parameters, before analytical and numerical tools are provided.

2. Dynamic mechanical investigation

In this section the multi-dimensional analysis of the power electronic circuit board is differentiated in experimental measurements to determine frequency responses and modal parameters, numerical model verification and discrete analytical modelling applying the theory of mass oscillations.

2.1. Experimental Reference

Experimental investigations are conducted by an electrodynamic shaker with zero-point position controls (Figure 2). The specimen is horizontally connected to the ceramic shaker head according to the real mounting position. The frequency response is discretely recorded by acceleration sensors, which are each positioned on the PCB and the exposed transformer. A third sensor is placed on the Aluminum mount to control the excitation system. As a first step, a noise analysis with 13.7 g_{RMS} input acceleration between 50 Hz and 6000 Hz determines the frequency response of the two measurement sensors. By randomizing frequencies potential resonances are excited at the same time and measurement times are significantly reduced. According to Figure 3 the frequency response indicates three resonances at (626 | 710 | 1777) Hz for the transformer device and (654 | 704 | 1764) Hz in case of the PCB. As the first two resonances are close together and coincide at higher excitation amplitudes, the characteristic curve shape of a two-masses-oscillator's frequency response is visible below 2000 Hz. While the resonance peaks are similar both for the PCB and the device, differences are obtained regarding the anti-resonance at 1551 Hz (device) and 980 Hz (circuit board). This is due to the fact that the elastic PCB resists less vibration-induced loads in comparison to the soldered transformer.

As a second step, the excitation signal is modelled harmonically to quantify the acceleration amplitudes. While random noise analysis resembles a test on frequency, harmonic signals are used to test amplitudes. Considering a load rate of 400 Hz/min and varying input signal (5, 10, 15 g), Figure 4

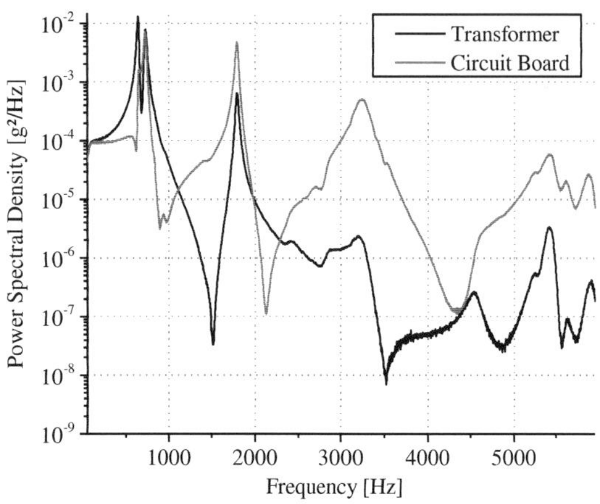

Figure 3. Random vibration response of the gate driver

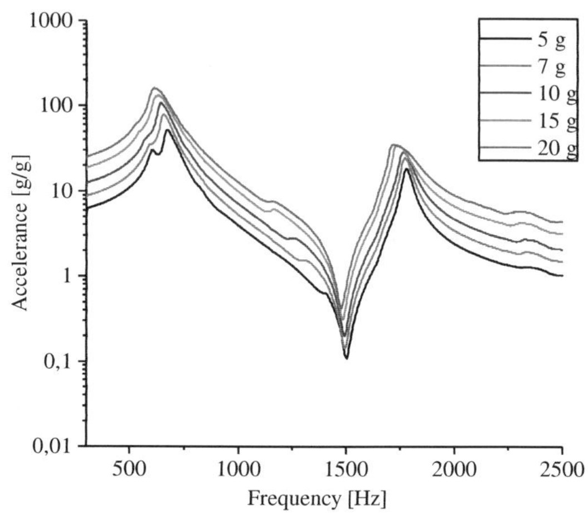

Figure 4. Sinusoidal analysis with varying amplitude

illustrates the amplitude responses, which are defined as dimensionless transfer accelerances (mm^2/mm^2) and referenced to the shaker head. The results show that the frequency response is not influenced by the excitation amplitude. Only at the second resonance, a slight but negligible shift to a lower frequency, i.e. digressive behaviour, can be obtained. Furthermore, beyond 10 g acceleration the double peak at the first resonance disappears. Here the device responds with tenfold amplitudes (100 m/s^2), which are likely to be excited by relevant automotive load profiles (Figure 1) and to cause the mentioned failure mechanism. The harmonic sweep were executed both forward (50 Hz → 2500 Hz) and backward (2500 Hz → 50 Hz) to identify nonlinear dynamic behaviour. Moreover the measurements indicate that the frequency response is independent of the excitation direction. Both forward and backward sweep curve identically, which is shown in Figure 5. Having proofed the independence of excitation amplitude and direction, linear dynamic behaviour can be assumed for the following analytical and numerical investigations.

2.2. Analytical Dynamics

Applying the theory of oscillations, the gate driver is analytically approached to increase theoretical knowledge and to proof that vibration resistance can be evaluated with minimal effort. Due to the experimental frequency response with two characteristic resonance peaks, the specimen is interpreted as a two-masses-oscillator (Figure 6). Both PCB and device are modelled as a point mass, which are connected with spring-damper elements with stiffness $k_{1/2}$ and damping constant $d_{1/2}$. The frequency response of a generic N mass oscillator is calculated as follows [10]:

$$\boldsymbol{H}(\Omega) = [-\Omega^2 \boldsymbol{M} + j\Omega \boldsymbol{D} + \boldsymbol{K}]^{-1} \tag{1}$$

In the case of a two-masses-oscillator matrix $\boldsymbol{H}$ is sized $2x2$. The imaginary part of the complex transfer function, which represents the ratio of response and excitation, contains phase information, while response amplitude is embedded in the real part. Matrix component $\boldsymbol{H_{11}}$ is the isolated response of the device, while $\boldsymbol{H_{12}}$ is the response of the device with respect to the PCB. The unknown components in the damping matrix are extracted from the experimental measurements by using the method of half-power points. The theory assumes that the dissipated oscillation energy is proportional to the square of the response amplitude:

$$A_{half} = \frac{A_{max}}{\sqrt{2}} \tag{2}$$

The so-called Lehr's damping ratio D is featured by the corresponding frequencies of the half-power-points:

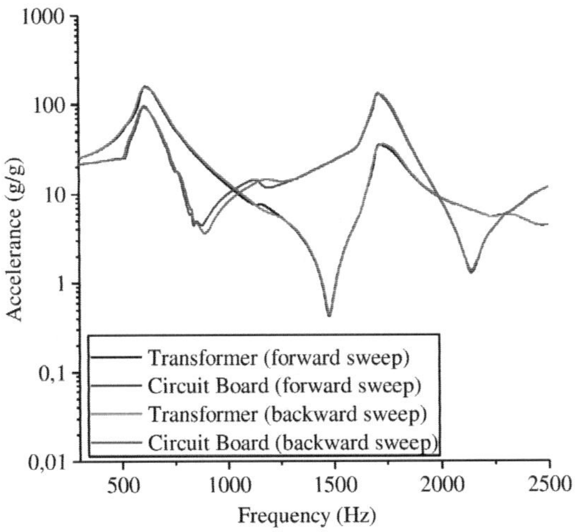

Figure 5. Frequency responses for positive and negative sweep

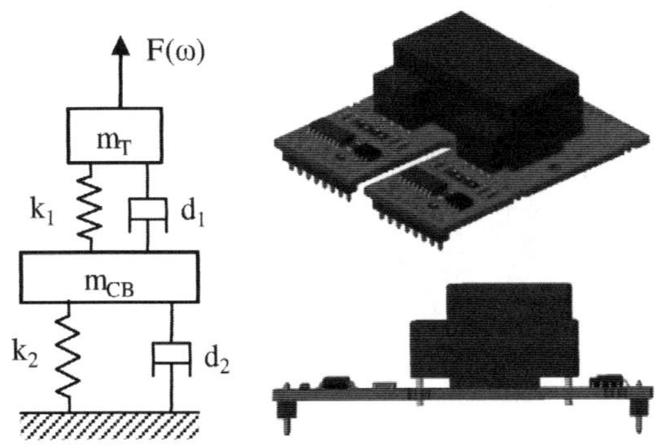

Figure 6. Analytically derived transfer functions

$$D = \frac{f_2 - f_1}{2f_0} \tag{3}$$

Having calculated the decay constant δ from this damping ratio and the excitation angular frequency ω_0

$$\delta = D\omega_0 \tag{4}$$

the damping constant is obtained by the following equation with mass m:

$$d = 2\delta m \tag{5}$$

Stiffness parameters k are empirically determined from the experiments by iteratively comparing analytical and experimental resonance frequencies. The absolute values $|\boldsymbol{H}(\Omega)|$ of these complex analytical response functions are shown in Figure 7.

Comparing the curvature of the matrix components with the measurements, it is evident, that the transfer function $\boldsymbol{H}_{11}$ is most similar to the experimental frequency response. Both the first resonance and the anti-resonance are present, only the second resonance is less sharply formed. The transfer function $\boldsymbol{H}_{22}$, which describes the excitation of the PCB, shows an anti-resonance, which is broadened to higher frequencies. Probably this is based on the relative large self-damping of the elastic FR-4 composite material. $\boldsymbol{H}_{12}$ and $\boldsymbol{H}_{21}$ indicate feedback effects, i.e. PCB and transformer influence each other, which is well dissolved by the analytical model. In general it can be stated that the dynamic behaviour of the first mass (transformer device) is directly transmitted to the second mass (PCB), which is due to the high stiffness factor.

2.3. Numerical Investigations

Using the finite element method, the experimental setup is numerically reproduced. The computational analysis is not performed in the time but frequency domain in order to minimize simulation time. However, this procedure requires the proof of linear dynamic behaviour, which was made in the previous experimental investigations. A verified finite element analysis (FEA) provides the possibility to conduct design studies by varying geometrical (e.g. PCB thickness, device position), material (e.g. PCB stiffness) or boundary conditions (e.g. mount). Furthermore, resonance shapes can be visualized and assigned to the identified frequencies. The numerical model is developed in the FEA software ANSYS V17 and shown in Figure 8.

The model considers PCB layers and electrical devices. Knowing mass and volume, the material properties of the devices are assumed to be isotropic and approximated as surrogates. SMD devices are modelled as mass points to maintain a uniform mesh, which neglects stiffening effects of the PCB.

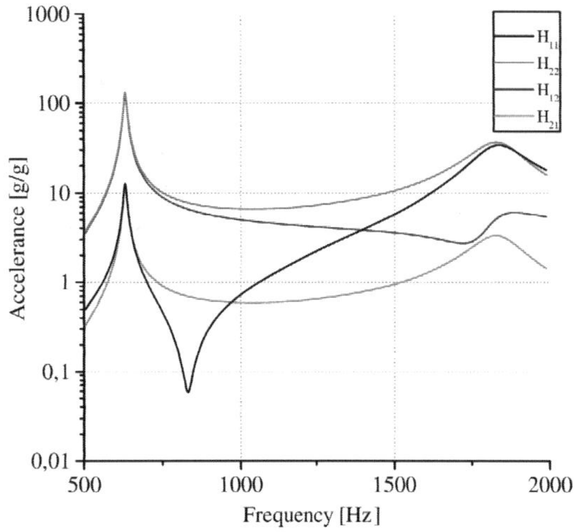

Figure 7. Analytical approach of the mass oscillator

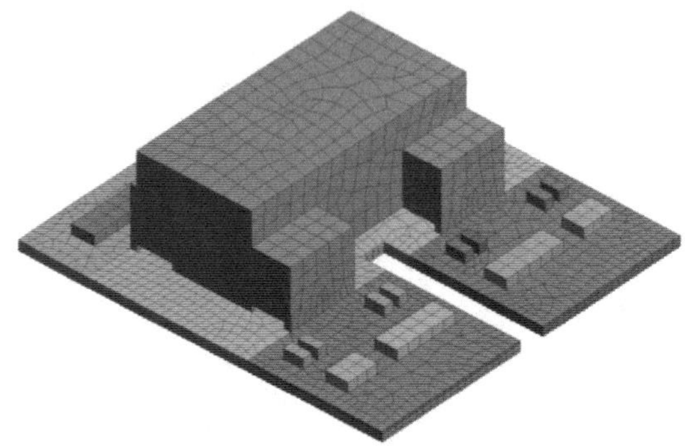

Figure 8. Finite element model of the gate driver

Using hexahedral finite elements with 11053 nodes, the component is meshed conformly, i.e. nodes are shared between material boundaries, and contact conditions do not need to be considered. Mass and stiffness configurations are taken into account according to the layer structure of the circuit board and the device positions. Solder joints are not explicitly modelled, since nonlinear material behaviour is automatically linearized in frequency response analysis. Element nodes of the underside pin headers are restricted in movement. Solder stresses are assessed at the contact surfaces of devices and circuit board. The FEA is calculated with modal superposition and the used material properties are listed in Table 1.

Table 1. Material properties for the numerical model

Component		Density [kg/m³]	Young's modulus [Pa]	Poisson's ratio
Transformer		1888	$1.6 \cdot 10^{9}$	0.402
Resistors		6955	$1.6 \cdot 10^{9}$	0.390
Capacitors		6851	$1.6 \cdot 10^{9}$	0.380
Circuit Board	FR-4	1850	$2.4 \cdot 10^{10}$	0.136
	Copper	8312	$1.1 \cdot 10^{11}$	0.340

Table 2 illustrates the dynamic shapes of the identified resonances. Hereby, the coloration only reflects relative movements because initially no load is defined. According to modal shape 1 and 2 the double peak consists of two bending shapes, where either the left or the right leg of the PCB is bent. Increasing the excitation amplitude closes the frequency difference and merges these two shapes. The second natural shape at 1756 Hz (modal shape 3) is bent symmetrically.

Table 2. Natural shapes of the driver board

Mode 1	Mode 2	Mode 3
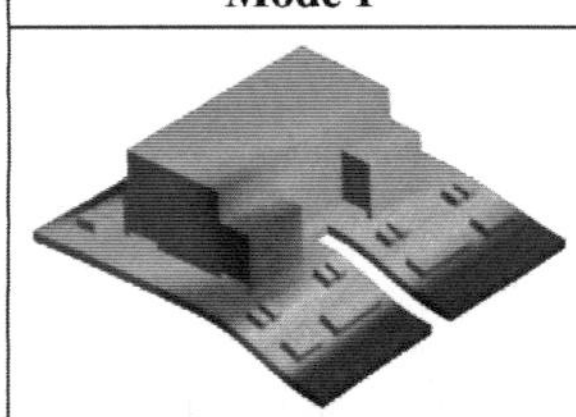	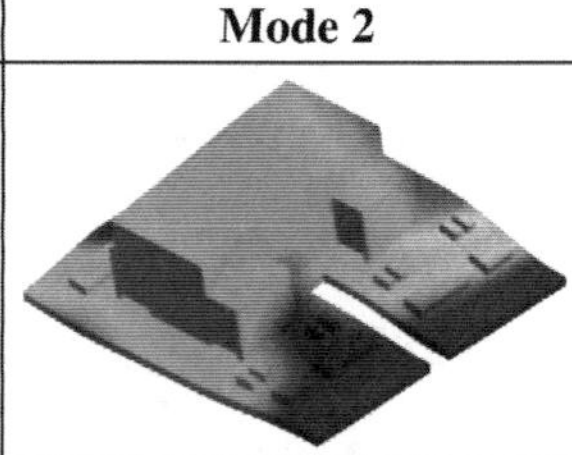	
625 Hz	741 Hz	1756 Hz
Asymmetric	Asymmetric	Symmetric

IOP Conf. Series: Journal of Physics: Conf. Series **1026** (2018) 012007 doi:10.1088/1742-6596/1026/1/012007

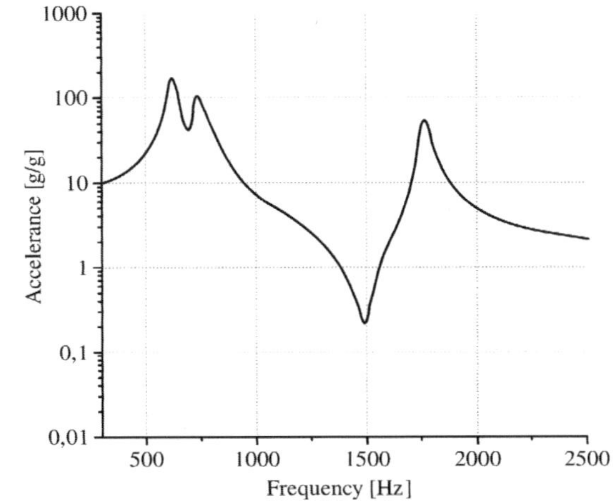

Figure 9. Numerical frequency response

Defining a sinusoidal load of 5 g at the pin headers, the numerical frequency responses are similar to the experimental results (Figure 9). The averaged responses are extracted in accordance with the experiment at the front edge of the PCB and the top surface of the transformer, while the global damping ratio is set to 2 %. Again, the characteristic response curve of a two-masses-oscillator is noticeable. The resonance frequencies and amplitudes match the experiment results very well.
Finally, a power spectral density analysis is conducted to assess the statistical stress response. The load is defined according to the DIN16750-3 vibration profile for the unmounted position and induced at the pin headers. The statistical noise signal is orientated orthogonal to the PCB. Apparently 3-σ stresses correlate with the circuit board's degree of bending and do not indicate plastic deformation (Table 3). However, high-cycle fatigue is characterized by low amplitudes and high frequency elastic deformation causing material failure in the long term. The maximum stress value in unsprung environment is at the transformer contacting and values 25.5 N/mm², which is 86 % higher than in sprung mount position. Applying *Steinberg*'s Three-Band-Technique [11] a damage of 24.2 % is expected after 365 days of load stress. In sprung condition this value reduces significantly to 1.1 %. This reliability knowledge can already evolve from the design process and be used to increase the system robustness. For instance, considering the low excitation signals (Figure 1) for middle frequencies, system performance can be improved by subtly adjusting mass and stiffness configuration and shifting resonances to lower amplitudes.

Table 3. 3-σ Stress configuration for sprung and unsprung mount position

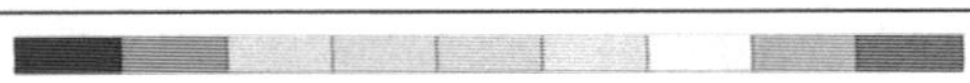 0 3 6 9 12 15 18 21 24 27 $\frac{N}{mm^2}$	
Sprung **Max.** 13.7 **Min.** 1.1 	Unsprung **Max.** 25.5 **Min.** 0.2 

3. Conclusion

In this paper a power electronic circuit board was dynamic-mechanically addressed to derive its degree of vibrational resistance. The described procedure is applicable to early evaluate system robustness of mechatronic products in the development phase. Using a multi-methodical approach of experimental measurements, numerical modelling and analytical description, natural frequencies and shapes were determined with consistent results. Through the interaction of these three methods, natural shapes could be assigned to the corresponding frequencies, material parameters were extracted and the analytical model of a mass oscillator was developed. The frequency response analysis with sinusoidal and random excitation signals were complemented with the numerical calculation of the resulting stress configuration.

Acknowledgement
This work was pursued within the projects *InTeLekt* funded by the German Federal Ministry of Education and Research (BMBF) according to the resolution of the German Federal Parliament (reference number 16EMO0023), and *LZE* ('Leistungszentrum Elektroniksysteme') supported by the Bavarian Ministry of Economic Affairs and Media, Energy and Technology.

References
[1] Guowei Z, Kashif R, Zhanzhi Z and Chengijng H 2011 Random vibration analysis of onboard PCBs *10th International Conference on Vibration Problems (ICOVP)*

[2] ISO16750-3 2003 International Standard: Road vehicles - Environmental conditions and testing for electrical and electronic equipment

[3] Yang Q, Pang H, Wang Z, Lim G, Yap F and Lin R 1999 Vibration reliability characterization of PBGA assemblies *Microelectronics Reliability*, vol 40, pp 1097-1107

[4] Dehbi A, Ousten Y, Danto Y and Wondrak W 2005 Vibration Lifetime Modelling of PCB Assemblies using Steinberg Model *Microelectronics Reliability*, vol 45, pp 1658–1661

[5] Amy R, Aglietti G and Richardson G 2010 Accuracy of simplified printed circuit board finite element models *Microelectronics Reliability*, vol 50, pp 86–97

[6] Lee Y, Wang B, Lai Y, Yeh C and Chen R 2008 Finite element model verification for packaged printed circuit board by experimental modal analysis *Microelectronics Reliability*, vol 48, pp 1837–1846

[7] Gu J, Barker D and Pecht M 2007 Prognostics implementation of electronics under vibration loading *Microelectronics Reliability*, vol 47, pp 1849–1856

[8] Amy R, Aglietti G and Richardson G 2009 Sensitivity analysis of simplified Printed Circuit Board finite element models *Microelectronics Reliability*, vol 49, pp 791–799

[9] Aytekint B and Ozgtiven H 2008 Vibration Analysis of a Simply Supported PCB with a Component - An Analytical Approach *10th Electronics Packaging Technology Conference*, pp 1178-1183

[10] Silva D 2005 Vibration and shock handbook, CRC Press, Boca Raton, Fla.

[11] Steinberg D 2000 Vibration analysis for electronic equipment, Wiley Interscience

IOP Conf. Series: Journal of Physics: Conf. Series **1026** (2018) 012008 doi:10.1088/1742-6596/1026/1/012008

A survey on multipliers, adders and adiabatic logic styles suitable for power reduction

S Jagadeesh Babu and R Sivakumar

Department of Electronics and Communication Engineering, RMK Engineering College, Kavaraipettai, Tamilnadu, India

Abstract: Low power circuits and designs are the need of the hour as they find applications in electronic components which have power efficient processing capabilities. High speed processors are power consuming and multipliers contribute to a maximum extent for this power consumption. Thus, to achieve low power designs, adiabatic logic is one of the noted technologies in this regard. Adiabatic logic can be implemented for different types of circuit designs and this paper concentrates on a survey of various multipliers and adders that can be modified into a low power multiplier using adiabatic logic.

1. Introduction

Multipliers and adders implemented using adiabatic logic are an ingredient to low power circuit design, but the dilemma in choosing a multiplier, adder and an adiabatic logic style from numerous multiplier, adder and adiabatic styles is a concern for researchers. This paper gives a brief survey about the multiplier, adder and adiabatic architectures commonly used and tries to get a solution to yield a low power design.

2. Multipliers

Multipliers are defined as the process of repeated addition and are an integral part of all processors. Multipliers are time consuming and power consuming units and if the power consumption is reduced, it will lead to subsequent reduction in the power consumption of the overall design. The three main parts of the multiplier are,

1. Partial product generation
2. Partial product accumulation
3. Final addition

1 1 0 1	Multiplicand
* 1 1	Multiplier
1 1 0 1	
1 1 0 1	Partial products
1 0 0 1 1 1	Final product

Generally partial product generation is the main area of concern when coming to implementing strategies regarding power reduction. Partial product accumulation is mainly shifting operation, and weightage in terms of low power implementation techniques is less. Finally, the addition part can also be altered for low power design.

2.1 Types of multipliers

The following multipliers are taken into consideration for analysing the multipliers suitable for adiabatic logic implementation:

1. Tree Multiplier
2. Array Multiplier
3. Vedic Multiplier
4. Carry Save Multiplier

Tree multiplier is a weight based multiplier design and has two phases of operation. In the first phase, AND operation of each bit takes place and the wires are awarded weights based on their position. In the second phase, full adder and half adder are used to reduce the partial products. Finally, a normal adder is used to do the normal addition for the last stage.

Array multiplier is similar to the conventional mathematical multiplier which is implemented using AND gates, shift registers and adders. It is a time consuming process, but the underlying principle is the simplest of all.

Vedic multipliers are based on the 16 vedic techniques mentioned in ancient Indian mathematical analysis. Present day usage of these techniques using modern EDA (Electronic Design Automation) tools has resulted in designing circuits with minimal hardware.

Carry save multiplier is based on rearranging the array multiplier design slightly in order to avoid carry propagation onto each row and finally adding the carry value to the results of each row.

3. Adders

Adders are an integral part of multipliers, and are the last step in multiplication. Major part of research has been undertaken in the partial product generation side, but considering power reduction in adders will add to the cause of generating low power designs.

3.1 Types of adders

The following adders are taken into consideration for analysing the addition process in the multipliers defined in the previous section.

1. Full Adder
2. Carry Look Ahead Adder

Full adder is a three bit adder which can be implemented using two exclusive OR gates, two AND gates and an OR gate. It is the most basic adder design and is the base for all other adders.

Carry look ahead adder is aimed at increasing the speed of higher bit addition by implementing parallelism. It eliminates the carry rippling effect by modifying the logic in adder design. Thus, n-bit adder can perform the addition of n bits simultaneously without waiting for the previous carry to propagate to the current stage of addition.

4. Adiabatic logic

Adiabatic logic works on the following two principles, "Never turn on a transistor when there is a voltage potential between the source and drain" and "Never turn off a transistor when there is current flowing through it."[1]

Switching power dissipation of CMOS (Complementary Metal Oxide Semiconductor) circuit with capacitive load has a lower limit of $C_L V_{dd}^2/2$ (where C_L: Load Capacitance, V_{dd}: Supply Voltage). Adiabatic switching circuits the switching power dissipation below this lower limit. It allows recycling of energy to reduce the total energy drawn from the power source. Thus, the power which may otherwise get wasted can be reused for the power requirement of the same system.

4.1 Challenges in adiabatic system

IOP Conf. Series: Journal of Physics: Conf. Series **1026** (2018) 012008 doi:10.1088/1742-6596/1026/1/012008

Adiabatic technique cannot be implemented for a single part of a single circuit as it is not recommendable in terms of design cost, manufacturing cost and time. So, part of the circuit which consumes large amount of power should be identified and possibility of applying adiabatic logic has to be checked. Also, the more the same type of circuit repeats, more is the gain in the power sector. A repetitive circuit which is bad in terms of power consumption and power dissipation and has a feasibility of applying adiabatic logic has to be first identified. Multiplier is one of the possible circuits as it is power consuming and also a common part in most of the processors. Another main problem that has to be addressed in adiabatic logic is increase in area.

4.2 Types of adiabatic logic styles

The following are some of the adiabatic logic styles commonly used in the low power principles,

1. ECRL (Efficient Charge Recovery Logic)
2. Clocked CMOS adiabatic logic

In ECRL, the power lost during capacitance discharge is reused by providing feedback mechanism to power devices. It performs precharge and evaluation simultaneously. It is a dual rail circuit.

Clocked adiabatic logic is also a dual rail logic to integrate all power control circuitry on the chip which leads to reduced expenses and good power control. Clocked adiabatic logic can also be operated from a dc power supply in a non energy-recovery mode compatible with standard CMOS logic.

5. State of the art

Adiabatic logic systems have been investigated by reasearchers and some of the works carried out are discussed in brief.

Wallace Dadda and Vedic Dadda multiplier have been implemented using ECRL (Efficient Charge Recovery Logic) adiabatic design and achieved 77% less power compared to conventional CMOS[2]. An array multiplier has been implemented using CEPAL (Complementary Energy Path Adiabatic Logic) in 32nm HSPICE tool and 47% reduction in power consumption was achieved[3]. A 8*8 tree multiplier has been implemented using ECRL (Efficient Charge Recovery Logic) adiabatic design in 22nm CADENCE tool and 61.5% less power consumption was observed compared to conventional CMOS[4]. A 16 bit logarithmic signal processor has been implemented using clocked adiabatic logic in 0.35μm and 10 times less power consumption than conventional CMOS was achieved[5]. 2*2, 4*4 and 8*8 vedic multipliers have been implemented using EEAL(Energy Efficient Adiabatic Logic) and the researchers achieved power saving of 64.3%, 56% and 41% respectively in 0.18μm CADENCE[6]. Adiabatic logic has been implemented in 8*8 tree multiplier and yielded 33% less power compared to conventional CMOS[7].

A full adder has been implemented in ECRL and SQAL (Secured-Quasi Adiabatic Logic) and the researchers achieved 69% less power in ECRL and 71.8% less power in SQAL in 45nm technology[8]. Adiabatic logic has been implemented in 4-bit carry look ahead adder and resulted in 25-30% less power compared to the conventional CMOS[9]. A semi-custom design has been implemented using PFAL (Positive Feedback Adiabatic Logic) and a power saving factor of 17 times at 10MHz and 7 times power saving at 100MHz was achieved compared to conventional CMOS[10]. NAND/NOR gates have been implemented using adiabatic logic and it has been reported that adiabatic logic is best suitable for less than 30MHz applications rather than 1MHz or lower applications[11]. A 0.8V CMOS adiabatic differential switch logic circuit has been implemented using bootstrap technique and resulted in 26% less power consumption and 52% smaller propagation delay[12]. A 8 bit carry look ahead adder has been implemented using Energy Recovery Adiabatic Logic Family and 2 to 5 times less dissipative circuit was achieved was achieved[13].

Adiabatic QSERL (Quasi Static Energy Recovery Logic) has been implemented in a 8*8 carry save multiplier using two complementary sinusoidal supply clocks and 34% energy saving has been reported[14]. Clocked CMOS adiabatic logic has been implemented using 1.8μm technology in inverters[15].

IOP Conf. Series: Journal of Physics: Conf. Series **1026** (2018) 012008 doi:10.1088/1742-6596/1026/1/012008

6. Conclusion

From a comprehensive review of the literature, it is evident that the types of multipliers and adders contribute to a lesser extent in power saving and the way in which type of adiabatic logic is used plays a major extent in power saving as seen in figure 1.

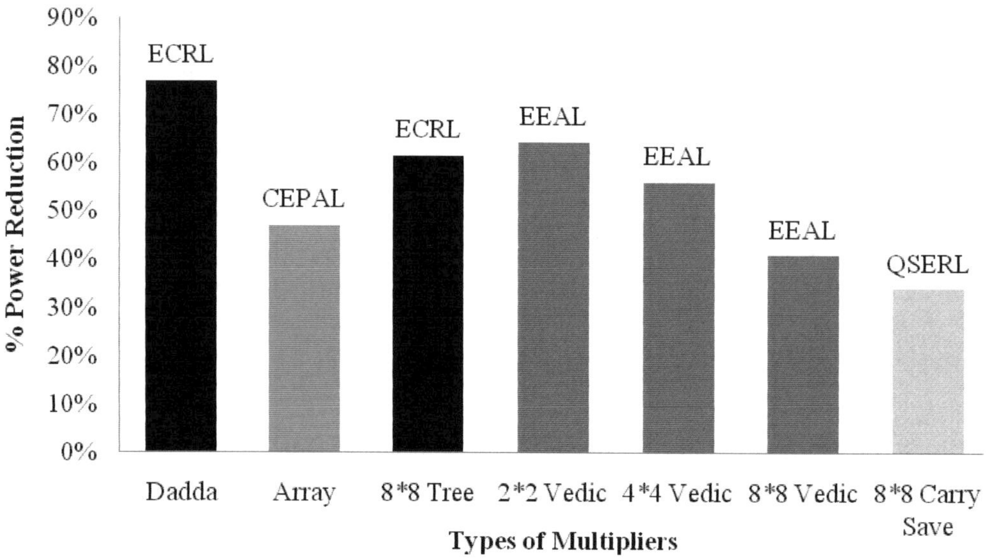

Figure 1. Power saving in multipliers by adiabatic logic.

ECRL (Efficient Charge Recovery Logic) and Clocked Adiabatic Logic are types of adiabatic logic which can be looked upon for constructing efficient low power designs using multipliers and adders.

7. References

[1] William C, et al, 1994, Low-power digital systems based on adiabatic switching principles, *IEEE Transactions on Very Large Scale Integration (VLSI) Systems*, **2** 4 pp 398-407.

[2] Ravish Aradhya H V, et al, 2017, Design and performance comparison of adiabatic 8-bit multipliers, *Distributed Computing, VLSI, Electrical Circuits and Robotics (DISCOVER), IEEE*, Mangalore, India.

[3] Suresh Kumar Pittala, Jhansi Rani A, 2016, Design of a novel array multiplier using adiabatic logic in 32nm CMOS technology, *IIOAB Journal*, **7** 9 pp 740-745.

[4] Manash Chanda, et al, 2016, Design and implementation of adiabatic multiplier in sub-threshold regime for ultra low power application, *International Conference On Communication And Signal Processing*, India.

[5] Gurtac Yemiscioglu, Peter Lee, 2015, Very-Large-scale integration implementation of a 16-bit clocked adiabatic logic logarithmic signal processor, *IET Computers & Digital Techniques*, **9** 5 pp 239-247.

[6] Manash Chanda, et al, 2013, Novel transistor level realization of ultra low power high speed adiabatic vedic multiplier, *International Multi-Conference on Automation, Computing, Communication, Control and Compressed Sensing (iMac4s),* Kottayam, India.

[7] Manash Chanda, et al, 2011, Design and analysis a/tree-multiplier using single clocked energy efficient adiabatic logic, *The 2011 IEEE Students' Technology Symposium*, IIT Kharagpur, India.

[8] Amalin Marina S, et al, 2016, Analysis of full adder using adiabatic charge recovery logic, *International Conference on Circuit, Power and Computing Technologies*, India.

[9] Manash Chanda, et al, 2015, Implementation of subthreshold adiabatic logic for ultralow-power application, *IEEE Transactions on Very Large Scale Integration (VLSI) System* **23** 12 pp 2782-2790.

[10] Antonio Blotti, Roberto Saletti, 2004, Ultralow-power adiabatic circuit semi-custom design, *IEEE Transactions on Very Large Scale Integration (VLSI) Systems*, **12** 11 pp 1248-1253.

[11] Massimo Alioto, et al, 2002, NAND/NOR adiabatic gates: power consumption evaluation and comparison versus the fan-in, *IEEE Transactions on Circuits and Systems-I: Fundamental Theory and Applications*, **49** 9 pp 1253-1262.

[12] Zhang Y, Chen H H and Kuo J B, 2002, 0.8V CMOS adiabatic differential switch logic circuit using bootstrap technique for low-voltage low-power VLSI, *IEEE Electronics Letters*, **38** 24 pp 1497-1499.

[13] Suhwan Kim, et al, 2001, True single-phase adiabatic circuitry, *IEEE Transactions on Very Large Scale Integration (VLSI) Systems*, **9** 1 pp 52-63.

[14] Yibin Ye, et.al, 2001, QSERL: Quasi-Static Energy Recovery Logic, *IEEE Journal of Solid-State Circuits*, **36** 2 pp 239-248.

[15] Dragan Maksimovic, et al, 2000, Clocked CMOS adiabatic logic with integrated single-phase power-clock supply, *IEEE Transactions on Very Large Scale Integration (VLSI) Systems*, **8** 4 pp 460-463.

IOP Conf. Series: Journal of Physics: Conf. Series **1026** (2018) 012009 doi:10.1088/1742-6596/1026/1/012009

Design of a noninvasive diabetes detector based on acetone recognition

Y C Jiang[1], M J Sun[2] and R J Xiong[1]

[1] School of Power and Energy, Northwestern Polytechnical University, Xi'an city, Shaanxi province, China
[2] School of Automation, Northwestern Polytechnical University, Xi'an city, Shaanxi province, China

Abstract. This paper presents the design project of a new type of diabetes detector system, consisting the unique gas sensor, the container, the conditioning circuit as well as the computer GUI developed by CVI. Noninvasiveness and the basis on acetone recognition are the most apparent characteristics of the system. The gas sample of the patients are collected and led to the gas container before reacting with the sensor and finally put out data on the screen at regular intervals. A controlled experiment is conducted to verify the precision of the PID sensor compared with the existing GC-MS method in the process of determining acetone concentration. The experiment shows that the system is effective and convenient in detecting low concentrated acetone exhaled by the patients, and an innovative product inspired by this may be prospective to be put into use in certain areas.

1. Introduction

The daily monitoring of diabetes mainly relies on the household blood glucose meter, and its accuracy and sensitivity are not very satisfactory. There are four main diagnostic methods for diabetes mellitus [1]: glycosylated hemoglobin diagnosis, oral glucose tolerance test (OGTT), VOC comprehensive diagnosis and gene diagnosis. At present, the existing blood glucose detection technology in the market mainly includes the professional testing equipment based on blood testing in hospital, as well as household handheld blood glucose detector. Both are invasive testing equipment which need regular needle detection, the operation is complex, and cost is high.

In recent years, the development of exhaled breath volatile organic compounds (VOC) [2] analysis technology provides the possibility of routine monitoring and early screening of diabetes. Exhaled acetone concentration is considered to be a reliable indicator of diabetes [3][4]. Many people around the world have studied the role of acetone in the diagnosis and treatment of diabetes, which is based on the fact that acetone is released from the skin, and its concentration is related to the h-hydroxy butyric acid.

A group of control experiments with cold trap and gas chromatography were carried out by Norio Yamane and his research team [5], they found that the skin acetone concentration in diabetic patients was significantly higher than that of control group (non-diabetes patients), and patients under different treatments show similar skin acetone content. The concentration of acetone in the skin is related to h-hydroxy butyric acid, blood glucose and glycosylated hemoglobin. Acetone concentration in the skin was significantly higher in patients with diabetic ketoacidosis (990ppb), but decreased to 80ppb after insulin treatment. On the conditions that the analysis of methods and tools are simplified, measurement of acetone concentration in the skin can be used as a screening test for ketoacidosis and can be used as the index of diabetes control and diabetic ketoacidosis and other ketogenic output.

Marco Righettoni et al developed a noninvasive respiratory monitoring detector [6] which is resisted to chemical corrosion, and in ideal conditions and real conditions are able to do rapid measurement of very low concentration of acetone. It also has high signal-to-noise ratio. The detector film is made of pure silicon and silicon hybrid WO3 nanoparticles in the gas phase, and placed directly on the cross electrode. The sensing characteristics, relative humidity and interference concentration. When the humidity was 90% and temperature was 400℃, there was a significant difference in acetone respiration between healthy people and patients with diabetes. There was a gap of 40% in the detector. So it shows that this new type of instrument that detects diabetes through breath analysis can provide a portable, cost-effective system to replace the cumbersome invasive diabetes detection system.

Acetone can be used as a promising biomarker for non-invasive diagnosis and detection of diabetes [7]. Researchers used the aluminum foil bag to collect the breath acetone of 25 patients with type 2 diabetes and 44 healthy people and analyzed through gas chromatography, discovering that acetone concentration found in T2D patients were significantly higher than those of healthy people and its characteristics curve indicates that the optimal cut-off value for diagnosis is 0.891 (the area under the curve is 0.999, sensitivity 100%, specificity 97.7%) when T2D patients were compared with healthy test-takers. It was also found that the ratio of acetone to total exhaled air and net exhaled air was 1:18 and 1:17. Therefore, breathing with high acetone percentage might be a useful indicator for accurate diagnosis and monitoring of diabetes.

In summary, acetone is closely related to diabetes [8], exhaled acetone concentration can be detected by gas chromatography, and a high probability exists to distinguish between diabetes and healthy people.

Compared with the traditional invasive and regular method of detecting blood sugar with expensive and big devices, the portable and convenient feature of the acetone recognition detector provides a better way to help diagnose, monitor and evaluate the treatment for diabetes [9]. Electronic nose instrument making more precise and accurate prediction of diabetes is also in continuous development, and the utilization of it is promising [10]. In this paper, acetone is used as the main indicating gas, assisted by other VOC together for diabetes monitoring and diagnosing. The ultimate aim is to design the different components of an effective recognition system.

2. Design of the acetone detector system

The design of the new system can be divided into three general modules: gas collection, reaction signal processing and data conversion. The second step are made up of three consecutive substeps while the other have two each. The gas gathered by the container is consumed by the sensor which generates signal into the computer for software output. The Figure 1 shows the detailed procedure of the system.

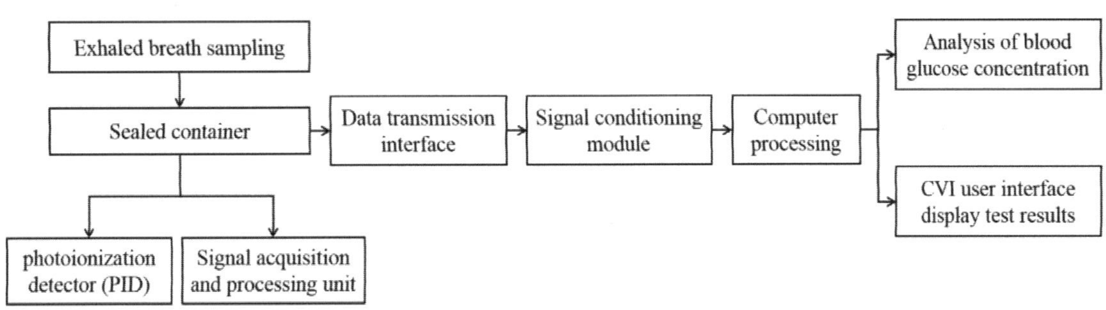

Figure 1. Flow chart of the detection system.

2.1 Type selection and curve correction of high precision gas sensor

The high demand for precision and efficiency requires the gas sensor to be sensitive and suitable enough for the normal testing environment. There are inevitable disturbing factors in the functioning process of the sensor that may greatly influence the testing results. So in order to solve the problem of cross sensitivity of sensors to non-target gases, the sensor array is used to increase the sensitivity to acetone.

According to the range of acetone content in the breath of diabetic patients and the convenience of the instrument, the electrochemical gas sensor of Baseline company is selected.

The main technical parameters are as follows:

Accuracy: ±1%FS;	response time: ≤10s;	recovery time: ≤5s;
Resolution: ±1ppm; ±1%LEL;	measurement range: 0-1500ppm;	humidity: ≤95%RH;
Working temperature: -20 to 60℃;	no frost.	

After testing 100 sets of standard concentration acetone samples under different environment temperature, we measured the sensor output curve of voltage-actual concentration shown in Figure 2.

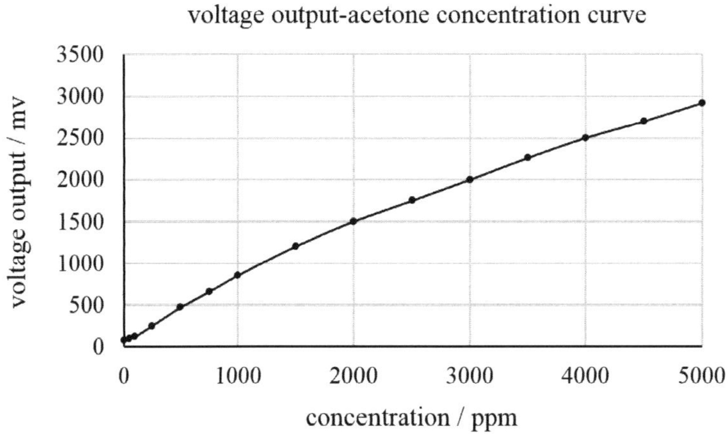

Figure 2. Characteristic curve of the gas sensor.

Conclusion can be made from the operation result that when the concentration is high, the output of PID is nonlinear, but repeatable, which can be corrected in software.

Because of the nonlinear of the curve, it is necessary to do the segmentation processing in the conversion process, after the test, the curve has high identification degree to the acetone gas of different concentration, with 200-1000 concentration of ppm resolution accuracy rate reaching 97%.

2.2 Design of autonomous container

In order to hold the gas exhaled by people, an autonomous container should be figure out and the gas channel should be carefully designed. The main purpose is to elevate the precision and convenience of the detection of samples.

The sealing device for experiment is designed as a cylindrical structure. The gas to be measured is led to the device bottom, so that it can fully react with the PID sensor and reduce the loss. The whole detection process is carried out in the device, so to improve the accuracy of the detection, the device needs to be a sealing one to prevent the impact of external environment, especially the air on detection.

Considering when the transparent plastics meet organic solvent like the acetone, organic solvent and the transparent plastic surface will produce a chemical reaction to generate other opaque material adhered on the surface of the transparent plastic, affecting the detection accuracy of the device, glass materials are used. Also, changing the gas channel from a silicone tube to fluorine hose can ensure measurement precision. At the mouth of the bottle, a matching frosted mouth is used to ensure the air

IOP Conf. Series: Journal of Physics: Conf. Series **1026** (2018) 012009 doi:10.1088/1742-6596/1026/1/012009

tightness of the device.

The gas to be measured is made up by acetone and the corresponding VOC. The acetone gas is injected by glass syringe and then sampled by headspace sampling, and finally imported into the reaction device by glass syringe and fluoride hose. Nitrogen is used as carrier gas to clean the reaction device and transport acetone gas, which is led by fluoride hose and vacuum pump into the reaction device.

The PID sensor is connected to a special round conditioning circuit, and then the data is imported into the computer by wire. The glass tube at the top of the bottle mouth and the exposed wire are sealed by glass glue, so as to ensure the tightness of the device, as is shown in Figure 3.

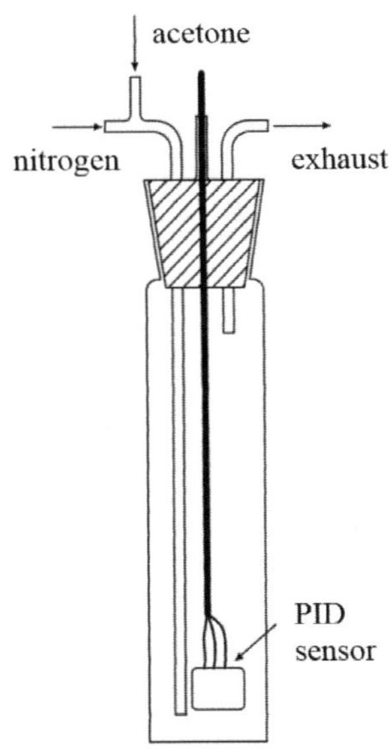

Figure 3. Cutaway view of the reaction bottle.

Two glass tubes function as the gas intake and exhaust channel of the cylinder device. The gas is mixed by nitrogen and acetone in a certain proportion. PID sensor is placed at the bottom of the glass bottle and connected with the wire, following which is the conditioning circuit.

2.3 Millivolt level signal conditioning circuit design

Using the gas exhaled by diabetes as collect object, the sensor transforms the gas concentration into the electrical signal. Normally the output voltage signal of PID sensor changes with the concentration of acetone. The voltage signal goes through the signal conditioning circuit and finally into the NI6009 data acquisition card. In the PC the data (the exhaled acetone by healthy people is close to the normal level between 8.9 and 30.8nmol, while acetone concentration range for T2D patients exhaled gas is 73.9 to 152.6nmol. The range for the acetone concentration of T1D patients is between 86.2 and 690nmol) was compared with setting threshold of the acetone concentration in patients with diabetes, and display the concentration of acetone in the CVI interface, so as to provide a reliable basis for whether the test-taker needs further complete blood glucose monitoring.

As the acetone concentration in the exhaled air is very small, the sensor can only output mV level voltage, special circuit is designed to make the sensor output signal more accurate and stable, as is shown in Figure 4.

IOP Conf. Series: Journal of Physics: Conf. Series **1026** (2018) 012009 doi:10.1088/1742-6596/1026/1/012009

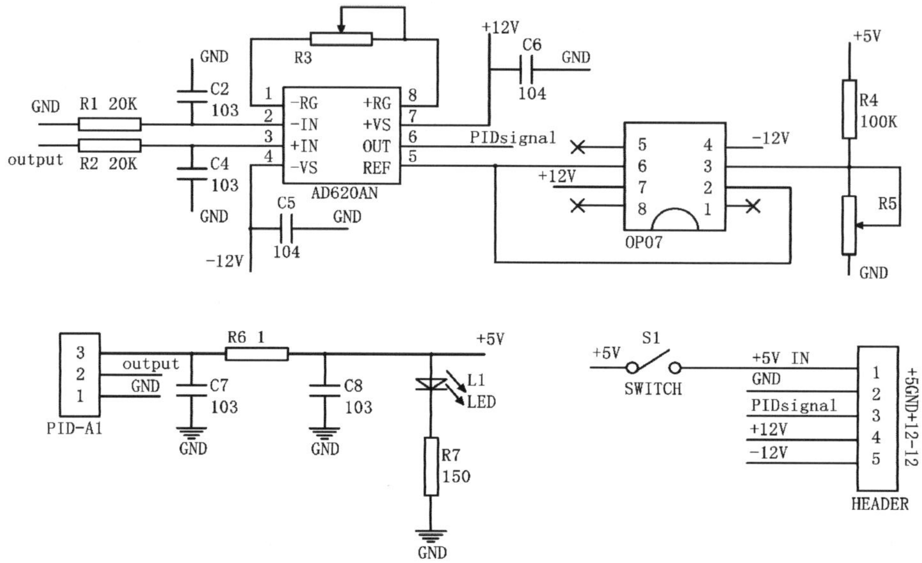

Figure 4. Design of signal conditioning circuit.

The core of the conditioning circuit is AD620 three stage amplifier chip, its magnification can reach 0-10000 times. We also designed the OP07 zeroing circuit to ensure that the reference voltage is 0, so that the true magnification of the circuit is more accurate.

2.4 GUI design

The output of data and its conversion into useful statistics should be accomplished by the computer and the commensurate software. The LABWINDOWS/CVI is chosen to do human-machine interaction.

Firstly, C language is used to edit the background code of the interface, then data is collected and analyzed by LABWINDOWS/CVI and PID sensor, which combines the advantage of professional detection tools with the virtual computer program. CVI software greatly integrated development environment, an interactive programming method, function panel and rich library function together to enhance the function of language, which provides an ideal software development environment for skilled developers to establish a data acquisition system.

On the other hand, NIDAQMX software has powerful functions, and its effective cooperation with CVI can enable the data collected by the acquisition card to be accurately input the computer, thus providing help for the later processing and detection conclusions.

At the same time, one advantage is that the data of the gas concentration detected can be displayed on the interface. If necessary, excel can be imported for data analysis, and get the diagnosis results.

The user interface is composed of different buttons, text boxes and graphs to realize the function. The basic information of the patients, the channel chosen and the sampling frequency are all portrayed as system information. The two graphs indicate the output voltage signal as well as the concentration calculated from the detection result, as is shown in Figure 5.

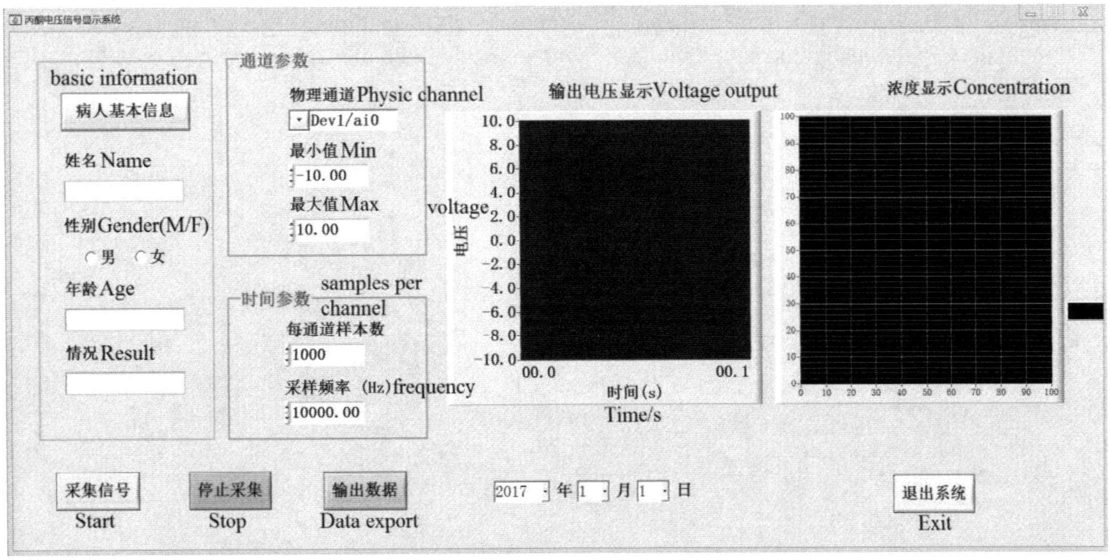

Figure 5. GUI design based on the LABWINDOWS/CVI and NIDAQMX.

3. Experiment simulation

To test the reliability of the PID gas sensor for exhaled acetone testing, we invited seven diabetics and a healthy breath of people to carry out clinical trials. No. 1 volunteer in the table 1 is healthy people, No. 2-8 are type 2 and type 1 diabetic patients in hospital, OHBA of the eight volunteers are measured in the hospital. Exhaled breath samples of the eight patients were collected using a 1.0-liter tedlar bag and each volunteer generated two bags of exhaled breaths, one of which was analyzed by gas chromatography-mass spectrometry (GC-MS) to detect acetone content breath sample [11], another gas sample used PID sensing for detection. GC-MS test results are given in Table 1.

Table 1. Basic information of 8 volunteers participating in exhaled breath test.

Number	State of illness	Age	Gender	Acetone content of Exhaled gas(GC-MS)/ppm
1	healthy	26	Female	0.486
2	Type 2 diabetes	61	Male	0.254
3	Type 2 diabetes	62	Female	0.490
4	Type 2 diabetes	57	Male	0.640
5	Type 2 diabetes	45	Male	1.292
6	Type 2 diabetes	30	Male	1.840
7	Type 2 diabetes	59	Male	2.940
8	Type 1 diabetes	43	Female	5.000

In the preliminary gas experiments, we recognized humidity, temperature, CO_2 and other influencing factors. Figure 6 gives the relationship between the concentration of acetone and the response measured by PID gas sensor, in which the condition that 35% RH pure acetone, 80% RH pure acetone, 35% RH humidity mixed gas (40000ppm CO_2 and acetone), 80% RH humidity mixed gas are given in it.

The relationship between the acetone concentration (GC-MS test results) and the response of the PID gas sensor in eight exhaled breath samples is indicated by the red pentagram. It can be seen that the corresponding points of the eight gas samples are well matched with the gas mixtures under the humidity of 35% RH. On the contrary, the stripline can be used to determine the actual acetone content in a sample of exhaled breath, thus achieving the purpose of eliminating CO_2 and interference gas. In fact, the gas cylinder design has taken into account the interference of other gas and ambient humidity on the PID sensor response.

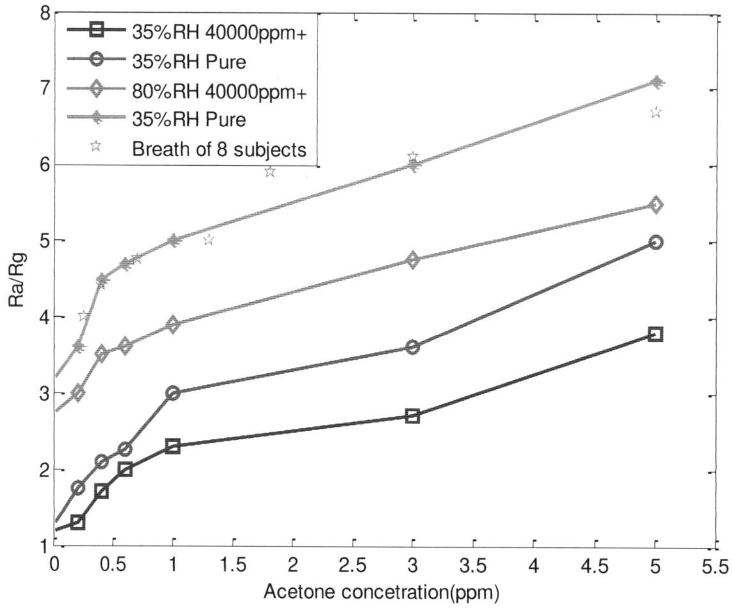

Figure 6. PID sensor response for acetone concentration. It indicates the relationship between acetone concentration test results by GC-MS and gas sensor of eight exhaled breath samples in Table 1.

4. Conclusion

This paper presents the design and experiment verification of a brand new type of diabetes detection system based on the acetone recognition. A PID sensor type is chosen, a cylinder container for mixed gas is designed and made, a conditioning circuit is established and a graphic user interface is edited by C programming and supported by CVI and NIDAQMX. The most obvious characteristic of the system is its noninvasiveness and high sensitivity. Then a controlled experiment is conducted to demonstrate that the PID sensor in the collecting device and the software work well together to generate the desired results, although precise in a finite range.

The test results indicate that the detection is merely precise under 35%RH, which means the adoption of the system should meet the relatively narrow scope compared with invasive detection. However, the advantage is obvious that the product based on the innovative system, if manufactured, can greatly reduce the infection probability and increase the chance of diagnosis of diabetes before it is too late.

More research is required both to support the feasibility of this type of system and to integrate these parts into a single entity. It is promising that the product will be useful not only in medical apparatus and instruments field, but also in daily monitoring of human health.

5. References

[1] Hoey H, Mlinac A, Tran C T, et al. 2012 5 (th) Annual symposium on self-monitoring of blood glucose (SMBG) applications and beyond, May 3-5, 2012, Dublin, Ireland. *Diabetes Technology &Therapeutics*, **14(12)**: 1155-73.

[2] Sulway M, Malins J 1970 Acetone in diabetic ketoacidosis *The Lancet*, **296(7676)**: 736-740

[3] Schubert J, Miekisch W, Geiger K, Noldge-Schomburg G 2004 Breath analysis in critically ill patients: potential and limitations *Expert Rev. Mol. Diagn.*, **4(5)**: 619–629.

[4] Minh T D C, Blake D R, Galassetti P R 2012 The clinical potential of exhaled breath analysis for diabetes mellitus *Diabetes Research and Clinical Practice*, **97(2)**: 195-205.

[5] Yamane N, Tsuda T, Nose K, Yamamoto A, Ishiguro H, Kondo T 2005 Relationship between skin acetone and blood beta-hydroxybutyrate concentrations in diabetes *Clinica Chimica Acta*, 3651-2.

[6] Righettoni M, Tricoli A, Pratsinis S E 2010 Si: WO3 Sensors for highly selective detection of acetone for easy diagnosis of diabetes by breath analysis *Analytical Chemistry*, **82(9)**: 3581-3587.

[7] Zhou M G, Liu Y, Li W W, Yuan X, Zhan X F, Li J, Duan Y X, Liu Y, Gao Z H, Cheng Y, Cheng S Q, Li H, Liang Y 2014 Investigation and identification of breath acetone as a potential biomarker for type 2 diabetes diagnosis *Chinese Science Bulletin*, **59(17)**: 1992-98.

[8] Wang C, Sahay P 2009 Breath analysis using laser spectroscopic techniques: breath biomarkers, spectral fingerprints, and detection limits *Sensors (Basel)*, **9(10)**: 8230-62.

[9] Minh T D, Oliver S R, Ngo J, et al. 2011 Noninvasive measurement of plasma glucose from exhaled breath in healthy and type 1 diabetic subjects *American Journal of Physiology-Endocrinology and Metabolism*, **300(6)**: E1166-75.

[10] Guo D, Zhang D 2010 A Novel Breath Analysis System Based on Electronic Olfaction *IEEE transactions on biomedical engineering*, **57(11)**:2753-62.

[11] Pauling L, Robinson A B, Teranishi R, Cary P 1971 Quantitative analysis of urine vapor and breath by gas-liquid partition chromatography *Proc Nat Acad Sci USA*, **68**: 2374-76.

Design and fabrication of a non-invasive, wireless system for monitoring needle insertion during epidural puncture

E Schena[1], C Massaroni[1], D Formica[2], A Mattei[3], V Piemonte[4], P Saccomandi[5], R Setola[6] and M Carassiti[3]

[1] Unit of Measurement and Biomedical Instrumentation, University Campus Bio-Medico di Roma, 00128, Rome, Italy
[2] Unit of Neurophysiology and Neuroengineering of Human-Technology Interaction, University Campus Bio-Medico di Roma, 00128, Rome, Italy
[3] Unit of Anesthesia, Intensive Care and Pain Management, University Campus Bio-Medico di Roma, 00128, Rome, Italy
[4] Unit of Chemical-physics Fundamentals in Chemical Engineering, University Campus Bio-Medico di Roma, 00128, Rome, Italy
[5] IHU (Institute of Image-Guided Surgery), Strasbourg, France
[6] Unit of Automatics, University Campus Bio-Medico di Roma, 00128, Rome, Italy

Abstract. Over the last decades epidural analgesia has gathered research interest and broad clinical acceptance. In this procedure, the detection of the epidural space is pivotal to avoid major complications. Although, some systems for supporting the anaesthetist in the epidural space detection are commercially available, this difficult procedure is often performed without any support. In previous articles, our research group described a new approach for a non-invasive detection of the epidural space; the assessment of the system was also performed both on a spinal column simulator and in *ex vivo* animal model (small pig). The mentioned system is based on a Force Sensing Resistor (FSR) that monitors the load exerted by the anaesthesiologist on the syringe plunger during the procedure. The resistance of the sensor is transduced into a voltage by means of a Wheatstone bridge (WB), then it is amplified, finally it is collected by a remote laptop. When the needle reaches the epidural space, the load applied by the anaesthesiologist decreases, so the consequent change of the system output may be used for the detection of the entrance within this space. The previous version of the system communicates to the laptop via USB. In this article, we described a new version of the system which communicates to the laptop via wireless. This solution aims at facilitating the use of the system in clinical settings. After the description of the measuring system, its preliminary assessment in patients undergoing epidural puncture will be reported.

1. Introduction

Epidural puncture has gathered interest in research and broad acceptance in clinical settings (e.g., during labour, and for relieving chronic back pain) [1],[2]. It has been shown that this procedure decreases patient morbidity after major surgery [2]. Although epidural analgesia is a very complex procedure with relevant risks of major complications (e.g., epidural haematoma, accidental dural puncture…) [3], it is usually performed without any technological support. During the procedure, the clinician pushes a syringe whose needle must reach the epidural space. This space is placed between the ligamentum flavum and the dura. The detection of the epidural space is usually performed by the anaesthetist, who feels a decrement of pressure (called loss of resistance, LOR) because the epidural space is softer than the ligamentum flavum. Since the epidural space is thin and the procedure is performed without any technological support, the risk of accidental dural puncture is high. In order to minimize this risk, several systems have been proposed, most of them based on the LOR detection, showing promising results [4],[5],[6]. In previous studies we have described the working principle and the fabrication of a new system for epidural space detection; it was also assessed by simulators and in *ex vivo* animal model [7],[8],[9]. The proposed system does not require any changes to the standard procedure since the sensing part consists on a very thin force sensor placed on the syringe plunger. This thin sensing element has been used by our group also in other applications related to the anaesthesiology field for force or pressure monitoring [10],[11].

The previous version of the system communicates to the laptop by means of an acquisition board via USB. The aim of this study is twofold: *i)* to design and develop a new version of the system which communicates to the laptop via wireless (Bluetooth Low Energy) standard. This solution aims at facilitating the use of the system in clinical settings; *ii)* the preliminary assessment of the system on three patients undergoing epidural procedures. The article is structured as follows: in Section 2 the working principle of the system, its main components, and the new design are described; Section 3 focuses on the description of the preliminary experiments in clinical settings, and shortly describe the data analysis; in Section 4 the results collected during epidural puncture are illustrated; in Section 5 results, conclusions and future works are discussed.

2. Working principle, design and fabrication of the system

2.1. Working principle

The measuring system here proposed consists of an FSR and a printed circuit board (PCB) to collect the analog output of the FSR, convert them into a digital signal and send data via Bluetooth connection (as show in figure 1).

The FSR (i.e., FSR 402, Interlink Electronics Inc., range of measurement from 0.2 N up to 20 N) is a polymer thick film that exhibits a decrease in resistance with increase in force applied to its active area. In this specific application, the FSR will be placed on the syringe plunger to sense the load exerted during the procedure by the operator.

The output of the FSR is an electrical resistance (R_{FSR}). For a force-to-voltage conversion, the FSR device is tied to a WB, as shown in figure 1.

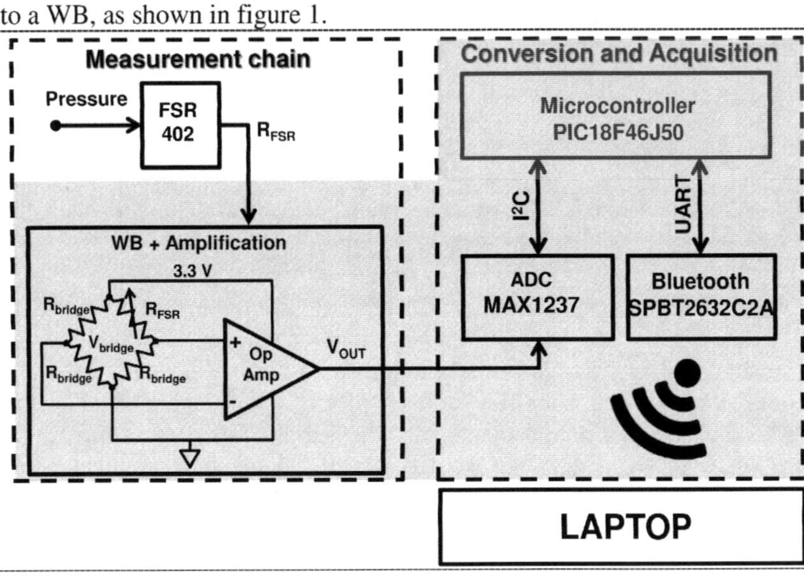

Figure 1. Schematic representation of the system. The electronic components of the printed circuit board are highlighted in green.

In general, the relationship between the output of an FSR (R_{FSR}) and the force (F) that acts on the active area of the sensor can be expressed as in (1):

$$R_{FSR}(F) = 10^{a \cdot \log_{10}(F) + b} \tag{1}$$

where *a* and *b* are two constant values that can be calculated by data provided by the manufacturer. By considering the plot R_{FSR} *vs* F reported in the datasheet of FSR 402 [12], the coefficients a=-0.839 and b=4.2309 have to be used to obtain the best fitting curve of the expression (1).

The output of the WB (V_{bridge}) as a function of the R_{FSR} is described by the following equation:

$$V_{bridge} = V_{supply} \cdot \frac{R_{bridge} - R_{FSR}}{2 \cdot (R_{bridge} + R_{FSR})} \tag{2}$$

Being the V_{supply} the bridge voltage supply (i.e., 3.3 V) and the R_{bridge} the value of the three resistances positioned on the WB (see figure 1).

By placing (1) in (2) together with the *a* and *b* coefficient values, the V_{bridge} can be expressed as a function of F as in (3):

$$V_{bridge} = V_{supply} \cdot \frac{R_{bridge} - (10^{0.839 \cdot \log_{10}(F) + 4.2309})}{2 \cdot (R_{bridge} + (10^{0.839 \cdot \log_{10}(F) + 4.2309}))} \tag{3}$$

The value of V_{bridge} strictly depends on the R_{bridge}. As shown in figure 2, according to (3), the sensitivity of the system increases with the R_{bridge} (1 kΩ, 10 kΩ, 50 kΩ, 68 kΩ, and 100 kΩ in figure 2); on the other hand, the V_{bridge} saturates at lower values of F.

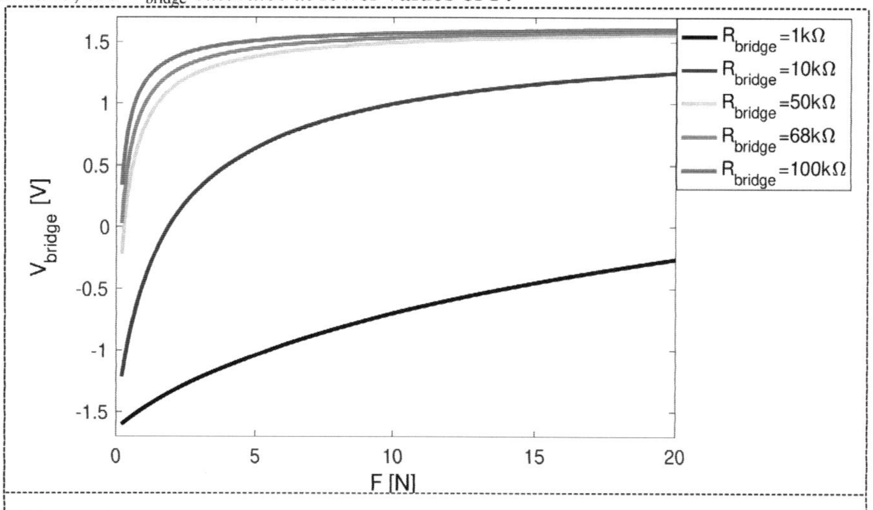

Figure 2. Output of the Wheatstone vs the force applied to the FSR using five different configurations.

In the specific application, the F can be considered uniformly applied on the syringe plunger: so, the pressure (P) exerted by the operator during the puncture may be expressed as the ratio between the F and the area in which F is applied. Since the epidural space can be detected by analyzing the passage from saturation to non-saturated condition of the system due to the LOR [7], [8], [9], and by considering the P values exerted during the passage through the ligamentum flavum and the epidural space, different values of R_{bridge} can be adopted. So, different configurations of the Wheatstone Bridges (in terms of R_{bridge} values) provides different sensitivities and measuring ranges to the LOR (see figure 2). In this study, we choose R_{bridge}= 10 kΩ: it allows collecting voltage changes from saturation (during the passage through the ligamentum flavum, F≥15 N [4]) to non-saturation (needle reaches the epidural space, F~2N).

2.2. Design and fabrication

The FSR used in this work was the FSR 402 (Interlink Electronics Inc). A custom PCB was designed in Eagle (Autodesk) and then manufactured in order to collect and transfer data to a laptop. The first stage of the PCB provides the physical connection to the FSR 402, the transduction of the R_{FSR} into a V_{bridge} by adopting one WB configuration (i.e., R_{bridge}= 10 kΩ) and the amplification of V_{bridge} to obtain a voltage output (V_{out}) with a differential amplifier. The second stage of the PCB allows the conversion and acquisition of the data: an analog-to-digital-converter (ADC) and a microcontroller are embedded into the PCB. In particular PCB includes: *i)* a 12-bit ADC unit (i.e., MAX1237, Maxim Integrated Inc.) with an integrated multiplexer, that converts the outputs of the WB; *ii)* a micro-controller (i.e., PIC18F46J50, Microchip Technology) that collects data from the ADC via an I2C digital bus, and sends the data through an UART port to a Bluetooth module (i.e., SPBT2632C2A, STMicroelectronics). This module provides the wireless communication to a remote laptop in MATLAB environment.

3. Preliminary experiments and data analysis

Preliminary experiments were performed using the proposed system with the configuration R_{Bridge}=10kΩ.This choice aims at figuring out the input-output relationship of this configuration which has intermediate metrological properties with respect to other analysed configurations (see figure 2): it has a sensitivity higher than the solution at 1 kΩ and lower than the one at 100 kΩ; while it has a measuring range lower than the solution at 1 kΩ, but bigger than the one at 100 kΩ. This preliminary analysis was performed on 3 patients undergoing epidural puncture, with age from 50 to 77 years. The enrollment started after the approval of the local medical ethics committee, and with written informed consent of the 3 patients. The 3 epidural punctures were performed by the same expert anaesthetist (M.C., with more than 5 years of experience), using the system by placing the FSR 402 on the syringe plunger. Figure 3 reports a picture of the typical scenario, where the positioning of the FSR on the syringe and the PCB, which embeds the Wheatstone bridge, the amplifier stage, the ADC unit, and the Bluetooth module are shown.

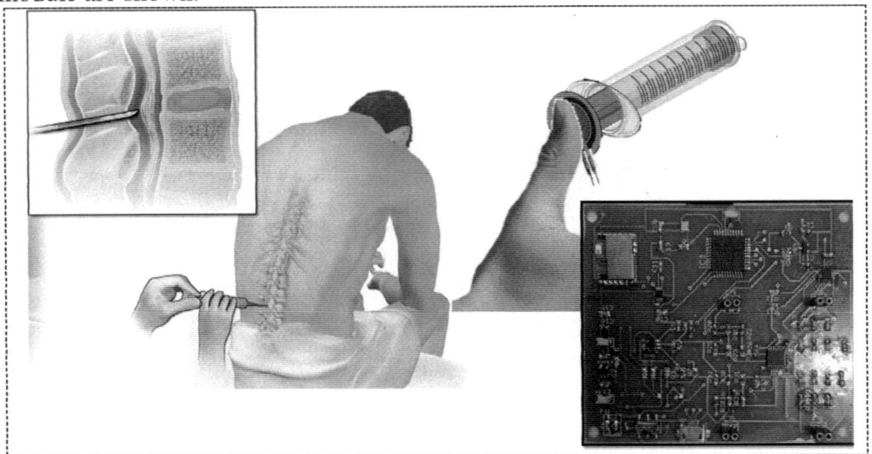

Figure 3. Schematic representation of the clinical scenario and of the proposed system.

The output of the system (V_{out}) was recorded during the whole procedure starting from the skin perforation until the needle reached the epidural space. The detection of the epidural space was made by considering the feedback of the expert anaesthesiologist.

The data analysis is devoted to the extraction of the system response from the data collected during epidural puncture procedure. Basically, it can be schematically divided in two main steps: *i)* the V_{out} value was collected during the entire procedure and was normalized in order to have an output (V_n)

ranging from 0 to 1 in all the trials; *ii)* the V_n change due to the LOR was individuated; *iii)* the starting point of the LOR was manually selected. This curve represents the system output over the LOR; *iv)* the previous step was repeated for the three trials.

4. Results

Figure 4 shows the curves obtained during the epidural puncture of the three patients after the fourth step described in the previous section.

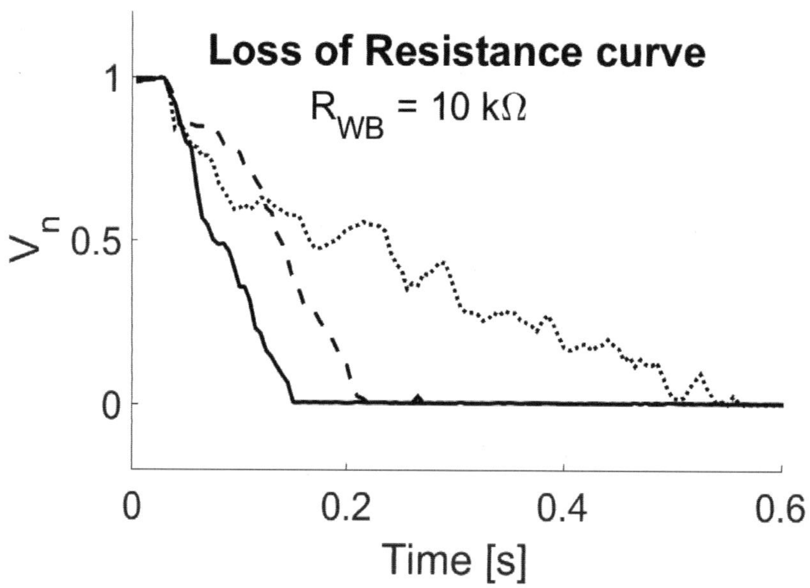

Figure 4. Trend of the normalized system output collected during the epidural puncture of the three patients during the passage from the ligamentum flavum to the epidural space.

The curve persists in saturation condition (V_n=1) up to the LOR, where the load exerted by the anaesthetist decreases. Related to this decrement there is a change of V_n up to a null value. This passage from V_n=1 to V_n=0 is the basis of the epidural space detection. In order to quantify this passage, the rise time, t_r, (defined as the time required for V_n to reach 0.10) was calculated for the three curves. It was 110 ms, 170 ms, and 450 ms. Since the anaesthetist pushes the syringe with a very low speed, a lag time shorter 500 ms (it fulfilled this requirement in all the three recorded procedures) is promising to consider the system available to support the clinician in the detection of the epidural space.

5. Discussion and conclusion

This article reports on design and assessment of non-invasive system for supporting the clinician in the detection of epidural puncture. Our group of research has already proposed new solutions for improving the management of critically ill patients [13],[14],[15] and has already assessed a previous version of the system on simulator and in *ex vivo* animal model [7],[8],[9]. In this article, we described a new version allowing the data collection via wireless communication to a remote laptop to facilitate the use of the system in clinical settings. The design of the system embeds four configurations by changing R_{bridge}, to adjust the system sensitivity and its measuring range. In this study, we have performed a preliminary assessment of the system on 3 patients using the configuration with R_{bridge}=10kΩ. Considering the theoretical predictions shown in figure 2, R_{bridge} value strongly influences the system response: the system sensitivity increases with R_{bridge} and the measuring range narrows with this parameter, the higher is the R_{bridge} value the shorter may be the t_r. Therefore, the use of a bigger value of R_{bridge} may improve the t_r value, but the risk is that the system will remain under saturation condition also after the LOR due to a narrower measuring range; on the other hand, the use of a smaller value of R_{bridge} may increase the t_r value, but will increase the measuring range. Future tests will focus on assessing the system on a bigger number of patients using different configurations to find the best trade-off between a shorter t_r and better static properties.

6. References

[1] Wheatley R G, Schug S A and Watson D 2001 Safety and efficacy of postoperative epidural analgesia *Br. J. Anaesth.* **87**, pp. 47–61.

[2] Rodgers A, Walker N, Schug S, McKee A, Kehlet H, van Zundert A, Sage D, Futter M, Saville G, Clark T and MacMahon S 2000 Reduction of postoperative mortality and morbidity with epidural or spinal anaesthesia: results from overview of randomised trials *B. M. J.*, **321**, p 7275.

[3] Christie I W and McCabe S 2007 Major complications of epidural analgesia after surgery: results of a six year survey *Anaesthesia* **62** pp 335–341

[4] Lechner T J M, van Wijk M G F and Maas A J J 2002 Clinical results with a new acoustic device to identify the epidural space *Anaesthesia* **57** pp. 768–772.

[5] Habib A S, George R B, Allen T K and Olufolabi A J 2008 A Pilot Study to Compare the Episure™ Autodetect™ Syringe with the Glass Syringe for Identification of the Epidural Space in Parturients *Anesth. Analg.* **106** pp. 541–543.

[6] Sawada A, Kii N, Yoshikawa Y, Yamakage M 2012 Epidrum®: A new device to identify the epidural space with an epidural Tuohy needle *J. Anesth.***26** pp. 292–295.

[7] Tesei M, Saccomandi P, Massaroni C, Quarta R, Carassiti M, Schena E, and Setola R 2016 A cost-effective, non-invasive system for pressure monitoring during epidural needle insertion: Design, development and bench tests In *38th Annual International Conference of the IEEE Engineering in Medicine and Biology Society (EMBC)*, pp. 194–197.

[8] Carassiti M, Mattei A, Quarta R, Massaroni C, Saccomandi P, Tesei M, Setola R, and Schena E 2017 A new, pressure guided management tool for epidural space detection: feasibility assessment on a simulator *Artif. Organs* 2017.

[9] Carassiti M, Quarta R, Mattei A, Tesei M, Saccomandi P, Massaroni C, Setola R and Schena E 2017 Ex vivo animal-model assessment of a non-invasive system for loss of resistance detection during epidural blockade In *39th Annual International Conference of the IEEE Engineering in Medicine and Biology Society (EMBC)*, pp. 759-762.

[10] Vallone N, Pizzo M C, Massaroni C, Saccomandi P, Silvestri S, Carassiti M, Mattei A, and Schena E 2015 Design and characterization of a measurement system for monitoring pressure exerted by bronchial blockers: In vitro trials In *37th Annual International Conference of the IEEE Engineering in Medicine and Biology Society (EMBC)* pp. 1691–1694.

[11] Carassiti M, Mattei A, Pizzo C M, Vallone N, Saccomandi P, and Schena E 2016 Bronchial blockers under pressure: in vitro model and ex vivo model *Br. J. Anaesth.* **117** pp. i92–i96.

[12] Datasheet of FSR 402, Interlink Electronics.

[13] Saccomandi P, Schena E, Silvestri S 2011 A novel target-type low pressure drop bidirectional optoelectronic air flow sensor for infant artificial ventilation: Measurement principle and static calibration *Rev. Sci. Instrum.*, **82(2)**, 024301.

[14] Schena E, Massaroni C, Saccomandi P, Cecchini S 2015 Flow measurement in mechanical ventilation: A review *Med. Eng. Phys.*, **37**, pp. 257-264.

[15] Schena E, Saccomandi P, Silvestri S 2013 A high sensitivity fiber optic macro-bend based gas flow rate transducer for low flow rates: Theory, working principle, and static calibration *Rev. Sci. Instrum.*, **84**.

IOP Conf. Series: Journal of Physics: Conf. Series **1026** (2018) 012011 doi:10.1088/1742-6596/1026/1/012011

Compact Model of Short-channel Cylindrical ballistic Gate-All-Around MOSFET Including the Source-to-drain Tunneling

H Cheng, Z Zhang, Z Yang and Z Liu

Department of Industrial Control Network and Systems, Shenyang Institute of Automation Chinese Academy of Sciences, No.114 Nanta St., Shenhe Dist., Shenyang, 110016 China

E-mail: liuzhifeng@sia.cn

Abstract. One compact model of drain current valid in the subthreshold region, for short-channel cylindrical gate-all-around metal-oxide-semiconductor field-effect transistors including the source-to-drain tunneling has been proposed. From a two-dimensional analysis, the drain-induced barrier lowering effect is modeled according to our previous work. In this paper, by introducing the profile of the energy subband level along the electron transport direction into the Wentzel-Kramers-Brillouin approximation, we can numerically derive the expression of the tunneling transmission coefficients. Then, the source-to-drain tunneling current in the subthreshold region is evaluated by using the Landauer formula. The results obtained are compared with non-equilibrium Green's function transport simulations with a good accuracy.

1. Introduction

As the dimension of semiconductor device scales down, undoped short-channel gate-all-around metal-oxide-semiconductor field-effect transistors (GAA MOSFETs) had been developed as one of the most important candidates for the next generation electronics [1-2]. Owing to their excellent gate controllability over the channel [3-5], such MOSFETs provide a better immunity to short-channel effects (SCEs), such as drain-induced barrier lowering (DIBL) and threshold voltage roll-off [1,6]. Since the gate length of the MOSFETs had reached sub-10 nm regime [7], it is widely investigated that the quantum transport has become significant of describing the properties in such devices [8]. Particularly, the quantum tunneling and ballistic transport of electrons across the source-to-drain potential barrier within GAA MOSFETs can be accurately simulated in several numerical simulators using non-equilibrium Green's function (NEGF) formalism [8-10]. However, the simulations require significantly long computation times, which are not suitable to the circuit level simulations.

In this work, we study the source-to-drain tunneling current in the GAA MOSFET shorter than 10nm, assuming the energy subband profile of an electron along the device channel can be analytically represented using one unknown parameter according to our previous work. Then, by considering the expression of the lowest subband level, the tunneling transmission coefficients can be numerically calculated using the Wentzel-Kramers-Brillouin (WKB) approximation. Therefore, we can propose a numerical compact model of the short-channel cylindrical GAA MOSFET including the source-to-drain tunneling in the subthreshold region from the Landauer formula. Furthermore, results obtained by this model are compared with those obtained by SILVACO NEGF simulation, successfully demonstrating a good agreement.

2. Calculation Theory

2.1. Model structure

The cylindrical GAA-MOSFET is considered with intrinsic silicon channel surrounded by the gate oxide and metal gate, as illustrated in figure 1. We assume that the source and drain are highly doped n-type silicon as ideal reservoirs of electrons. Thus, we also consider that the electrons flow into the source and drain without reflections back to the channel. In the following formulations, cylindrical coordinate with length z, radius r, and angular component φ is used. In addition, we also assume that the transverse and longitudinal effective masses are fixed at $m_t = 0.19m_0$ and $m_l = 0.91m_0$, respectively, in the (100)-oriented channel. The effective mass in the confinement cross section is represented as the expression $2(mt^{-1} + ml^{-1})^{-1}$ [11].

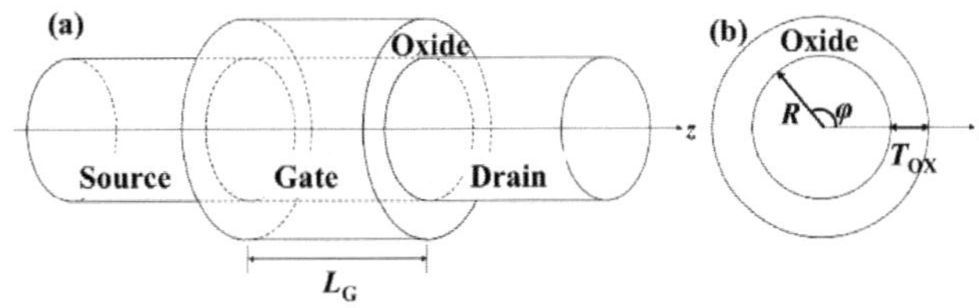

Figure 1. Schematic drawing of cylindrical GAA-MOSFET model considered in this work. (a) The device structure and (b) wire cross section, where the channel length, radius, and gate oxide thickness are denoted as L_G, R and T_{OX}, respectively.

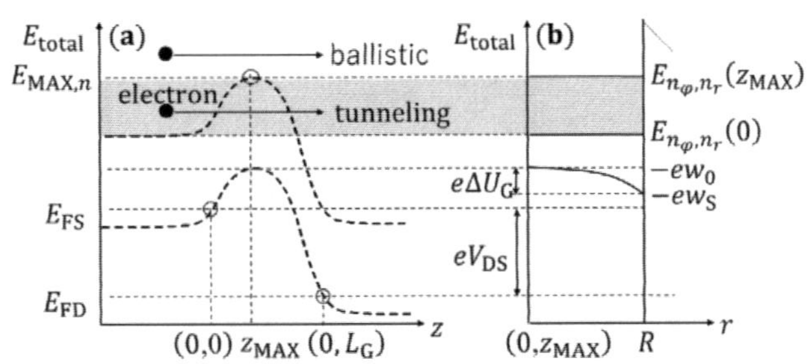

Figure 2. (a) Representation of conduction band edge and higher energy subabnd profile along the z-direction of channel. Electrons whose energy subband level is higher/lower than the potential barrier in the channel contribute to the ballistic/tunneling current. Turning points z_{MAX}, z_1 and z_2 have literal expressions. (b) Schematics of confinement potential energy distribution along r-component at z_{MAX} in the cross section.

2.2. Landauer formula

As shown in figure 2(a), the profiles of the conduction band edge and higher energy subband along z-direction are illustrated. The total energy of an electron is denoted as E_{total}, and the source and drain Fermi energies are denoted as $E_{F,S}$ and $E_{F,D}$, respectively. The difference in Fermi level between the source and drain is given by eV_{DS}, where V_{DS} is defined as voltage across the source and drain. The maximum energy level of nth-subband along the z-axis at the barrier top z_{MAX} is denoted as $E_{MAX,n}$, corresponding to the discrete subband level measured from $E_{F,S}$ expressed by $E_{n\varphi,nr}(z_{MAX})$, where $E_{n\varphi,nr}(z)$ represents the confinement energy level along z-direction with n_φ and n_r represented as the angular and radial quantum numbers, respectively. Then, the drain current can be represented using the Landauer formula as follows [12-14]:

$$I_{\mathrm{DS}} = \frac{e}{\pi\eta}\sum_{n_v}\sum_{n_\varphi,n_r}\int_0^\infty dE_z T(E_z)\times\left[f(E_{\mathrm{F,S}},E_{\mathrm{total}})-f(E_{\mathrm{F,D}},E_{\mathrm{total}})\right] \tag{1}$$

$$E_{\mathrm{total}} = E_{n_\varphi,n_r}(z)+E_z \tag{2}$$

where $T(E_z)$ is the transmission coefficients for electrons traveling through the source-to-drain energy barrier at E_z, E_z is the electron kinetic energy along z-direction, $f(E_{\mathrm{F,S}},E_{\mathrm{total}})$ denotes the Fermi-Dirac distribution function in the source reservoirs. Moreover, the source-to-drain tunneling and ballistic currents are considered involving electrons with the energies $E_{\mathrm{total}} < E_{\mathrm{MAX},n}$ and $E_{\mathrm{total}} > E_{\mathrm{MAX},n}$, respectively.

Figure 2(b) illustrates the potential profile along r-direction in the cross section at z_{MAX}. Defining the electrostatic potential at the center and surface of the cylindrical channel as w_0 and w_{S}, the conduction band edges at the center and surface are given by $-ew_0$ and $-ew_{\mathrm{S}}$, respectively. In addition, the potential is measured from $E_{\mathrm{F,S}}$. The difference between $-ew_0$ and $-ew_{\mathrm{S}}$ is described using an unknown parameter ΔU_{G}. According to our previous work, we use $\Delta U_{\mathrm{G}}(z)$ instead of ΔU_{G} to describe the z-direction dependence [13,14]. Therefore, the electron subband level $E_{n\varphi,nr}(z)$ can be rewritten in terms of $\Delta U_{\mathrm{G}}(z)$ as follows:

$$E_{n_\phi,nr}(z) = E^{\mathrm{q}0}_{n_\phi,n_r} - eV^*_{\mathrm{GS}} + e\left(H_{n_\phi,n_r} + \frac{4\pi\varepsilon_{\mathrm{CH}}}{C_{\mathrm{OX}}}\right)\Delta U_{\mathrm{G}}(z), \tag{3}$$

where $V_{\mathrm{GS}}^* = V_{\mathrm{GS}} - \varphi_{\mathrm{GC}} + w_{\mathrm{FB}}$, V_{GS} is the gate-source voltage, φ_{GC} is the work function difference between the gate and channel material, w_{FB} is the conduction band edge with respect to $E_{\mathrm{F,S}}/(-e)$ under the flat-band condition, C_{OX} is the gate oxide capacitance per unit length, $\varepsilon_{\mathrm{CH}}$ is the dielectric constant of the channel, $E_{n\varphi,nr}{}^{\mathrm{q}0}$ is the confinement energy level in one-dimensional infinite square well potential with respect to $-ew_{\mathrm{S}}$. The matrix element $H_{n\varphi,nr}$ is given by [13]

$$H_{n_\varphi,n_r} = \int_0^R \Psi^{0*}_{n_\varphi,n_r}(r)\left(1-\frac{r^2}{R^2}\right)\Psi^0_{n_\varphi,n_r}(r)rdr, \tag{4}$$

where $\Psi_{n\varphi,nr}{}^0(r)$ denotes unperturbed electron wave function in terms of confinement. Furthermore, $\Delta U_{\mathrm{G}}(z)$ is derived based on our previous study [14]:

$$\Delta U_{\mathrm{G}}(z) = A\exp(\gamma\cdot z)+B\exp(-\gamma\cdot z), \tag{5}$$

where coefficients A and B are expressed as

$$A = K\frac{(V_{\mathrm{bi}}-V^*_{\mathrm{GS}})\cdot[1-\exp(-\gamma\cdot L_{\mathrm{G}})]+V_{\mathrm{DS}}}{2\sinh(\gamma\cdot L_{\mathrm{G}})}, \tag{6}$$

$$B = K\frac{(V_{\mathrm{bi}}-V^*_{\mathrm{GS}})\cdot[\exp(\gamma\cdot L_{\mathrm{G}})-1]-V_{\mathrm{DS}}}{2\sinh(\gamma\cdot L_{\mathrm{G}})}, \tag{7}$$

with

$$K = \frac{\gamma^2R^2}{4\left(1-\frac{\gamma^2R^2}{4}\right)^2+\gamma^2R^2}. \tag{8}$$

Here, we define V_{bi} as the built-in potential at source-channel and channel-drain junctions, γ is the scaling parameter depending on device structure given by

$$\gamma = \frac{2}{R}\left(\frac{4\pi\varepsilon_{\text{CH}}}{C_{\text{OX}}}+1\right)^{-1/2}. \tag{9}$$

Moreover, by solving the derivative of (3) equals to zero, z_{MAX} is then represented by

$$z_{\text{MAX}} = \frac{1}{2\gamma}\ln\left(\frac{B}{A}\right). \tag{10}$$

2.3. Transmission coefficient calculation

The source-to-drain tunneling current cannot be neglected in a device when the channel length is shorter than 10 nm [15]. In the subthreshold region, we assume the electron transmission coefficients for each subband $T_{n\varphi,nr}(E_z)$ can be calculated using the WKB approximation as follows [15,16]:

$$T_{n_\varphi,n_r}(E_z) = \exp\left[-2\int_{z_1}^{z_2}\kappa(z,E_z)dz\right], \tag{11}$$

where $\kappa(z,E_z)$ is the electron wave vector inside the potential barrier at E_z along the z-direction, which is represented as follows:

$$\kappa(z,E_z) = \frac{1}{\hbar}\left[2m_t\left(E_{n_\phi,n}(z)-E_z\right)\right]^{1/2}. \tag{12}$$

By substituting (3) into (11), $T_{n\varphi,nr}(E_z)$ can be numerically obtained. Then, z_1 and z_2 are defined as the turning points, at which the electron energy corresponds to E_z, respectively. Thus, resulting that z_1 and z_2 read as the same form obtained in [11]

$$z_{1,2} = \frac{1}{\gamma}\ln\left[\frac{-C\pm\Lambda^{1/2}}{2A}\right], \tag{13}$$

where the quantities C and Λ are given by

$$C = \frac{E^{q0}_{n_\varphi,n_r} - eV^*_{\text{GS}} - E_z}{e\left(H_{n_\varphi,n_r} + \dfrac{4\pi\varepsilon_{\text{CH}}}{C_{\text{OX}}}\right)}, \tag{14}$$

$$\Lambda = C^2 - 4AB. \tag{15}$$

Note that this method requires a numerical subroutine to evaluate the transmission coefficients, it makes this model become numerical. However, by substituting (11) into (1), we can derive the expression of the source-to-drain tunneling current in the subthreshold region.

3. Results and Discussion

In the subthreshold region, since most carriers are only exited up to the lowest subband in the channel, the tunneling current can be expressed by only considering the lowest the subband (n_φ=0, n_r=1). Figure 3 shows comparisons of the lowest subband profile along the channel between NEGF simulation [17] (green open circle lines) and our model (red solid lines). The dashed lines show interfaces between source/channel and drain/channel. Three lines correspond to the results with V_{GS} = 0, 0.2 and 0.5V, respectively. In order to achieve good agreements, channel length used in the compact model has been expanded into the source and the drain regions, respectively in [14]. It demonstrates that the potential profiles calculated using our compact model successfully capture the shape of the potential calculated by NEGF simulator in the subthreshold operating region. Figure 4 illustrates electron kinetic energy dependencies of the transmission coefficients for different gate bias. Four lines correspond to the results with V_{GS} = 0, 0.1, 0.2 and 0.3V. Note that the transmission coefficients must be 1 when E_{total} of an electron becomes larger than $E_{0,1}(z_{MAX})$ under different gate bias condition correspondingly.

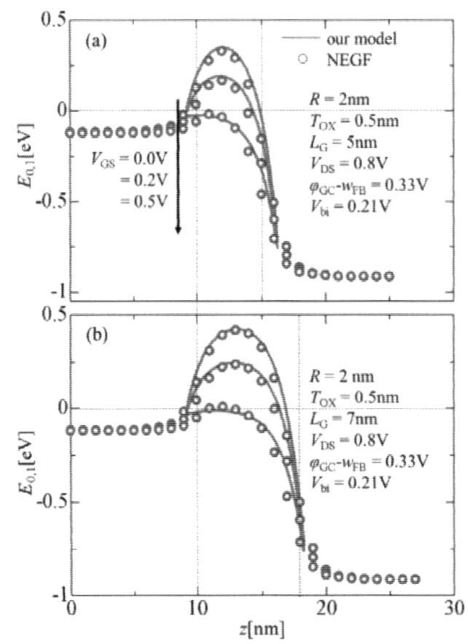

Figure 3. Comparison results of the lowest subband prole plot between the compact model and NEGF simulation under different gate bias conditions.

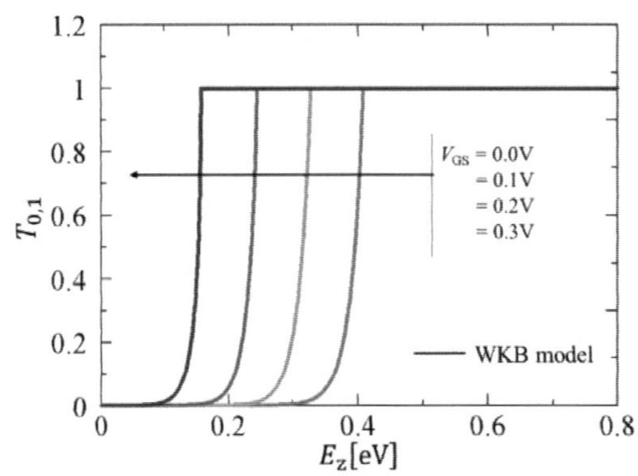

Figure 4. Electron kinetic energy dependence of the transmission coefficient calculated by the compact model with different gate bias.

In figure 5, for different channel length setting to 5 and 7 nm from the top, the subthreshold drain current vs. gate voltage characteristics are shown, demonstrating good accuracy with the WKB compact model. The green open circle lines and red solid lines show the results obtained by NEGF simulation and our WKB model, respectively. Note that the WKB model current decreases as V_{GS} gets large. That is considered when the potential barrier height in the channel decreases, the amount of electrons tunneling through the source-to-drain potential barrier reduces. Figure 6 shows the comparison results of the subthreshold drain current between models considering the WKB tunneling (red solid lines) and our ballistic model (blue dashed lines). However, note that $T_{0,1}$ is assumed to be 1 for $E_{total} > E_{MAX,0,1}$ and 0 for $E_{total} < E_{MAX,0,1}$ when considering the ballistic model. As expected, the subthreshold drain current evaluated using the ballistic model is lower than the WKB tunneling current. It demonstrates that the source-to-drain tunneling becomes dominant and contributes significantly to the current in the subthreshold region when the channel length of the GAA MOSFET decreases below 10 nm.

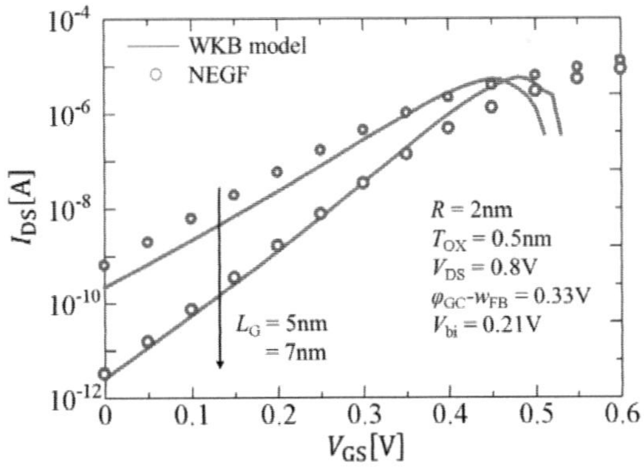

Figure 5. Comparison results of current characteristics in the subthreshold region between NEGF simulation and this WKB model.

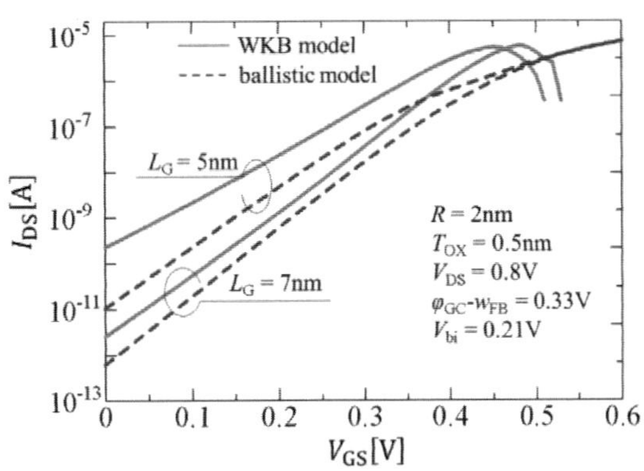

Figure 6. Comparison results of current characteristics in the subthreshold region between the models considering ballistic transport and this WKB model, respectively.

4. Conclusions

We have presented a compact model of short-channel cylindrical GAA MOSFET including the source-to-drain tunneling. By using the WKB approximation to evaluate the transmission coefficients for the source-to-drain tunneling, we calculate the subthreshold drain current considering the lowest subband from the Landauer formula. In addition, this compact model has been tested against NEGF simulation, and showed a good accuracy in the subthreshold region. Furthermore, we intend to extend this tunneling current model of the GAA MOSFET using the WKB approximation with an analytical expression in our future study. Then, we can propose an analytic compact model of ballistic GAA MOSFETs for all operating regions.

5. Acknowledgment

This work has been supported by National Science and Technology Major Project of the Ministry of Science and Technology of China (Grant No. 2015ZX03003010).

6. References

[1] Cousin B, Reyboz M, Rozeau O, Jaud M -A, Ernst T and Jomaah J, 2011, *Solid-State Electron.*, **56**, **1**, 40-46.

[2] Colinge J -P, 2004, *Solid-State Electronics*, **48**, **6**, 897-905.
[3] Hosseini R and Teimuorzadeh N, 2013, *Physical Review & Research International*, **3**, 568-576.
[4] Colinge J -P, 2008, FinFETs and Other Multi-Gate Transistors, *Spinger*.
[5] Abdi M A, Djeal F, Arar D and Hafiane M L, 2008, *J. Mater. Sci.: Mater. Electron.*, **19**, **1**, 248-253.
[6] Suk S D, Li M, Yeoh Y Y, Yeo K H, Cho K H, Ku I K, Cho H, Jang W, Kim D –H, Park D and Lee W –S, 2007, *Electron Devices Meeting, IEDM 2007. IEEE International*, 891-894.
[7] Bangsaruntip S, Balakrishnan K, Cheng S -L, Chang J, Brink M, Lauer I, Bruce R L, Engelmann S U, Pyzyna A, Cohen G M, Gignac L M, Breslin C M, Newbury J S, Klaus D P, Majumdar A, Sleight J W and Guillorn M A, 2013, *Electron Devices Meeting, IEDM 2013. IEEE International.*, 20.2.1-4.
[8] Hosseini R, Fathipour M and Faez R, 2012, *Int. J. Phys. Sci.*, **7**, **28**, 5054-5061.
[9] Adnan M M R, Hafiz M. S B, Tasneem N and Khosru Q D M, 2016, *Electron Devices and Solid-State Circuits (EDSSC), 2016 IEEE International Conference on*, 62-65.
[10] Khan T, Iztihad H M, Sufian A, Alam M N K, Mollah M N and Islam M R, 2015, *International Conference on Electrical Engineering and Information Communication Technology,* 1-4.
[11] Autran J L, Munteanu D, Tintori O, Aubert M and Decarre E, 2004, *2004 NSTI Nanotechnology Conference and Trade Show, NSTI Nanotech 2004*, **2**, 171-174.
[12] Lundstrom M, 1997, *IEEE Electron Device Lett.*, **18**, **7**, 361363.
[13] Cheng H, Uno S, Numata T and Nakazato K, 2013, *Jpn. J. Appl. Phys.*, **52**, 04CN03**.**
[14] Cheng H, Uno S and Nakazato K, 2015, *J Comput Electron*, **14**, **1**, 321-328.
[15] Chaudhry A and Roy J N, 2010, *J. electron. sci. and technol.*, **8**, **4**, 346-350.
[16] Griffiths D J, Introduction to quantum mechanics, 2nd edition, *Pearson Eduction Inc*.
[17] Atlas User's Manual, 2011, *Silvaco, Inc. CA. www.silvaco.com*, **Charpter 13**, 754-795.

IOP Conf. Series: Journal of Physics: Conf. Series **1026** (2018) 012012 doi:10.1088/1742-6596/1026/1/012012

Design of an Inductor-less Direct AC-AC Converter Realizing 1/4x and 4x Conversion

K Eguchi[1], F Asadi[2], K Kuwahara[1] and I Oota[3]

1 Department of Information Electronics, Fukuoka Institute of Technology, Fukuoka, Japan
2 Department of Mechatronics Engineering, Kocaeli University, Kocaeli, Turkey
3 Department of Electronics Engineering and Computer Science, National Institute of Technology, Kumamoto College, Kumamoto, Japan
E-mail: eguti@fit.ac.jp

Abstract. In this paper, an inductor-less direct AC-AC converter realizing 1/4x and 4x conversion is designed by using switched-capacitor (SC) techniques. Unlike conventional SC AC-AC converters, the proposed AC-AC converter is synthesized with two converter blocks. By converting an AC input twice by these converter blocks, the 1/4x step-down / 4x step-up conversion is achieved by the proposed converter. The proposed converter has a simple circuit configuration, where each converter block consists of only four bidirectional switches and two capacitors. Hence, the proposed converter can realize smaller circuit components and higher input power factor than conventional converters. The effectiveness of the proposed converter was confirmed by simulation program with integrated circuit emphasis (SPICE) simulations and theoretical analysis. The SPICE simulation revealed that the proposed converter outperforms the conventional AC-AC converter using a flying capacitor.

1. Introduction

As an alternative appliance of autotransformers, an AC-AC converter has been developed by using switched-capacitor (SC) techniques in these last few decades. The SC AC-AC converter can be designed without magnetic components. Therefore, small size and light weight can be realized by the SC ac-ac converter, though the autotransformer suffers from its heavy weight and large volume for a magnetic core and winding.

In past studies, many researchers have undertaken the development of the SC AC-AC converter. In 2006 and 2009, Terada et al. and Eguchi et al. designed the SC AC-AC converter by utilizing a ring-type converter [1, 2]. However, these ring-type-based AC-AC converters require many circuit components and complex circuit control. To simplify the circuit configuration, Lazzarin et al. and Andersen et al. developed the direct SC AC-AC converter with a flying capacitor [3, 4]. In [3, 4], the direct AC-AC converter achieves the conversion ratio of 1/2x and 2x. Following this study, You et al. expanded Lazzarin's converter to realize the conversion ratio of 1/4 and 4x [5]. However, these direct AC-AC converters [3-5] suffer from low power efficiency and low input power factor. To solve these problems, Eguchi et al. proposed the nesting-type AC-AC converter [6]. By using multi-conversion topology, Eguchi's converter can achieve higher power efficiency and higher input power factor than the converters reported in [3-5]. However, Eguchi's converter requires many circuit components.

IOP Conf. Series: Journal of Physics: Conf. Series **1026** (2018) 012012 doi:10.1088/1742-6596/1026/1/012012

In this paper, an inductor-less direct AC-AC converter realizing 1/4x and 4x conversion is designed by using SC techniques. Unlike the conventional SC AC-AC converters, the proposed AC-AC converter is synthesized with two converter blocks. By converting an AC input twice by two converter blocks, the proposed converter offers the 1/4x stepped-down / 4x stepped-up voltage to an output load. The proposed converter has a simple circuit configuration, where each converter block consists of only four bidirectional switches and two capacitors. Hence, the proposed converter can realize smaller circuit components than conventional converters. Furthermore, high input power factor can be achieved by reducing circuit components from the conventional converter. The characteristics of the proposed converter are evaluated through simulation program with integrated circuit emphasis (SPICE) simulations and theoretical analysis.

2. Circuit Configuration

2.1. Conventional converter

The circuit configuration of the conventional AC-AC converter [3, 4] is depicted in figure 1, where S_1 and S_2 are bidirectional switches which are driven by non-overlapped two-phase clock pulses with constant switching frequency and duty cycle. By controlling S_1 and S_2, the conventional converter of figure 1 provides the 1/2x stepped-down / 2x stepped-up voltage to an output load.

The operation principle of figure 1 (a) is as follows. As figure 1 (a) shows, the input ac voltage is divided by main capacitors C_2 and C_3. By changing the connection of the flying capacitor C_1, the voltages of C_2 and C_3 are averaged. Therefore, the conventional converter achieves the 1/2x step-down conversion. Of course, as shown in figure 1 (b), the conventional converter can achieve the 2x step-up conversion by swapping the input and output terminals. Furthermore, by increasing the number of stages as described in [5], the conventional converter can realize high gain. The output voltage of the conventional converter can be expressed as

$$\frac{v_{out}}{v_{in}} = \begin{cases} 1/(N+1) & \text{(Stepup)} \\ N+1 & \text{(Stepdown)} \end{cases} \tag{1}$$

where N is the number of stages. The number of circuit components for the conventional converter is $2(N+1)$ switches and $2N+1$ capacitors. Concretely, in the conversion ratio of 1/4x and 4x, the conventional converter requires 8 bidirectional switches and 7 capacitors [5]. However, in the conventional converter, the number of circuit components increases linearly according to the step-up/step-down gain. For this reason, the conventional converter suffers from the reduction of input power factor and power efficiency.

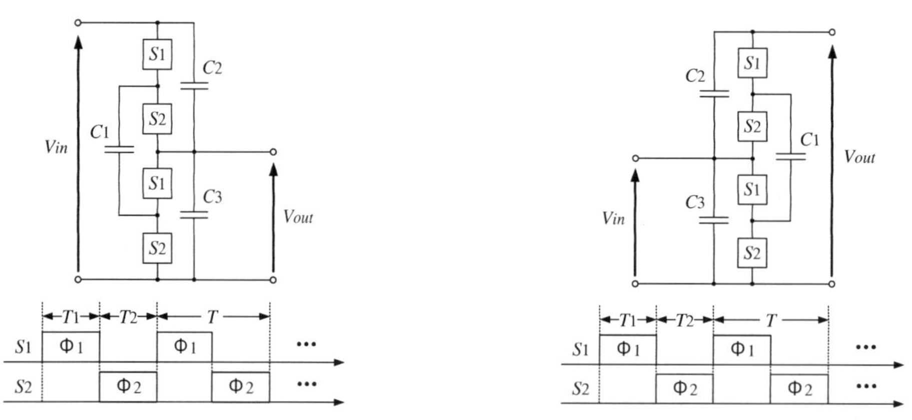

Figure 1. Conventional AC-AC converter: (a) 1/2x step-down gain and (b) 2x step-up gain.

IOP Conf. Series: Journal of Physics: Conf. Series **1026** (2018) 012012 doi:10.1088/1742-6596/1026/1/012012

2.2. Proposed converter

The circuit configuration of the proposed AC-AC converter is shown in figure 2. As figure 2 shows, the proposed converter consists of two converter blocks. figure 3 illustrates the instantaneous equivalent circuits of figure 2, where the bidirectional switch is modelled by an ideal lossless switch and an on-resistance R_{on}. In each converter block, the I/O terminal is connected alternately to one of capacitors to realize 1/2x / 2x conversion. In other words, electric charges in C_1 and C_2 are averaged by changing the connection of I/O terminal. Therefore, unlike the conventional converter of figure 1, the number of capacitors connected to the I/O terminals is constant. By connecting two converter blocks in series, the proposed converter shown in figure 2 achieves the conversion ratio of 1/4x / 4x. In general case, the output voltage of the proposed converter is given by

$$\frac{v_{out}}{v_{in}} = \begin{cases} \prod_{i=1}^{M} \frac{1}{2} & \text{(Stepup)} \\ \prod_{i=1}^{M} 2 & \text{(Stepdown)} \end{cases}, \tag{2}$$

where M is the number of converter blocks. The number of circuit components for the proposed converter is $4M$ switches and $2M$ capacitors. Owing to the proposed topology, the proposed converter can reduce the number of circuit components. As figure 2 shows, the proposed converter can be synthesized with 8 bidirectional switches and 4 capacitors in the conversion ratio of 1/4x and 4x.

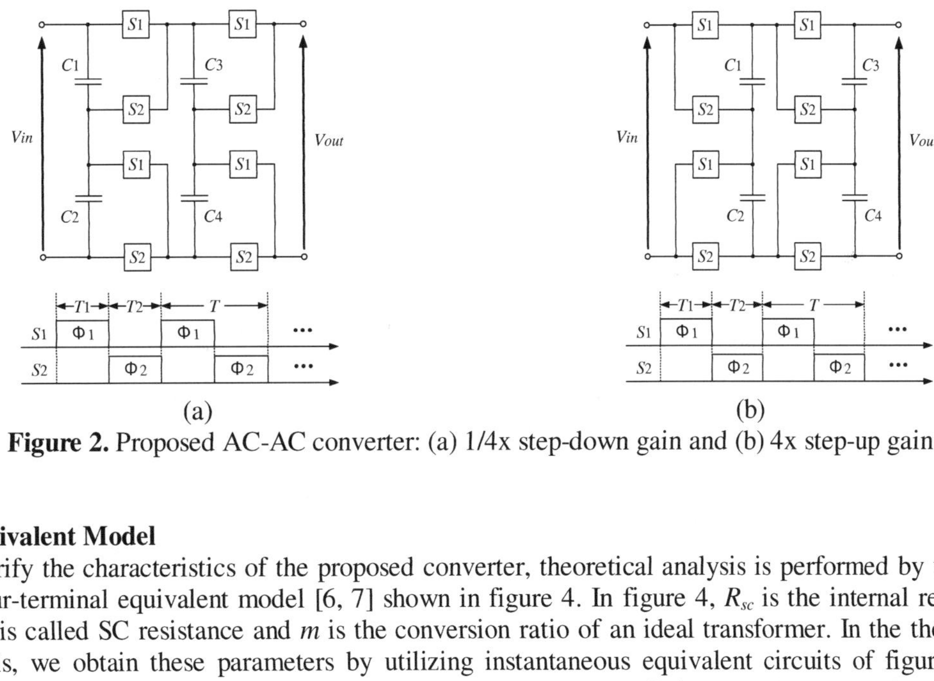

Figure 2. Proposed AC-AC converter: (a) 1/4x step-down gain and (b) 4x step-up gain.

3. Equivalent Model

To clarify the characteristics of the proposed converter, theoretical analysis is performed by utilizing the four-terminal equivalent model [6, 7] shown in figure 4. In figure 4, R_{sc} is the internal resistance which is called SC resistance and m is the conversion ratio of an ideal transformer. In the theoretical analysis, we obtain these parameters by utilizing instantaneous equivalent circuits of figure 3 and derive the theoretical formulas to estimate the maximum power efficiency and output voltage, where the AC input is assumed as a pulse waveform. To save space, only the theoretical analysis for the conversion ratio of 1/4x is discussed in this paper.

In a steady state, we consider the differential value of electric charges Δq_{Ti}^{k} in C_k (k=1, ..., 4). In the steady state, the overall change in electric charges is zero. Thus, the differential value Δq_{Ti}^{k} satisfies

$$\Delta q_{T_1}^{k} + \Delta q_{T_2}^{k} = 0, \tag{3}$$

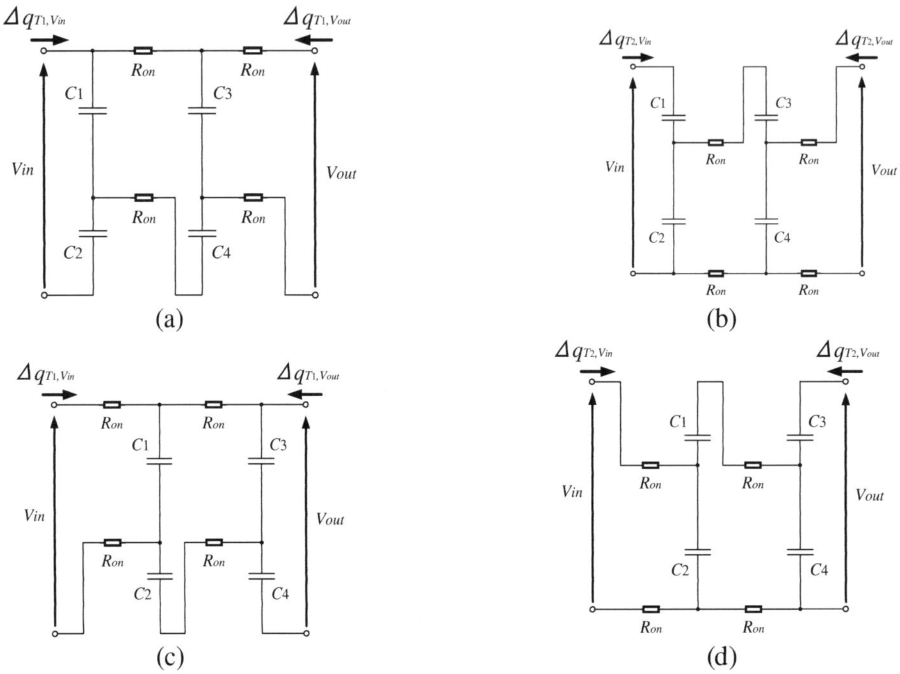

Figure 3. Instantaneous equivalent circuit: (a) Figure 2 (a) in state-T_1, (b) Figure 2 (a) in state-T_2, (c) Figure 2 (b) in state-T_1, and (d) Figure 2 (b) in state-T_2.

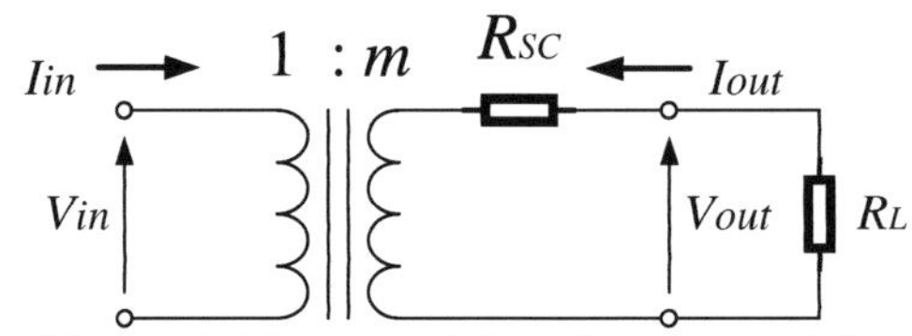

Figure 4. Four termial equivalent model.

where $\Delta q_{Ti}^{\ k}$ (i=1, 2) denotes the electric charge of the k-th capacitor in State-T_i. The interval of T_1 and T_2 satisfies

$$T = T_1 + T_2 \quad \text{and} \quad T_1 = T_2 = \frac{T}{2}, \tag{4}$$

where T is a period of the clock pulse.
In State-T_1 of figure 3 (a), the differential values of electric charges in the input and the output, $\Delta q_{T1,vin}$ and $\Delta q_{T1,vout}$, are expressed as

$$\Delta q_{T_1,v_{in}} = \Delta q_{T_1}^{1} + \Delta q_{T_1}^{3} - \Delta q_{T_1,v_{out}}, \tag{5}$$

$$\Delta q_{T_1}^{1} = \Delta q_{T_1}^{2} - \Delta q_{T_1}^{4}, \tag{6}$$

and

$$\Delta q_{T_1,v_{out}} = \Delta q_{T_1}^{3} - \Delta q_{T_1}^{4}. \tag{7}$$

On the other hand, in State- T_2 of figure 3 (b), the differential values of electric charges in the input and the output, $\Delta q_{T2,vin}$ and $\Delta q_{T2,vout}$, are expressed as

$$\Delta q_{T_2,v_{in}} = \Delta q_{T_2}^1, \tag{8}$$

$$\Delta q_{T_2}^1 = \Delta q_{T_2}^2 + \Delta q_{T_2}^3, \tag{9}$$

and

$$\Delta q_{T_2,v_{out}} = \Delta q_{T_2}^4 - \Delta q_{T_2}^3. \tag{10}$$

Here, the I/O currents, I_{in} and I_{out}, are expressed by the overall change in electric charges. Using equations (5) - (10), we have I_{in} and I_{out} as

$$I_{in} = \frac{\Delta q_{v_{in}}}{T} = \frac{\Delta q_{T_1,v_{in}} + \Delta q_{T_2,v_{in}}}{T}, \tag{11}$$

and

$$I_{out} = \frac{\Delta q_{v_{out}}}{T} = \frac{\Delta q_{T_1,v_{out}} + \Delta q_{T_2,v_{out}}}{T}. \tag{12}$$

In equations (11) and (12), Δq_{vin} and Δq_{vout} are electric charges in V_{in} and V_{out}, respectively. Substituting equations (3) - (10) into equations (11) and (12) yields the following relation between the input current and the output current:

$$i_{in} = -\frac{1}{4} i_{out}, \tag{13}$$

where

$$\Delta q_{vin} = -\frac{1}{4}\Delta q_{vout}. \tag{14}$$

Therefore, we have m=1/4.

Next, the total consumed energy in one period is discussed to derive the SC resistance R_{SC}. In figures 3 (a) and (b), the consumed energy, W_{T1} and W_{T2}, is expressed as

$$W_{T_1} = \frac{\left(\Delta q_{T_1}^4\right)^2}{T_1} R_{on} + \frac{\left(\Delta q_{T_1,v_{in}} - \Delta q_{T_1}^1\right)^2}{T_1} R_{on} + \frac{\left(\Delta q_{T_1,v_{out}}\right)^2}{T_1} 2R_{on} \tag{15}$$

$$W_{T_2} = \frac{\left(\Delta q_{T_2}^3\right)^2}{T_2} R_{on} + \frac{\left(\Delta q_{T_2,v_{in}} - \Delta q_{T_2}^2\right)^2}{T_2} R_{on} + \frac{\left(\Delta q_{T_2,v_{out}}\right)^2}{T_2} 2R_{on} \tag{16}$$

Substituting equations (3) - (10) into equations (15) and (16) yields

$$W_T = W_{T1} + W_{T2} = \left(\frac{5R_{on}}{2}\right)\frac{\left(\Delta q_{v_{out}}\right)^2}{T}, \tag{17}$$

where W_T is the total consumed energy in one period. Therefore, we have the SC resistance R_{SC} as

$$R_{SC} = \frac{5}{2} R_{on}, \tag{18}$$

because the consumed energy of figure 4 can be defined as

$$W_T = R_{SC}\frac{\left(\Delta q_{v_{out}}\right)^2}{T}. \tag{19}$$

By combining equations (13) and (18), the proposed converter is presented by

$$\begin{bmatrix} V_{in} \\ I_{in} \end{bmatrix} = \begin{bmatrix} 4 & 0 \\ 0 & 1/4 \end{bmatrix} \begin{bmatrix} 1 & (5/2)R_{on} \\ 0 & 1 \end{bmatrix} \begin{bmatrix} V_{out} \\ -I_{out} \end{bmatrix}, \tag{20}$$

because the four-terminal model of figure 4 can be expressed by the K-matrix. Finally, the equation (20) gives the maximum efficiency and the maximum output voltage as

$$\eta_{\max} = \frac{R_L}{(5/2)R_{on} + R_L} \tag{21}$$

and

$$v_{out.\max} = \left\{ \frac{R_L}{(5/2)R_{on} + R_L} \right\} \left(\frac{v_{in}}{4} \right). \tag{22}$$

Of course, the characteristics of the step-up mode can be analyzed by the same method. As equations (21) and (22) show, the SC resistance is one of the most important factors to improve circuit characteristics.

Tables 1 and 2 show the comparison of characteristics between the proposed converter and the conventional converter [5]. As table 1 shows, the number of circuit components for the proposed converter is the smaller than that of the conventional converter [5]. Hence, the proposed converter can realize simple circuit configuration. On the other hand, the internal resistance of the proposed converter is smaller than that of the conventional converter [5]. Therefore, the proposed converter can achieve higher power efficiency than the conventional converter [5].

Table 1. Comparison of the number of circuit components.

	Gain	Switch	Capacitor
Proposed converter	1/4 x or 4x	8	4
Conventional converter [5]	1/4 x or 4x	8	7

Table 2. Comparison of the internal resistances.

	Gain	R_{SC}
Proposed converter	1/4 x	$(5/2)R_{on}$
	4 x	$40R_{on}$
Conventional converter [5]	1/4 x	$3R_{on}$
	4 x	$48R_{on}$

4. Simulations

In SPICE simulations, the characteristics of the proposed converter were compared with that of the conventional ac-ac converter [5]. The SPICE simulations were conducted under conditions that V_{in} = 220V@50Hz, $C_1 = \ldots = C_4 = 33\mu F$, $C_{out} = 1nF$, $R_{on} = 0.83\Omega$, $T = 10\mu s$, and $T_1 = T_2 = 5\mu s$, where C_{out} denotes the output capacitance of the ac-ac converter.

Figure 5 shows the simulated output voltage as a function of time. In figure 5 (a) and (b), about 55V@50Hz and 875@50Hz outputs are obtained by converting the 220@50Hz input.

Figure 6 shows the simulated power efficiency as a function of the output power. In figure 6 (a), about 8% power efficiency is improved by the proposed converter when the output power is 0.25kW. On the other hand, in figure 6 (b), about 12% power efficiency is improved by the proposed converter when the output power is 4kW. Figure 7 shows the input power factor as a function of the output power. In figure 7 (a), more than 0.4 input power factor is improved by the proposed converter when the output power is 0.25kW. On the other hand, in figure 7 (b), more than 0.2 input power factor is improved by the proposed converter when the output power is 4kW.

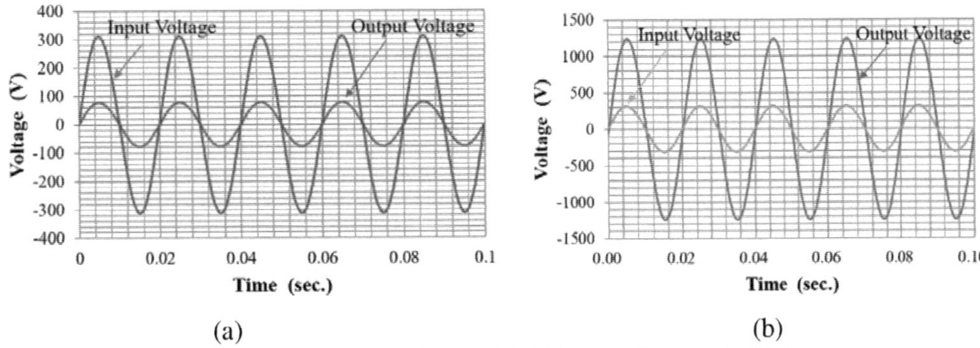

Figure 5. Simulated output voltage: (a) 1/4x step-down and (b) 4x step-up.

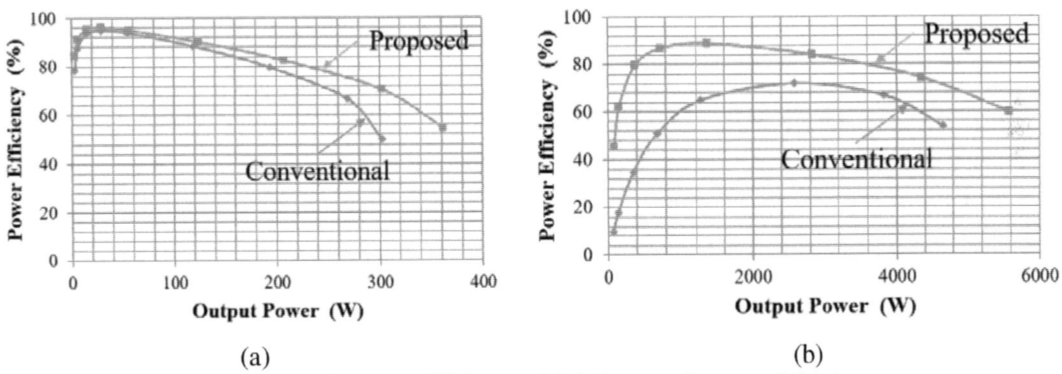

Figure 6. Simulated power efficiency: (a) 1/4x step-down and (b) 4x step-up.

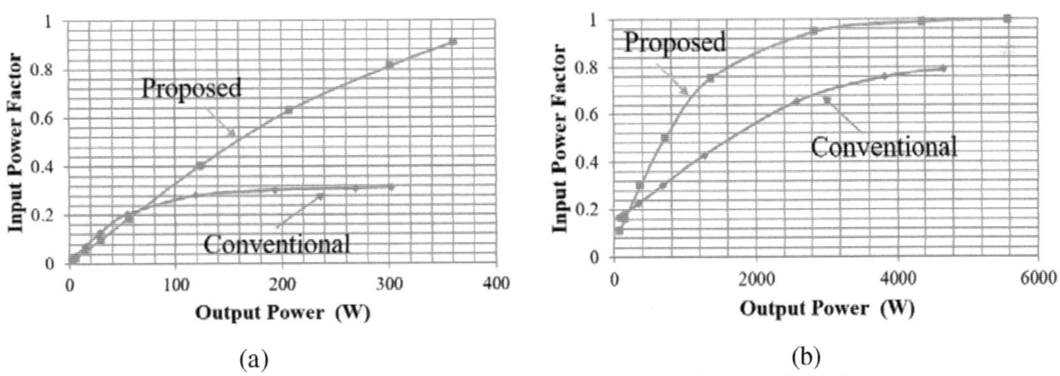

Figure 7. Simulated input power factor: (a) 1/4x step-down and (b) 4x step-up.

5. Conclusion

An inductor-less direct AC-AC converter realizing 1/4x and 4x conversion has been proposed in this paper. The results of this work are as follows:

1. The proposed converter consisting of only 8 switches and 4 capacitors can achieve step-down / step-up conversion.

2. When the output power was 0.25kW, the proposed 1/4x step-down converter improved about 8% power efficiency from the conventional converter. On the other hand, when the output power was 4kW, the proposed 4x step-up converter improved about 12% power efficiency from the conventional converter.
3. When the output power was 0.25kW, the proposed 1/4x step-down converter improved more than 0.4 input power factor from the conventional converter. On the other hand, when the output power was 4kW, the proposed 4x step-up converter improved more than 0.2 input power factor from the conventional converter.

The experimental evaluation is left to a future study.

6. References

[1] Terada S, Oota I, Eguchi K and Ueno F 2006 *Proc. of the 37th IEEE Power Electronics Specialists Conference* 1884-1887.
[2] Eguchi K, Oota I, Terada S and Inoue T 2009 *International Journal of Innovative Computing, Information and Control* **5** 2927-2938.
[3] Andersen R L, Lazzarin T B and Barbi I 2013 *IEEE Trans. on Power Electronics* **28** 3329-3340.
[4] Lazzarin T B, Andersen R L and Barbi I 2015 *IEEE Trans. on Industrial Electronics* **62** 735-745.
[5] You J and Hui C 2014 *Proc. of the 2014 17th Int. Conf. on Electrical Machines and Systems* 3205-3207.
[6] Eguchi K, Do W, Oota I and Sasaki H 2017 *ICIC Express Letters, Part B: Applications* **8** 1191-1198.
[7] Eguchi K, Do W, Kittipanyangam S, Abe K and Oota I 2016 *International Journal of Innovative Computing, Information and Control* **12** 1411-1421.

IOP Conf. Series: Journal of Physics: Conf. Series **1026** (2018) 012013 doi:10.1088/1742-6596/1026/1/012013

The analysis of IEEE 802.11 DCF protocol based on LEO satellite Wi-Fi network

Wen He[1], **Xiaofeng Tao**[1], **Defeng Ren**[1]

[1] China Academy of Space Technology (Xi'an), China

Abstract. Aiming at the problems of traditional IEEE 802.11 DCF protocol applied to LEO satellite Wi-Fi network, the influence of propagation delay, hidden terminal and user collision on the throughput of MAC layer is fully analysed, and a specific scheme is proposed to adjust the ACK Timeout and the minimum contention window value. The improved two-dimensional Markov model has been used to analyse the DCF protocol throughput theoretically. By using MATLAB simulation, it is verified that the throughput performance is directly related to the transmission probability, and throughput performance is negative correlated with the number of nodes and transmission distance.

1. Introduction

Due to remote locations, sparsely populated areas, and low returns, many rural areas in developing countries in the world have no access to the communication network. They are in short supply of a low-cost solution to network connectivity. In recent years, some people have proposed to combine satellite communications with Wi-Fi technology and utilize LEO satellites to achieve wide-area coverage of Wi-Fi, making Wi-Fi more widely available and more flexible for more users. However, most wireless LANs currently on the market use the IEEE 802.11 standard, and the media access control layer mainly uses Distributed Coordination Function (DCF). However, the LEO satellite Wi-Fi network has the problems of longer propagation delay, hidden terminal, and more user collision. Some technologies of the conventional IEEE 802.11 DCF protocol have shown incompatibility in this network environment, resulting in a drastic throughput reduce [1].

This paper briefly introduces DCF protocol and analyses the existing problems in the LEO satellite Wi-Fi network. According to the characteristics of the LEO satellite channel environment, the corresponding improvements are simply proposed. A modified Markov chain analysis model established. The theoretical analysis results are verified by MATLAB.

2. A brief introduction to DCF protocol

DCF protocol is divided into basic CSMA/CA mode and RTS / CTS mode [2]. The RTS / CTS mode solves the problem of short-range "hidden terminal" by introducing RTS frames and CTS frames.

In the CSMA/CA mode, when the site has data to be transmitted, the busy/free state of the channel must be determined first. Continue to listen when listening to the channel is busy; once idle, and reach the DIFS duration, the data can be sent to the destination node. A random back-off process is performed before sending data, binary exponential back-off algorithm, as shown in Figure 1.

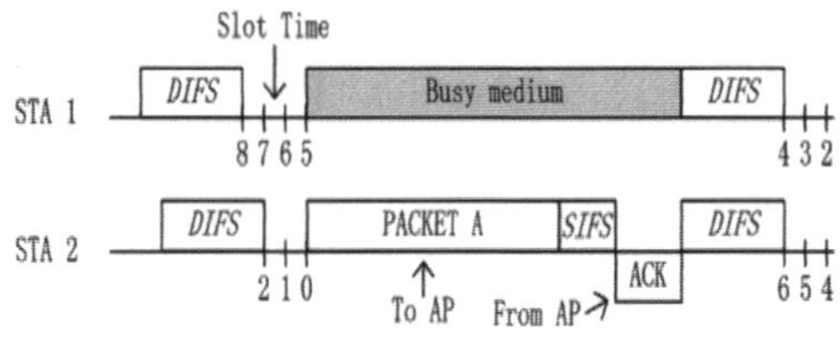

Figure 1. Random back-off process.

Each node in the retreat process needs to maintain a window parameter. When a node fails to transmit, the contention window of the node will double. When increasing to the maximum, continuous retransmission will keep the maximum value unchanged until the node sends successfully, or the limit of maximum retransmission times is reached, the contention window will be reset to the minimum.

3. The analysis of LEO satellite Wi-Fi MAC protocol

Compared with the quasi-static channel environment of the ground Wi-Fi, The LEO satellite link has a long propagation time and a large area. The direct application of traditional Wi-Fi to the satellite scene can seriously affect the performance of the MAC protocol. Figure 2 shows the LEO satellite Wi-Fi system model used in this study. With LEO satellite as the access point AP, it can realize two-way communication with terrestrial base stations via Wi-Fi. The base station can form an extended terrestrial wireless link with a terminal such as a mobile phone.

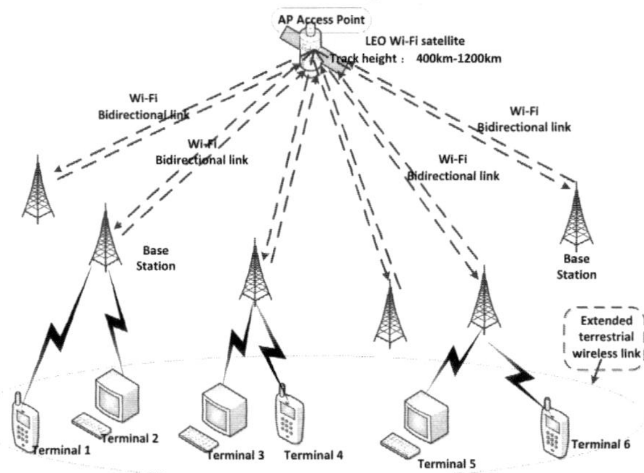

Figure 2. LEO satellite Wi-Fi system model.

3.1. Effect of propagation delay on MAC protocol

802.11 standard uses the Positive Acknowledgment mechanism. All transmitted frames must be acknowledged. The original DCF protocol ACK Timeout is set to 44us, the cell radius does not exceed 1.2km. Relative to the terrestrial Wi-Fi network, The LEO satellite link environment has a long propagation delay. If the ACK Timeout is set too short, it may cause wrong judgment of the transmission situation, which causes the data frame to be retransmitted continuously and may eventually be discarded, thereby reducing the system throughput. Figure 3 shows the interaction of MAC layer data frame in CSMA / CA mode.

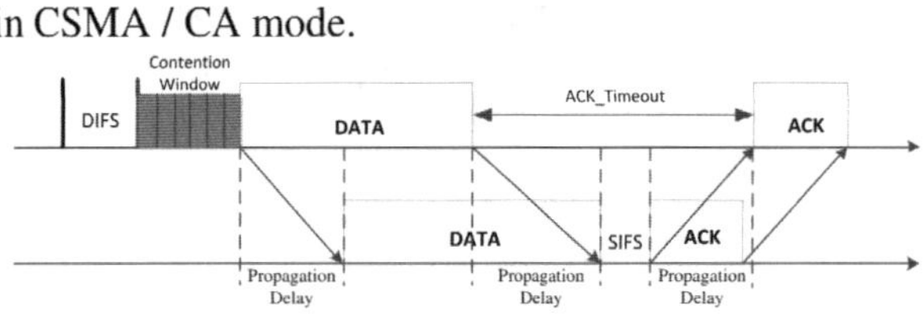

Figure 3. The interaction of MAC layer data frame.

Therefore, considering setting a reasonable ACK Timeout based on propagation delay is a better way to improve the system performance. As shown below:

$$ACK\ Timeout = 2PropTime + SIFS + SlotTime \tag{1}$$

3.2. The problem of "hidden terminal"

LEO satellites with orbital altitude of 400km can cover hundreds of thousands of square kilometres. Due to the limited transmitting power of the terrestrial terminals, the terminals that are far apart on the ground can hardly monitor the state of each other, and almost all the terminals become hidden nodes. The RTS / CTS mode is adopted to reduce the adverse effects of "hidden terminals", but it is only suitable for the short-distance indoor environment. In the long-distance link environment, due to the longer propagation delay, and the relatively small value of the slot time, it is possible that other stations receive a station RTS frame before the back-off counter has reached zero, the station will begin data transmission, then there will be collisions, so that the throughput greatly reduced.

Therefore, TDMA can be considered as a basic multiple access method, using a combination of fixed and dynamic scheduling algorithms [3], TDMA access mode for fixed allocation, and contention based CSMA / CA scheme for dynamic allocation. The two will be fused, complement improve each other.

3.3. A serious user conflicts

Due to the LEO satellite-to-ground link environment, the coverage area is large, so the traffic volume is serious. In particular, in the case of hidden terminals, multi-user conflicts will be aggravated. Once a collision occurs, the data frame needs to be retransmitted. However, a longer propagation delay takes longer time and even discards data frames, which seriously affects the system throughput.

Aiming at the shortcomings of DCF protocol, we can improve it from both back-off mechanism algorithm and window parameter [4]. On the one hand, the back-off window algorithm is dynamically adjusted according to the number of active nodes or the network load; on the other hand, $CW_{\min}$ and $CW_{\max}$ in the DCF protocol may also be adjusted according to actual network conditions.

The following gives a solution to the problem of "user conflicts".

In a certain transmission period, the number of collision packets N_c is obtained through the carrier sense retransmission counter, and the number of successful packets N_s is monitored according to the number of received ACKs. The success rate SR is defined as the following:

$$SR = \frac{N_c}{N_s} \tag{2}$$

We can determine the network's busy state by “threshold”.

$$CW_{\min} = \begin{cases} 0.5CW_{\min}, & \text{SR>threshold} \\ 2CW_{\min}, & \text{SR<threshold} \end{cases} \tag{3}$$

The algorithm dynamically adjusts the contention window through network conditions to avoid the disadvantages of long delay under light load and high conflict rate under heavy load, and reduce the unfair access between nodes under the BEB algorithm, thereby improving the system throughput.

4. 802.11 DCF Analysis Model of LEO Satellite link

In allusion to the adjustment and optimization of 802.11 DCF protocol, Bianchi first proposed the Markov Chain model [5]. However, the model is proposed for the short-range indoor environment of the ground without considering the propagation delay and hidden terminal problems, and can't be directly applied to the LEO satellite Wi-Fi link environment. Therefore, this paper presents a modified Markov analysis model for LEO satellite Wi-Fi network based on the literature [6]. The model considers the influence of hidden nodes and long distance.

4.1. Markov Chain model

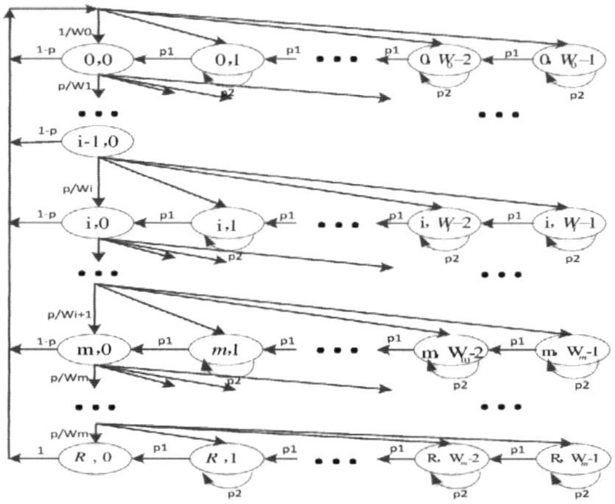

Figure 4. An improvement of Markov chain model.

By Markov state transition diagram can get the following transition probability:

$$P\{i,\ j \mid i\text{-}1,0\} = \frac{p}{W_i} \ ,\ i \in [0,\mathrm{m}]\ , j \in [0,\mathrm{W}_i\text{-}1] \tag{4}$$

$$P\{0,\ j \mid i,0\} = \frac{1-p}{W_0} \ ,\ i \in [0,\mathrm{m}\text{-}1]\ , j \in [0,\mathrm{W}_0\text{-}1] \tag{5}$$

$$P\{i,\ j \mid i\text{-}1,0\} = \frac{p}{W_m} \ ,\ i \in [\mathrm{m}+1,\mathrm{R}]\ , j \in [0,\mathrm{W}_m\text{-}1] \tag{6}$$

$$P\{0,\ j \mid R,0\} = \frac{1}{W_0} \ ,\ \mathrm{j} \in [0,\mathrm{W}_0\text{-}1] \tag{7}$$

If the active node in the channel is the access point, and the nodes in the ground are visible to each other, when the station detects that the wireless channel is idle and the probability of the back-off counter reduces by 1 is,

$$p_1 = P\{i,\ j | \ i,\ j+1\} = 1-p \ ,\ i \in [0,\mathrm{R}]\ ,\ j \in [0,\mathrm{W}_i\text{-}2] \tag{8}$$

If the station detects the wireless channel is busy, the probability of its back-off counter frozen is

$$p_2 = P\{i,\ j | \ i,\ j\} = p \ ,\ i \in [0,\mathrm{R}]\ ,\ j \in [0,\mathrm{W}_i\text{-}1] \tag{9}$$

If the active node in the channel is a base station, it is a hidden node with other nodes, and the station can't detect the busyness of the channel (except the access point AP). Regardless of any situation, the probability of the back-off counter reduces by 1 for each slot time is

$$p_1 = P\{i,\ j | \ i,\ j+1\} = 1 \ ,\ i \in [0,\mathrm{R}]\ ,\ j \in [0,\mathrm{W}_i\text{-}2] \tag{10}$$

$$p_2 = 0 \tag{11}$$

Where,

$$W_i = \min(2^i W_0 - 1, 2^m W_0 - 1) \tag{12}$$

$$W_0 = CW_{min} + 1 \tag{13}$$

$b_{i,j} = \lim_{t\to\infty} P\{s(\mathrm{t}) = i\ ,\ \mathrm{b(t)}=j\}\ ,\ i \in [0,\mathrm{m}]\ , j \in [0,\mathrm{W}_i\text{-}1]$ represents the steady-state distribution probability of Markov chain.

It is assumed that there are n nodes in the LEO satellite Wi-Fi network. The transmission probability τ and collision probability p of each node are the same. A represents an access point

AP, B represents a terrestrial base station $STA(i)$, and C represents another terrestrial base station $STA(j)$ $(i \neq j)$. In the LEO satellite Wi-Fi network, due to the long propagation delay and the hidden terminal problem, the collision between nodes may occur between different slottime.

$$\tau = \frac{1}{1 + \frac{1-p}{1-p^{R+1}} \sum_{i=0}^{R} p^i (1 + \frac{W_i}{2})} \tag{14}$$

$$p = 1 - \prod_{C=1, C \neq STA(i)}^{n-1} (1 - \mu_{BAC}) \tag{15}$$

μ_{BAC} represents the probability of another base station C colliding with it when a base station B transmits a message to the access point A.

$$\mu_{BAC} = \sum_{i=0}^{R} \sum_{j=0}^{CW_i} b_{X,i,j} P\{\lambda_{B,A,C,j}\} \tag{16}$$

Where,
$$P\{\lambda_{B,A,C,j}\} = N_{BAC,j} \left(\prod_{y=1, y \neq B,C}^{n} \sum_{l=0}^{R} \sum_{m=j}^{W_j - 1} b_{y,l,m} \right) \left(1 - \sum_{a=0}^{R} \sum_{b=0}^{W_a - 1} \min(\frac{j}{W_a}, 1) \cdot b_{C,a,b}\right) \tag{17}$$

$$b_{C,i,j} = \frac{W_i - j}{W_i} p^i \tau \frac{1-p}{1-p^{R+1}} \tag{18}$$

$$N_{BAC,j} = \begin{cases} 1 & int(NT_{BAC}) > j \\ (NT_{BAC} - j) & int(NT_{BAC}) = j \\ 0 & int(NT_{BAC}) < j \end{cases} \tag{19}$$

Let $T_{B,C}$ represent the time interval that another station C occur collision when the station B is transmitting data, just as Figure 5.

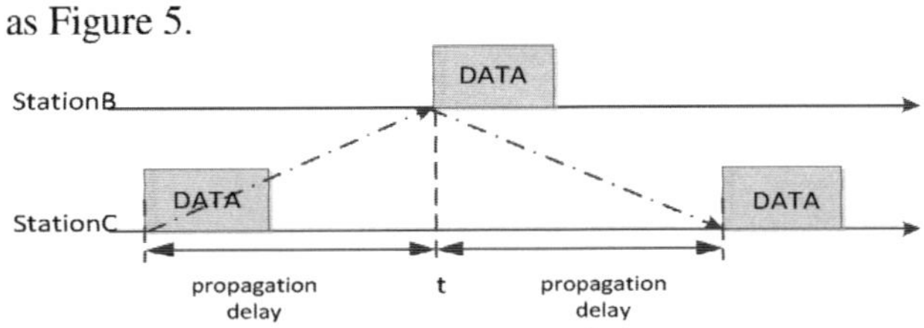

Figure 5. The analysis of $T_{B,C}$.

Thus,
$$T_{B,C} = (t - \delta_{B,C},\ t + \delta_{B,C}) \tag{20}$$

Where $\delta_{B,C}$ represents the transmission distance between station B and C.

$$NT_{BAC} = \frac{2\delta_{B,C}}{Slot\ Time} \tag{21}$$

4.2. A calculation of the throughput

The throughput is an important parameter to measure the performance of the network, it refers to the amount of data transmitted successfully in a unit of time. The throughput of this system is defined as

$$S = \frac{p_s p_{tr} E[p]}{(1 - p_{tr})\sigma + p_s p_{tr} T_s + p_{tr}(1 - p_s) T_c} \tag{22}$$

Where p_{tr} is the probability of at least one transmission in a system slot time, p_s is the probability that only one transmission is transmitted successfully, σ is the length of a slot time, T_s is the average length that transmit one frame successfully, T_c is the average duration of a transmission collision occurs. It is classified as the conflict length of the sending stations ($T_{c,i}$) and the conflict length of other stations ($T_{c,not\text{-}i}$). $E[\mathrm{p}]$ is the average payload size of the packets

$$p_{tr} = 1-(1-\tau)^n \tag{23}$$

$$p_s = \frac{n\tau(1-\tau)^{n-1}}{1-(1-\tau)^n} \tag{24}$$

In the basic access mode,

$$T_s = H + E[p] + 2 \cdot PropTime + SIFS + T_{ACK} + DIFS \tag{25}$$

$$T_c = \frac{\tau}{p_{tr}} T_{c,i} + (1-\frac{\tau}{p_{tr}}) T_{c,not\text{-}i} \tag{26}$$

$$T_{c,i} = H + E[\mathrm{p}] + ACK\ Timeout + DIFS \tag{27}$$

$$ACK\ Timeout = 2 \cdot PropTime + SIFS + \sigma \tag{28}$$

$$T_{c,not\text{-}i} = H + E[p] + PropTime + EIFS \tag{29}$$

$$H = PHY\_\mathrm{hdr} + MAC\_\mathrm{hdr} \tag{30}$$

By substituting equations (24)-(30) into equation (23), it can be seen that the throughput is determined by the transmission probability. By the differentiation of equation (23) on the transmission probability, it can be seen that the optimal transmission probability for maximizing the throughput depends on the number of active nodes and the transmission distance.

4.3. The simulation verification

In order to verify the relationship between the throughput and the transmission probability, the number of nodes and the transmission distance, the above theoretical analysis results are verified by MATLAB simulation based on the 802.11b protocol standard.

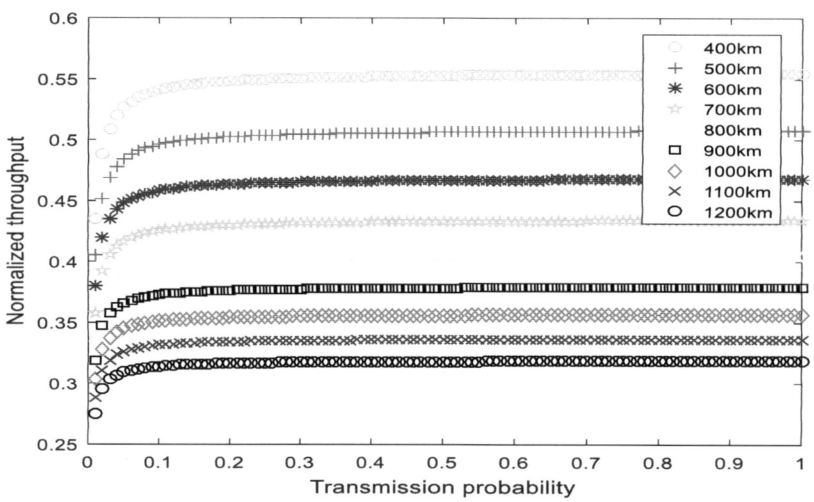

Figure 6. With different transmission distances, the throughput varies with the transmission probability(the number of nodes is 1).

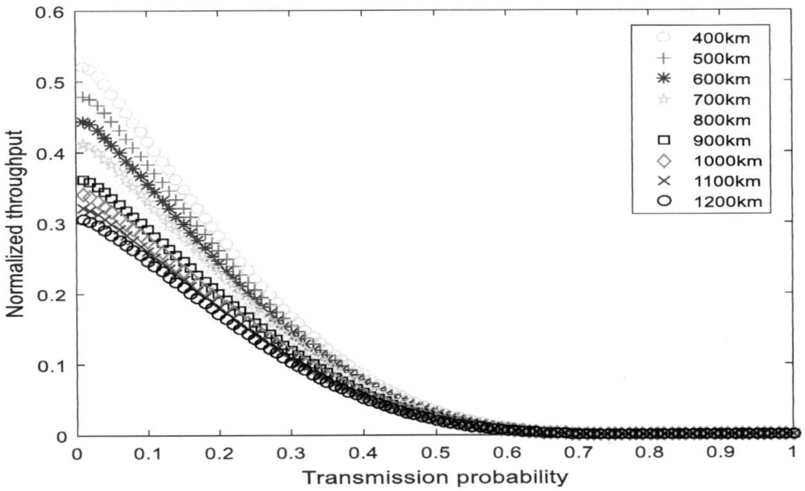

Figure 7. With different transmission distances, the throughput varies with the transmission probability(the number of nodes is 7).

As can be seen from the figure above, the throughput increases with the increase of transmission probability τ for a single user, and when τ reaches 0.2, it tends to be stable, and the shorter the transmission distance is, the larger the throughput is. When multi-user, the throughput decreases rapidly with the increase of transmission probability. When τ approaches 0.6, the throughput tends to zero.

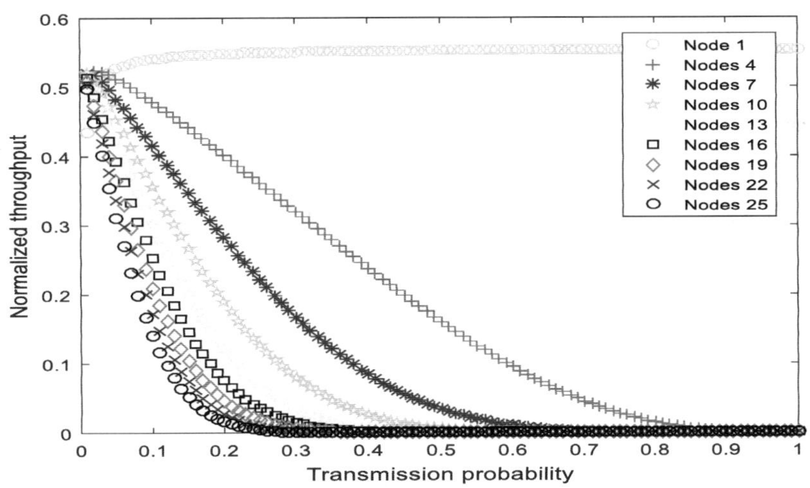

Figure 8. With different number of nodes, the throughput changes with the transmission probability(Transmission distance of 400km).

As can be seen from the figure above, the throughput is not sensitive to the sending probability for a single user. When multi-user, the throughput decreases rapidly with the increase of sending probability, and the more the number of users, the lower the throughput.

The simulation results show that in the LEO satellite Wi-Fi network, the throughput is directly related to the transmission probability, and the more the number of users, the longer the transmission distance, the poorer the throughput performance will be.

5. Conclusions

Based on the study of the characteristics of the long propagation, "hidden terminal" and the user conflict of LEO satellite-to-ground link, this paper improves the ACK Timeout and $CW_{\min}$ parameter for the traditional DCF protocol, and a two-dimensional Markov model for LEO satellite link is proposed. The simulation results show that the throughput performance is directly determined by the transmission probability and decreases with the increase of number of users and transmission distance. The follow-up study will propose specific solutions and simulation verification for multi-user wireless communication in the LEO satellite link environment.

References

[1] Wang N 2012 Experimental Study of 2P MAC Protocol in Long Distance Wireless Mesh Networks *Tianjin University*.

[2] MATTBEW GAST S. *802.11 Wireless Network Authoritative Guide* Second Edition. Nanjing: Southeast University Press, 2007.

[3] Tan Fy, Ye L, Yu Fs, Liu Zb 2011 Multiple access protocol with combination of TDMA and CSMA in Ad hoc network *Computer Engineering and Design* 32(11):3656-3659+3681.

[4] Wang P 2015 Improvement and performance analysis for DCF mechanism in IEEE 802.11 MAC layer *Southwest Jiao tong University*.

[5] Wang Jj, Li D, Liu Sy 2017 New improved back-off algorithm and simulation for 802.11 DCF *Electronic Design Engineering* 25(10):148-153.

[6] Simóreigadas F J, Ramoslópez F J, Seoanepascual J 2010 Modeling and Optimizing IEEE 802.11 DCF for Long-Distance Links *IEEE Transactions on Mobile Computing* 9(6):881-896.

IOP Conf. Series: Journal of Physics: Conf. Series **1026** (2018) 012014 doi:10.1088/1742-6596/1026/1/012014

The Design of Broadband Multi-target Folding Jammer Based on a Periodic Non-uniform LFM Local Oscillator

Qi Zhang[1], Jun Zhu[1], Bin Tang[1] , Huan Yang[2]

1. School of Electronic Engineering (EE),University of Electronic Science and Technology of China, (UESTC),Chengdu, China
2. Institute of Electronic Engineering,China Academy of Engineering Physics,Mianyang, China

E-mail: 809768504@qq.com

Abstract. This paper proposes a novel broadband multi-target folding jammer based on a periodic non-uniform linear frequency modulation (LFM) local oscillator (LO). It can achieve broadband receiving and transmitting with a small count of equipment. This paper also proposes a de-chirp function (DF) which is used to estimate the Nyquist zone index of the received signals and a new multi-beamforming method. The estimation value can be used to guide the transmitting frequency of the jamming signals and the new multi-beamforming method can be used to achieve the **accurate** multi-target jamming. The scale experiments show that the proposed folding jammer is feasible to achieve in practice and the method of DF works better than the cubic phase function (CPF) for guiding the frequency of jamming signals. The multi-beamforming simulation shows that the designated areas are covered by the relevant jamming.

1. Introduction

Owing to the limits of the conventional jammer [1-3] on the instantaneous receiving and transmitting bandwidth, jamming the broadband signals in real time needs a combination of multi-stage LO and frequency sweep mode [4]. This will make the jammer complicated. In order to reduce the complexity and cost of the jammer, this paper proposes a novel broadband folding jammer which can be used to jam the broadband signals with a small amount of equipment.

Increasing the receiving and transmitting bandwidth is difficult for existing technology [5]. A feasible technology which is proposed in recent years for broadband receiving and transmitting is based on compressed sensing technology [6]. Fudge et al. [7] proposed a simplified structure for the compressed sensing system. But the related research of Fudge is limited to the wavelet analysis [8] for the processing of receiving signals. The performance of this system needs to be improved. The folding transmitting structure has not been discussed before.

Inspired by the DF method of LFM signal [9], this paper proposes a novel periodic non-uniform LFM LO which is different from the LO in [7]. Based on this design, the paper proposes a Nyquist zone index estimation algorithm which has better performance than the CPF [10]. The estimation value can be used to guide the jamming signals accurately on the appointed frequency. This paper also proposes an available structure of the transmitting LO which is different from the receiving LO for the jammer.

At last, the paper proposes a new multi-beamforming method for transmitting of jamming signals. This method lead into the parameter of distance dimension [11-12]. Therefore, the jamming signals can cover the areas which have the fixed distance and direction.

In summary, we propose a novel broadband multi-target folding jammer with a small count of equipment in this paper. We demonstrate the complete structure of the folding jammer and analyze the signal processing flow in Section 2. Next, we propose a robust Nyquist zone index estimation algorithm based on the LFM LO and a new multi-beamforming method in Section 3. In Section 4, we conduct scale experiments to demonstrate that the folding jammer is feasible to achieve and the proposed estimation algorithm has better performance than the CPF under this structure, then the multi-beamforming simulation proves that the designated areas are covered by the relevant jamming signals. In Section 5, we conclude this paper.

2. System structure

2.1. The structure of the folding jammer

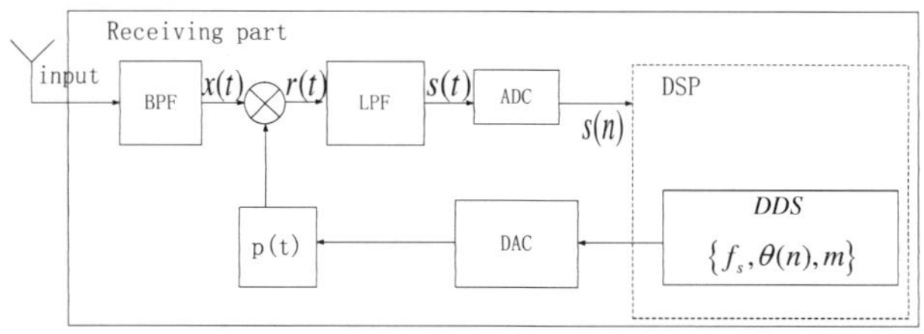

Figure 1. The receiving structure of the folding jammer.

The receiving structure of the folding jammer is shown in Fig.1. The input signal is received in the broadband filter BPF. After filtering the out-of-band noise, the complex signal $x(t)$ is obtained. $x(t)$ is then mixed with the periodic non-uniform LFM LO $p(t)$ to obtain the signal $r(t)$. The $p(t)$ is generated by the digital-to-analog converter (DAC) and the direct digital synthesizer (DDS) in digital signal processor (DSP). $r(t)$ passes through the complex low-pass filter LPF and then the signal $s(t)$ is obtained. $s(t)$ is then sampled by the analog-to-digital converter (ADC) to obtain the baseband digital signal $s(n)$.

The transmitting structure of the folding jammer is shown in Fig.2. The jamming signal $s'(t)$ is mixed with the periodic non-uniform LO $p'(t)$ to obtain the signal $x'(t)$. The $p'(t)$ is also generated by the DAC and the DDS in DSP. $x'(t)$ then passes through the filter banks to obtain the output signal of jammer.

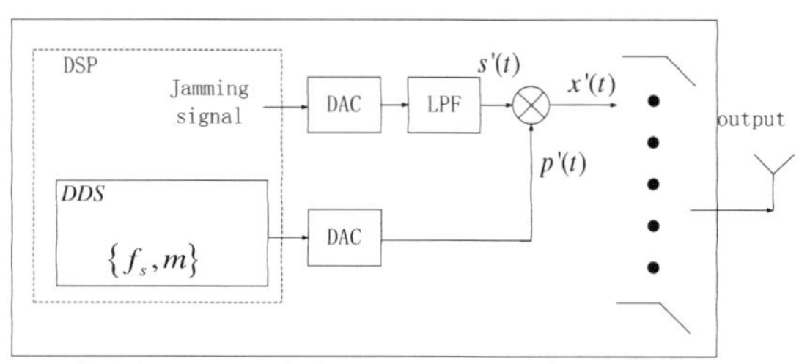

Figure 2. The transmitting structure of the folding jammer.

2.2. Signal processing in the folding jammer

In the receiving part of the jammer, the signal model received can be expressed as follows

$$x(t) = \exp\left[j\left(2\pi f_0 t + \varphi_0\right)\right] + v(t) \tag{1}$$

where f_0 and φ_0 are carrier frequency and initial phase of the signal. $v(t)$ is the zero-mean white Gaussian noise whose variance is σ_v^2 and it is independent of the signal.

The periodic non-uniform LFM LO can be simplified as

$$p(t)=\sum_{m=0}^{M-1}\sum_{h}\mu(t-hT)\exp\left[j\left(2\pi f_s t/2+2\pi m\left(f_s t+\theta(t)\right)\right)\right] \quad (2)$$

where $\theta(t)=f_{LFM}(t-hT)+0.5K_0(t-hT)^2$ is a periodic instantaneous phase modulation, $h=0,1,\cdots$. f_{LFM}, K_0 and T are the initial frequency, chirp rate and repetition period of the LFM LO. f_s is the carrier frequency of the LO, m represents the Nyquist zone, M is the number of the Nyquist zone, the $\mu(t)$ can be expressed as

$$\mu(t)=\begin{cases}1, 0\le t<T\\ 0, otherwise\end{cases} \quad (3)$$

After the deducing and arrangement, the receiving signal of DSP $s(n)$ can be expressed as

$$\begin{aligned}s(n)=&\sum_h \mu(t-hN_T)\bullet\\ &\exp\left[j\left(2\pi f_0' nT_s-2\pi m_H f_{LFM}(n-hN_T)T_s\right)\right]\bullet\\ &\exp\left[j\left(\varphi_0-m_H\pi K_0\left(nT_s-hN_T T_s\right)^2\right)\right]+v'(n)\end{aligned} \quad (4)$$

where $f_0'=f_0-f_s/2-m_H f_s$, N_T is the number of sampling points for in time T, $T_s=1/f_s$. $v'(n)$ is also the zero-mean white Gaussian noise. $m_H\in\{0,1.....,M-1\}$ is the index of the Nyquist zone,
In the transmitting part of the jammer, we assume the jamming signal model as

$$s'(t)=\exp\left[j\left(2\pi f_c t+\varphi_c\right)\right]+n(t) \quad (5)$$

where f_c and φ_c are carrier frequency and initial phase of the jamming signal. $n(t)$ is the zero-mean white Gaussian noise whose variance is σ_v^2 and it is independent of the signal.
The transmitting LO can be simplified as

$$p'(t)=\sum_{m=0}^{M-1}\exp\left[jm\left(2\pi f_s t\right)\right] \quad (6)$$

After the jamming signal is mixed with the transmitting LO, the mixing signal $x'(t)$ can be expressed as

$$x'(t)=\exp\left[j((2\pi f_c\pm 2\pi f_s m)t+\varphi_c)\right]+n''(t) \quad (7)$$

In (7), $n''(t)$ is also the zero-mean white Gaussian noise. Finally $x'(t)$ passes through the filter banks to obtain the output jamming signal on the appointed frequency based on the value of m_H.

3. Algorithm Analysis

3.1. Nyquist zone index estimation based on the DF

The method defines the DF as follows

$$DF(d,K)=\frac{1}{N_T-d}\sum_{n=d}^{N_T-1}s(n)s^*(n-d)\exp\left(jKnT_s\right) \quad (8)$$

where d is the time delay. The DF will produces the auto term of the signal. When $K=2\pi mK_0 dT_s\left(m=0,1,\cdots,M-1\right)$, the auto term can be denoted as

$$DF\left(d,2\pi mK_0dT_s\right)= \frac{1}{N_T-d}\exp\left[j\left(2\pi f_0^{''}dT_s+m_H\pi K_0\left(dT_s\right)^2\right)\right]\cdot \sum_{n=d}^{N_T-1}\exp j2\pi\left(m-m_H\right)K_0dnT_s^2 \tag{9}$$

In (9), $f_0^{''}=f_0^{'}-m_Hf_{LFM}$ and we have

$$DF\left(d,2\pi mK_0dT_s\right)|_{m=m_H}=\exp\left[j\left(2\pi f_0^{''}dT_s+m_H\pi K_0\left(dT_s\right)^2\right)\right] \tag{10}$$

$$DF\left(d,2\pi mK_0dT_s\right)|_{m\neq m_H}=0 \tag{11}$$

When $m=m_H$, $DF\left(d,2\pi mK_0dT_s\right)$ is a non-zero value. When $m\neq m_H$, $DF\left(d,2\pi mK_0dT_s\right)$ is zero. Therefore, $\hat{m}_H$ can be obtained when the value of $DF\left(d,2\pi mK_0dT_s\right)$ comes to non-zero value by scanning the value of m. $\hat{m}_H$ is the estimation value of m.

After obtaining the estimation value of m_H, we can choose the output of the corresponding filter in the filter banks. The filter banks is described in Fig.3.

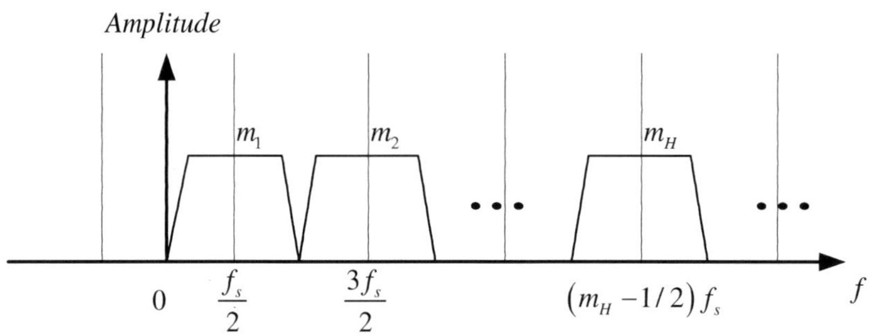

Figure 3. The structure of the filter banks.

3.2. Multi-beamforming algorithm

For the conventional beamforming algorithm, array transmitting signal model can be expressed as

$$\mathbf{S}(f_0)=\left[S_1(f_0),S_2(f_0),\cdots,S_N(f_0)\right] \tag{12}$$

In (12), $\left\{S_i\left(f_0\right)\right\}_{i=1}^N$ is the transmitting signal for each array element. f_0 is the transmitting frequency of signal. N is the number of array element.

In this paper, the array transmitting signal model has been modified with a frequency offset. It can be denoted as

$$\begin{bmatrix} S_1(f_0) & S_2(f_0) & \cdots & S_N(f_0) \\ S_1(f_0+f_T) & S_2(f_0+f_T) & \cdots & S_N(f_0+f_T) \\ \cdots & \cdots & \cdots & \cdots \\ S_1(f_0+(M_z-1)f_T) & S_2(f_0+(M_z-1)f_T) & \cdots & S_N(f_0+(M_z-1)f_T) \end{bmatrix}_{M\times N} \tag{13}$$

where M_z is the number of the frequency offset, f_T is the frequency offset, $f_0\gg f_T$. Each column represents the transmitting signal of each array element. Then the arrived signals of far field from one array element can be expressed as

$$\begin{aligned} P_F &= \sum_{m=1}^{M} w_m \left(\exp\left(-j2\pi\left(\left(f_0+(m-1)f_T\right)(t-t_0)\right)\right)\right)/R \\ &= \frac{1}{R}\exp\left(-j2\pi f_0 (t-t_0)\right) \\ &\quad \cdot \sum_{m=1}^{M} w_m \exp\left(-j2\pi\left(\left((m-1)f_T\right)(t-t_0)\right)\right) \\ &= F(t)\mathbf{w}_d^{\mathrm{H}}\mathbf{A}_d \end{aligned} \tag{14}$$

In (14), $F(t)=\frac{1}{R}\exp\left(-j2\pi f_0 (t-t_0)\right), t_0 = R/c$, c is the light speed, R is the distance. $\mathbf{w}_d$ is the weight vector for distance dimension, $\mathbf{w}_d = \left[w_{d_1}, w_{d_2}, \cdots, w_{d_M}\right]^{\mathrm{T}}_{M\times 1}$, $\mathbf{A}_d$ is the steering vector for direction dimension. $\mathbf{A}_d$ can be expressed as

$$\mathbf{A}_d = \begin{bmatrix} 1, \exp\left(-j2\pi\left(f_T (t-t_0)\right)\right), \cdots, \\ \exp\left(-j2\pi\left(\left((m-1)f_T\right)(t-t_0)\right)\right) \end{bmatrix}^{\mathrm{T}}_{M\times 1} \tag{15}$$

The arrived signals of far field from all array elements can be expressed as

$$\begin{aligned} P &= \sum_{n=1}^{N}\sum_{m=1}^{M} w_m w_n \left(\exp\left(-j2\pi\left(\left(f_0+(m-1)f_T\right)(t-t_\phi)\right)\right)\right)/R_\phi \\ &= \sum_{n=1}^{N}\sum_{m=1}^{M} \frac{1}{R_\phi}\exp\left(-j2\pi f_0 (t-t_\phi)\right) \\ &\quad \cdot w_m w_n \left(\exp\left(-j2\pi\left(\left(f_0+(m-1)f_T\right)(t-t_\phi)\right)\right)\right) \\ &= \sum_{n=1}^{N}\sum_{m=1}^{M} F_\phi(t) w_m w_n \exp\left(-j2\pi\left(\left((m-1)f_T\right)(t-t_\phi)\right)\right) \\ &\quad \cdot \exp\left(j2\pi f_0 \frac{(n-1)d\sin\theta}{c}\right)\exp\left(j2\pi\left(\frac{mnf_T d\sin\theta}{c}\right)\right) \end{aligned} \tag{16}$$

In (16), $F_\phi(t)=\frac{1}{R_\phi}\exp\left(-j2\pi f_0 (t-t_\phi)\right)$, $t_\phi = \frac{R_\phi}{c}$, $R_\phi = R + nd\sin\theta$. We can see that the energy of beam is associated with the distance parameter and direction parameter.The conventional beam is only associated with the direction parameter.

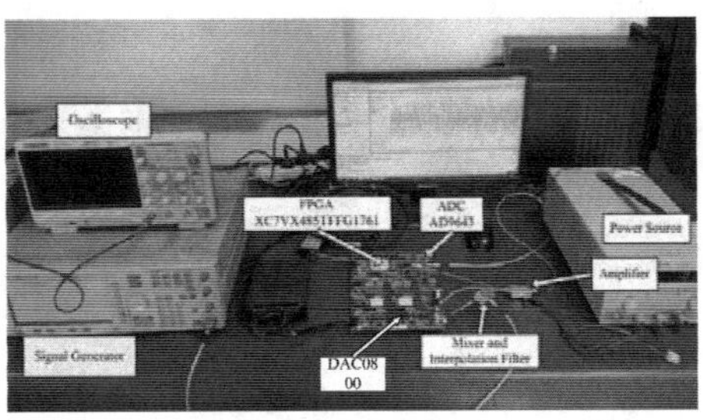

Figure 4. The folding jammer laboratory scale experiment.

4. Simulation Analysis

We construct a laboratory scale experiment as shown in Fig.4 to verify our analyses. The field-programmable gate array (FPGA) (XC7VX485TFFG1761) generates the receiving and transmitting LO signals. The signal generator generates the analog input signals. The non-uniform receiving LO signal and the analog input signal are mixed and filtered in the mixer-filter and its output is sampled by a uniform sampling ADC (AD9643). The model of DSP is the TI TS201. The model of

IOP Conf. Series: Journal of Physics: Conf. Series **1026** (2018) 012014 doi:10.1088/1742-6596/1026/1/012014

DAC is DAC0800. Considering the FPGA processing capability, the receiving and transmitting LO carrier frequency are 31.25MHz. The sampling frequency of the ADC/DAC are 70MHz.
The parameters of the input analog signals are shown in Table 1.
In Table 1, MP represents the mono-pulse signal, PRI is the pulse repetition interval and PW is the pulse width. These two parameters are arbitrarily variable. The Signal1 is a narrowband signal and The Signal2 is a wideband signal.

Table 1. Parameters of analog input signals

Signal Parameter	Signal 1	Signal 2
Modulation type	MP	LFM
Carrier frequency	41.25MHz	66.5MHz
Signal bandwidth	0MHz	6MHz
PRI	8μs	10μs
PW	4μs	5μs

4.1. The results of the folded receiving signals
Fig.5 and Fig.6 show the time-frequency distribution of the signals which are sampled by ADC. As is shown in Fig.5, the frequency of Signal 1 has been folded to the 10MHz. As is shown in Fig.6, the initial frequency of the LFM pulse signal has been folded to the 4MHz. These results satisfy the folding theory which is described above. Therefore, the design of the receiving part is achieved in practice.

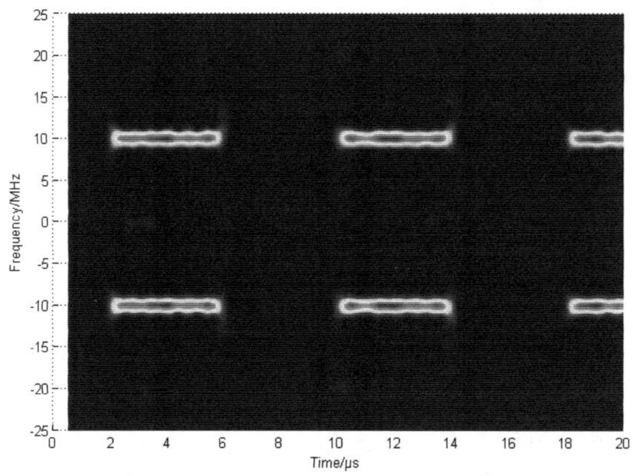

Figure 5. The time-frequency distribution of Signal 1.

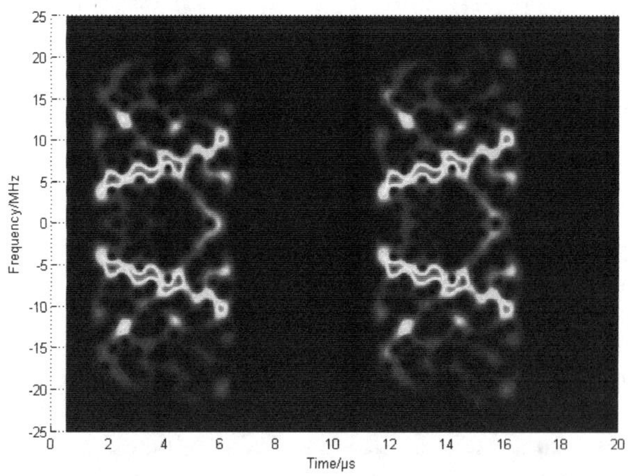

Figure 6. The time-frequency distribution of Signal 2.

4.2. Comparative` experiment between DF and CPF

We set the SNR of the input signals as $-10\sim15dB$. 500 times hardware tests are used for each SNR value in DSP. Then we export the each test result.

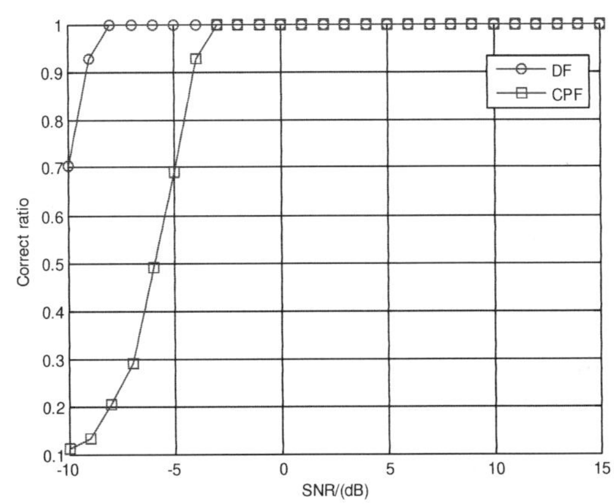

Figure 7. The correct ratio of the Nyquist zone index estimation.

Fig.7 compares the correct ratio of Nyquist zone index estimation using the DF with that using CPF by hardware tests. The correct ratio represents the correct values in these 500 times tests. The proposed DF method works better than the CPF method when SNR<-3dB. When SNR>-8dB, the correct ratio of Nyquist zone index estimation using the DF can reach 100%. So the CPF method is more sensitive to noise than the DF method. The performance of the proposed method is better in test.

4.3. The results of the jamming signals

After obtaining the estimation value of the Nyquist zone index, we can transmit the jamming signal to the corresponding zone by folding transmitting. We set the jamming signal as the broadband noise jamming signal whose bandwidth is 10MHz. In this paper, we use the ADC whose sampling frequency is 150MHz to sample the transmitting signal of the folding jammer again. Then we can use the sampling data to analyze the jamming signals.
As is shown in Fig.8 and Fig.9, the jamming signals are folded to the appointed frequency accurately. The results prove that the design of transmitting part is also achieved in practice.

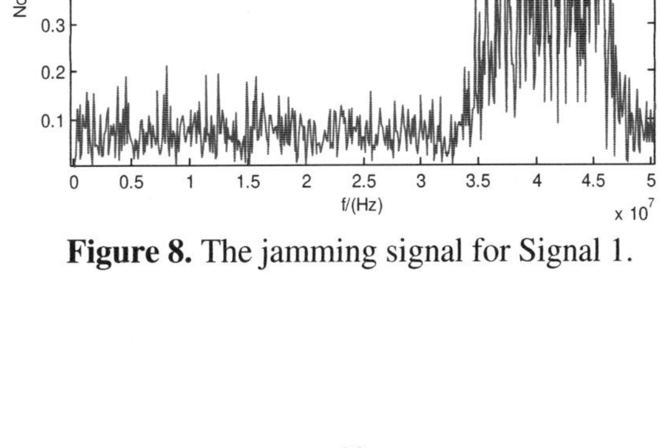

Figure 8. The jamming signal for Signal 1.

IOP Conf. Series: Journal of Physics: Conf. Series **1026** (2018) 012014 doi:10.1088/1742-6596/1026/1/012014

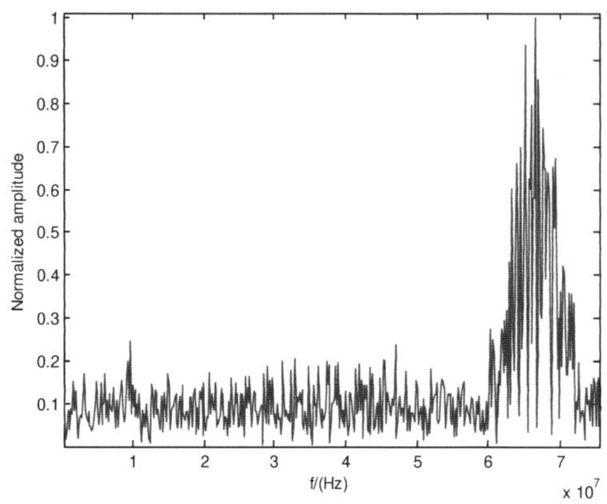

Figure 9. The jamming signal for Signal 2.

4.4. The results of the beamforming method

In this correspondence, we consider an 31-element uniform linear array operating at a carrier frequency f_0 = 10 GHz with a fixed inter-element spacing $d = c_0 / (2f_0)$. Each element transmits the signal in a linearly increased carrier frequencies, where the frequency step is f_T = 3 kHz. In the simulation we use the following parameters: T_p = 1/(10 f_T) seconds, t_s = 3/(10 f_T) f_T 4/(10 seconds, R = r_s = 20 km, θ= 30°, ε = 0.5, and U − L = 0.05. These parameters are described in complete detail in [13].

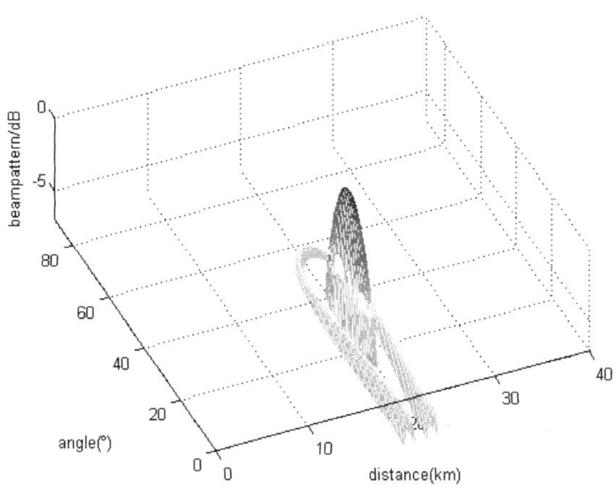

Figure 10. The beam pattern of jamming.

As is shown in Fig.10, the main lobe of beam is formed at the fixed distance and direction. The energy of beam is very weak at the other areas. Therefore, the concerned area is covered by jamming, but the others is not. This can improve the accuracy of the jamming, and develop the stealth of the jammer. If we transmit multi-jamming, the each beam for multi-jamming could be added in digital domain, and then transmitted together. So multi-beam is the superposition of single beam.

5. Conclusion

The broadband folding jammer presented in this paper not only can achieve the receiving of broadband signals with a small amount of equipment, but also can efficiently obtain Nyquist zone index of broadband signals to guide the jamming signals. The proposed DF method works better than the classical method based on the design of LFM LO. The scale experiments prove that the proposed folding jammer is feasible to achieve. The transmitting beamforming method uses the distance parameter firstly. The simulation shows that the areas which has the fixed direction and distance are covered by the energy of jamming, and the jamming energy on the other areas are so weak. Therefore, the broadband folding jammer has many advantages, it can has broader applications in the higher frequency jamming, ultra-broadband jamming and multi-target jamming field.

6. Acknowledgment
This work is supported by National Natural Science Foundation of China (61571088).

7. References
[1] M. Soumekh:" SAR-ECCM using phase-perturbed LFM chirp signals and DRFM repeat jammer penalization". IEEE Transactions on Aerospace and Electronic Systems, vol. 42, 2006, pp.191-205.

[2] S Gezici, S Bayram, M N Kurt, M Reza :"Optimal Jammer Placement in Wireless Localization Systems". IEEE Transaction on Signal Processing, vol. 64, 2016, pp.4534-4549.

[3] S. Djukanovic, M. Dakovic, T. Thayaparan, L. Stankovic:" Method for non-stationary jammer suppression in noise radar systems". IET Signal Processing, vol. 4, 2010, pp.305-313.

[4] G Torrisi, O Leonardi, G Sorbello et al :" A novel microwave frequency sweep interferometer for ECR plasma electron density measurements". 46th European Microwave Conference, pp. 811, 2016.

[5] Namgoong W. A:" A channelized digital ultrawideband receiver". IEEE Transactions on Wireless Communications, vol. 2, 2003, pp.502-510.

[6] Donoho D L:" Compressed sensing". IEEE Transaction on information Theory, vol. 52, 2006, pp.1289-1306.

[7] Fudge G L, Bland R E, Chivers M A, et al:" A Nyquist folding analog-to-information receiver [C]". 42nd Asilomar Conference on Signals, Systems and Computers, Pacific Grove, California, USA, pp.541-545, 2008.

[8] O O Odejide, C M Akujuobi, A Annamalai et al :" Application of analytic wavelet transform for signal detection in Nyquist folding analog-to-information receiver." IEEE International Conference on Communications, pp.1-5, 2009.

[9] P Bruce, A G Nathan :" Structured de-chirp for compressive sampling of LFM waveforms". IEEE 7th Sensor Array and Multichannel Signal Processing Workshop (SAM), pp. 37, 2012.

[10] P O Shea :" A new technique for instantaneous frequency rate estimation". IEEE Signal Processing Letters, vol. 9, 2002, pp.251-252.

[11] Antonik P, Wicks M C, Griffiths H D, et al. Frequency diverse array radars [C]. Proceedings of IEEE Radar Conference, Verona, pp.215-217, 2006.

[12] Antonik P, Wicks M C, Griffiths H D, et al. Multi-mission multi-mode waveform diversity [C]. Proceedings of IEEE Radar Conference, Verona, pp.580-582, 2006.

[13] GAO K D, WANG W Q, CHEN H, et al. Transmit beamspace design for multi-carrier frequency diverse array sensor [J]. IEEE Sensors Journal, vol. 16, 2016, pp.5709-5714.

Design of Automatic Storage System Monitoring Software Based on OPC Communication

Zhang Peng, Zhang Teng and Zhou Guangcheng

School of Mechanical and Automotive Engineering, Qilu University of Technology (Shandong Academy of Sciences), Jinan, China

E-mail: sqzhangpeng@163.com

Abstract. In order to solve the problem of heterogeneous network interconnection between PLC control network and monitoring software network in automatic storage system, OPC communication technology is used to build the network interconnection model. The integration of heterogeneous networks is achievable, and C# is used to design automatic storage system monitoring software. After taking test, the data transmission between PLC control network and computer control network is reliable. The monitoring software runs smoothly, and is applied successfully to one automatic storage simulation system.

1. Introduction

Because of the different equipment manufacturers, the traditional automatic storage system uses their own communication protocol and form industrial control network which are based on PLC, Industrial Ethernet, DCS, field bus, wireless network, etc. The heterogeneous networks of monitoring platform are difficult to interrelate and share data, which can't analyze data effectively and provide comprehensive decision support. Application of OPC (OLE For Process Control) communication technology [1] can solve the interconnection between heterogeneous networks, and provide a shared platform for all data in the system.

OPC is an OLE industry standard for process control, which is managed by the International Organization's OPC Foundation. It defines how to use the Microsoft operating system to exchange real-time data between computers. It aims to develop an inter-operable interface standard, which is based on Microsoft COM, DCOM, and ActiveX technology. And it provides greater inter-operable ability of automated control, system field devices, and offices [2].

In this paper, one automatic storage system is taken as an example, and the monitoring software based on OPC communication technology is developed. In order to ensure the operation of the monitoring software work smoothly, it is necessary to establish the communication of the monitoring software and the PLC controller program, and the field devices.

2. System Equipment Structure

Hardware equipment of the automatic storage system includes IPC, stacker, conveyor belt, hoister, transplanting machine, shuttle car, PLC and other auxiliary equipment. IPC is used to support the installation of the monitoring software. Stacker is used to access the goods. Conveyor belt and hoister are used to transport the goods. Transplanting machine is used to change the transmission line at each connection of the transmission line. PLC includes stacker PLC control cabinet (slave station) and scene equipment PLC control cabinet (master station). Stacker control cabinet is used to control the stacker's action according to the task. Scene equipment control cabinet is used to control the conveyer belt, hoister, shuttle cars, transplanting machine and other field equipment to work [3]. The hardware equipment framework of control system is shown in Fig.1.

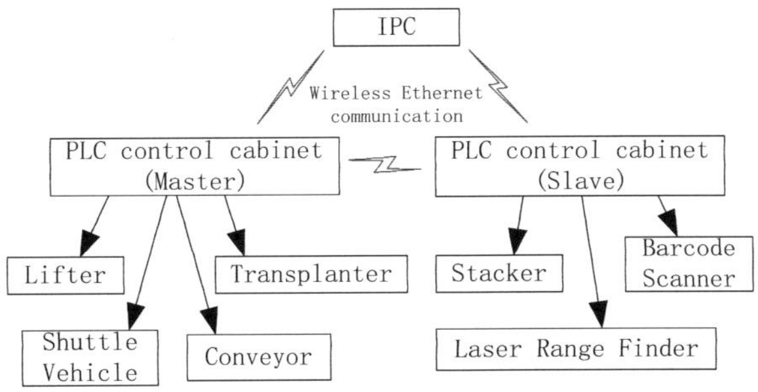

Figure 1. Hardware equipment framework of control system

3. Software Design Scheme

As an important part of the automatic storage system, the monitoring software is responsible for the control strategy dispatch, data processing and calculation, information transmission and so on [4]. It includes three parts: database, PLC control, and OPC communication interface. OPC communication interface includes OPC server communication configuration and computer OPC server program interface development. OPC server communication configuration is used to ensure the connection of PLC program and the interface program. OPC server program interface completes the reliable controlling of monitoring software to the field devices.

According to the requirement of automatic storage system, the design of the monitoring software is summarized as the design of the four main support structures that are the equipment operation scheduling block, the PLC control block, the program interface block and the database information processing block [5][6]. Monitoring software design framework is shown in Figure 2.

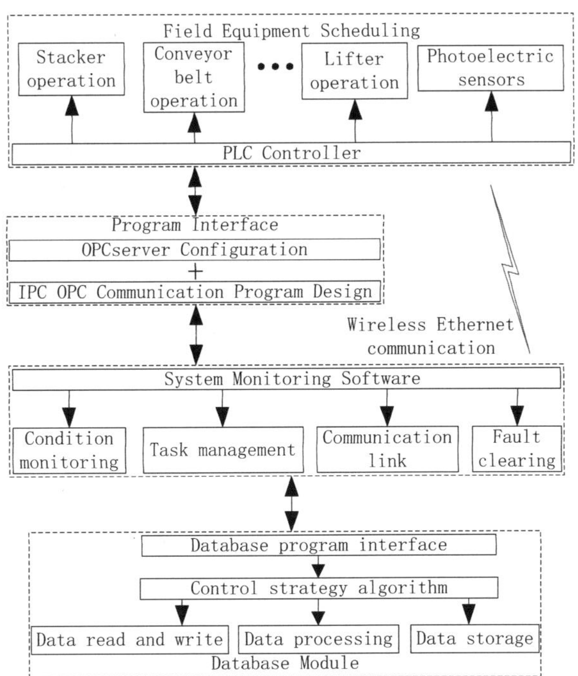

Figure 2. Monitoring Software Design Architecture

4. OPC Program Interface Design

The software of Simatic Net and STEP7 are used to build OPC server. C# is used to develop OPC server interface program, to realize the communication between the IPC and the S7-300PLC.

4.1 OPC Communication Configuration

In order to realize the communication between OPC and PLC, OPC server must be configured. In this paper, wireless Ethernet is used to realize the communication between the IPC and the PLC. STEP7 and Simatic Net software are required. The hardware devices include one IPC and two sets of Siemens S7-300PLC. Configuration steps are as follows:

a) Open Simatic Net software →configure a PC site→add an OPC server→create Ethernet(1) communication network→set IP address to 192.168.1.105→ compile and save.

b) Open STEP7 PLC software and create a new project of PLC controller→configure Ethernet(1) communication network→set IP address 192.168.1.102→download into the PLC controller.

c) Open 'Station Configuration Editor' of the Simatic Net software→click 'Import Station'→find the XDB file under the XDBs folder→import it. The OPC server configuration is completed.

4.2 OPC Interface Program

By configuring the OPC server and developing the OPC server program interface, the variable structure consists of three levels, which is shown in Figure3. The first level is the OPC server type, the server type is OPC.SimaticNET. The second level is the group, which can be put all the same variables into. The third level is item that is a series of variables link to the OPC server, corresponding to the I/Q of PLC, etc.

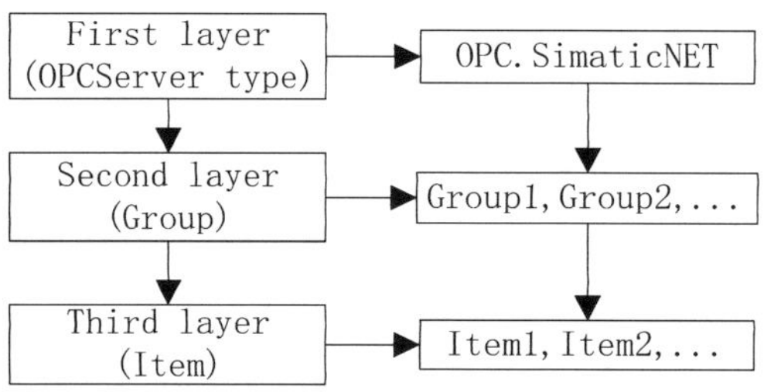

Figure 3. Variable structure diagram

OPC server configuration completes the connection between the IPC and the PLC. In order to realize the real-time communication and control of the monitoring software and the PLC, it is necessary to develop the OPC server interface program. This paper uses C# to develop the interface program. The main steps and the program code are as follows:

(1) Declare variables and list the main variables as follows:

Private OpcRcw.Da.IOPCServer ServerObj; private Object GroupObj = null;

private int pSvrGroupHandle = 0;

private OpcRcw.Da.IOPCSyncIO IOPCSyncObj = null.

(2) Create an OPC server and establish the connection to the OPC server: OPC server is created by the founctions of public OPC Server() and public bool Connect(), and the connection with the OPC server is established. The return value is used to show the connection result.

(3) Add group into the OPC server: OPC server group object is created by the function of public bool AddGroup(), and show the result with the return value.

(4) Items are added into the group object. Through the function of public bool AddItems(), more items add into the group object of the server.

IOP Conf. Series: Journal of Physics: Conf. Series **1026** (2018) 012015 doi:10.1088/1742-6596/1026/1/012015

(5) Data synchronization read and write programs are developed by variable value write and read functions of public bool SyncWrite() and public bool SyncRead().

5. Monitoring software Design

Monitoring software interface is shown in Fig4. This software has the functions of connecting and disconnecting with the database, PLC, field devices, system fault clearing, system working mode switching, initializing device interface and whole initialization. Among them, the function of initialization equipment interface is to complete the preparation of field equipment such as conveyor interface and stacker interface to ensure the normal operation of the monitoring software after receiving the task. The initialization equipment interface program code is as follows:

```
private void
tsbInitDevice_Click(objectsender,EventArgs e)
{
if (ds.Tables[0].Rows.Count > 0)
{
AddListSystemEvent;
foreach (DataRow row in ds.Tables[0].Rows)
{
Stacker stk = new Stacker(this, row);
stackers.Add(row.ItemArray[0].ToString(), stk);
frm.Text=string.Format(stackers[row.ItemArray[0].
ToString()].deviceId);
     this.Controls.Add(stackers[row.ItemArray[0].ToString()].pic);
stackers[row.ItemArray[0].ToString()].pic.BringToFront();
stackers[row.ItemArray[0].ToString()].pic.DoubleClick
=newSystem.EventHandler(this.pic_StackerDoubleClick);
}}
if (ds.Tables[0].Rows.Count > 0)
{
 AddListSystemEvent();
foreach (DataRow row in ds.Tables[0].Rows)
{ cr= new Convyer(this, row);
convyers.Add(row.ItemArray[0].ToString(), cr);
frm.Text=string.Format(convyers[row.ItemArray[0].
ToString()].deviceId);
this.Controls.Add(convyers[row.ItemArray[0].ToString()].pic);
convyers[row.ItemArray[0].ToString()].pic.BringToFront();
convyers[row.ItemArray[0].ToString()].pic.DoubleClick
+=newSystem.EventHandler(this.pic_ConvyerDoubleClick);
}}
```

During the execution of the task, we can intuitively monitor the running status of the system in real-time from the image display area of the monitoring software. Stacker control flow chart is shown in Fig.5.

6. Real-time Analysis of Data Transmission

In order to verify the real-time data transmission in the system, the data transmission response time between the communication stations is tested and analyzed.

IOP Conf. Series: Journal of Physics: Conf. Series **1026** (2018) 012015 doi:10.1088/1742-6596/1026/1/012015

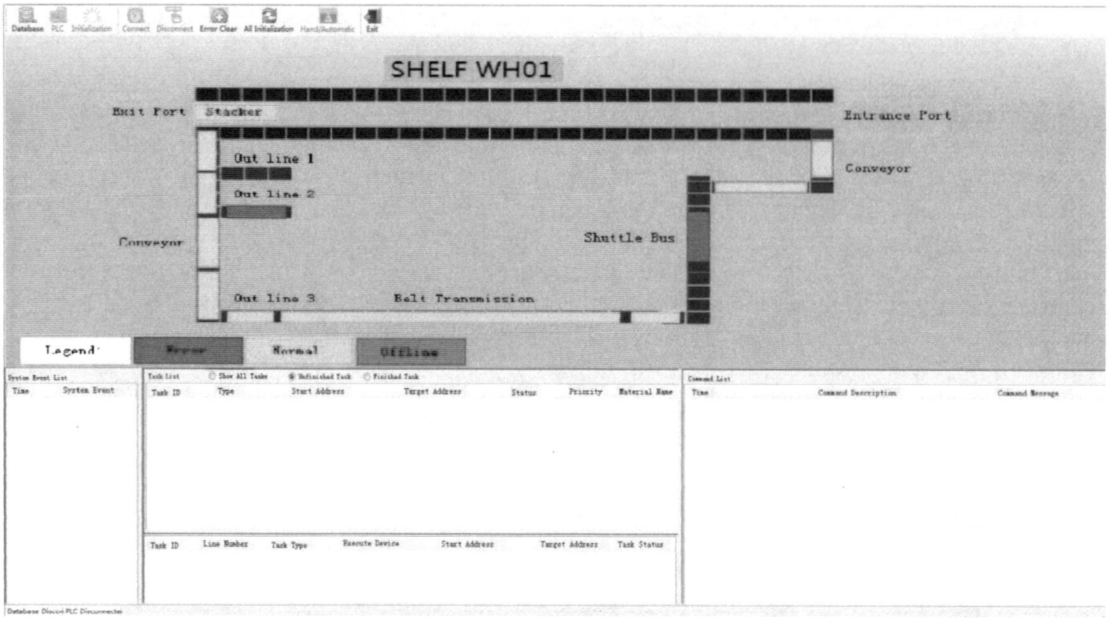

Figure 4. Monitor software

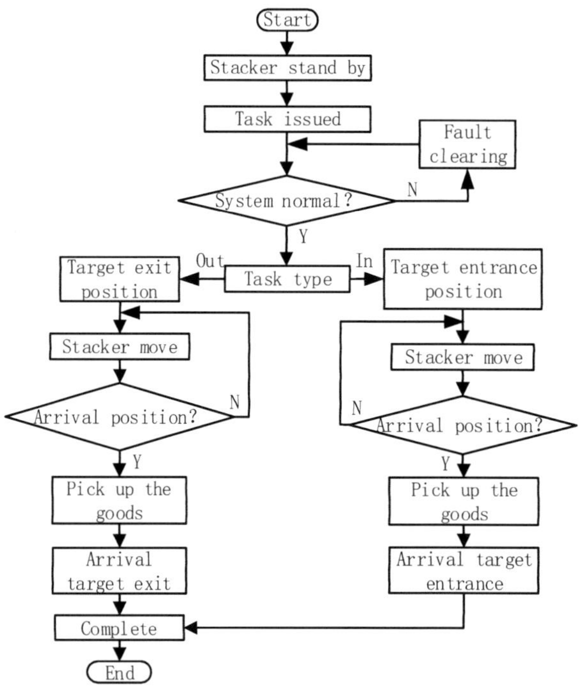

Figure 5. Stacker control flow

Data transmission response time acquisition and analysis are carried out by sending different packets. The monitor computer as packet transmission site send different packets to the management computer site, main site PLC, and stacker crane PLC, respectively.

From Table1, the average response time of data transmission are 5ms and 2ms respectively. From the collected response time data, in the case of wireless network communication mode, the data transmission response time are kept within 15ms between different communication sites. The average response time from the monitor computer to the management computer are 2ms and 4ms, respectively. The average response time of data transmission from the monitor computer to the main site PLC are 3ms and 4ms, respectively. The average response time of the data transmission from the monitor computer to the stacker crane PLC is 5ms and 2ms respectively.

Table 1. Wireless communication mode.

Site Name	Monitor Computer	Management Computer	Main Site PLC	Stacker Crane PLC
IP Address	（192.168. 1.105）	（192.168.1 .104）	（192.168.1 .100）	（192.168. 1.103）
Send 10 Packets （32byte） Response Time（ms）		Max=3ms Min=2ms Ave=2ms	Max=5ms Min =2ms Ave=3ms	Max=6ms Min=2ms Ave=5ms
Send 10 Packets （64byte） Response Time（ms）		Max=7ms Min=3ms Ave=4ms	Max=8ms Min=2ms Ave=4ms	Max=8ms Min=1ms Ave=2ms

The data transmission in wired Ethernet communication mode has be tested. The data transmission response time are less than 1ms. The wireless network communication response time are more than wired communication, which are still able to meet the system communication requirements. It should be noted that, in the larger and longer transmission distance of the automatic storage system, in order to ensure the stability of wireless network communication and real-time transmission, we can use the way to increase the wireless relay, or the wireless AP installed to a higher position as far as possible to reduce the interference of the metal shelf on the wireless signal.

7. Conclusions

The OPC communication technology can solve effectively the interconnection problem between heterogeneous networks. And C# is used to develop system monitoring software based OPC of automatic storage simulation system successfully. Experiments show that the connection between heterogeneous networks is stable.

References

[1] Wang Jie, Gao Kunlun, Wang Wanzhao. “Backstage Control of DCS in Thermal Power Plant Based on OPC Communication Technology,” Electric Power Automation Equipment, vol.4, pp.142-147, April 2013.

[2] Gao Kunlun. “Research and Application of OPC Communication Technology,”Zheng Zhou: Zhengzhou University, pp. 6-9, 2012.

[3] Zhao Lihui, Huo Chunbao. “Design of automatic sorting and storage system in laboratories,” Experimental Technology and Management, vol.30, pp.130-133, March, 2013.

[4] Chen Weike. “Design on Monitoring System of Storage,” Tian Jin: Hebei University of Technology, pp.5-8, 2014.

[5] Meng Lei, Zang Huadong. “The Warehouse Control System and Configuration Control Based on PLC,” Techniques of Automation and Applications, vol.29, pp.69-72, March, 2010.

[6] Li Ning, Tang Shaoqin, Bian Juange. “Design of monitoring system for stereoscopic warehouse of production line based on configuration software,” Automation and Instrumentation, vol.7, pp.61-64, April, 2014.

IOP Conf. Series: Journal of Physics: Conf. Series **1026** (2018) 012016 doi:10.1088/1742-6596/1026/1/012016

Design and implementation of GPS baseband correlator

Qingming Yi, Min Shi, Zhiji Lin and Chong Luo

College of Information Science and Technology, Jinan University, Guangzhou, China

E-mail: tyqm@jnu.edu.cn

Abstract. Correlator design is the core part of the GPS baseband processor design. In this paper, the idea of the modular is used to design the local NCO, pseudo-code generator, mixer, integral-clear filter, phase detector and loop filter. Each module is completed by Verilog and validated on the FPGA. An alternative true value operation is ultilized on the mixer to replace the multiplier operation of the input signal in this paper, thereby reducing the computational complexity. By initializing G2 register of the PN generator, it can remove the design of the phase selector. FPGA verification results show that the improved GPS baseband correlator can reduce resource consumption, accelerate the processing speed, and correctly demodulate the navigation data from input GPS signals.

1. Introduction

Global Positioning System (GPS) is a satellite system that provides continuous accurate and reliable location, speed and time information for all types of users and receiver platforms. The satellite system has become a significant support system and infrastructure of major in national economic security, information security and national security, which is the embodiment of comprehensive national strength [1]. Therefore, the research and development of GPS baseband processor has become a hot area of research. Correlator is a core part of the baseband processor, whose efficiency and performance affect the speed of positioning and the accuracy of the entire GPS receiver [2].

Therefore, this paper researches and implements the GPS baseband processor correlator. The bit number of mixer operations is less relatively because truth-operations can reduce the amount of computation. It can remove the design of the phase selector by initializing G2 register of the PN generator. FPGA simulation results suggest that the system has characteristics of occupation less hardware resource and wide range of the phase discrimination.

2. Structure and principle of correlator

As the core of the baseband processor, correlator detects correlation cumulative value to determine whether the processing is successful during the capture. It also stabilizes the tracking GPS signal during the tracking processing, and demodulates the navigation data meanwhile [3]. The correlator structure block diagram of baseband processor is shown as Figure 1.

GPS digital IF signal can be expressed as [4]:

$$s(t) = AC(t)D(t)\cos(w_c t + \varphi) + n_0 \tag{1}$$

IOP Conf. Series: Journal of Physics: Conf. Series **1026** (2018) 012016 doi:10.1088/1742-6596/1026/1/012016

Where A is the amplitude of the GPS signal, C(t) is the C/A code, D(t) is the navigation message, W_c is the carrier frequency of the intermediate frequency, φ is the initial phase of the carrier, n_0 is the additive white Gaussian noise.

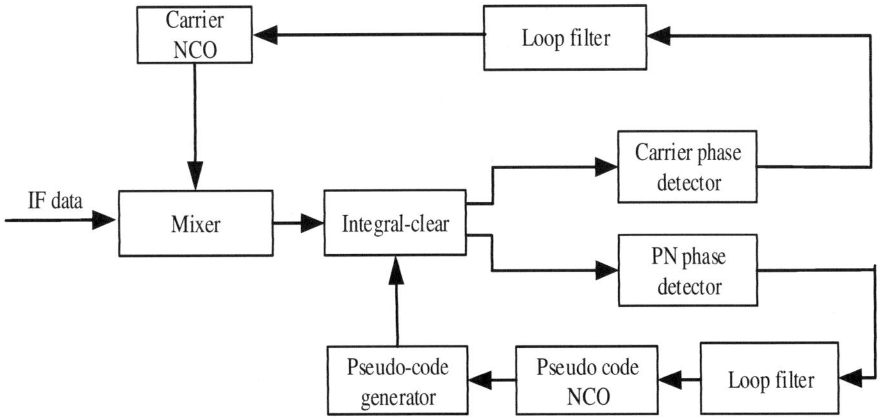

Figure 1. Block diagram of the baseband processor correlator

In-phase component I and quadrature component Q generated by the local carrier NCO are expressed as:

$$I(t) = \cos(w_c t + \varphi(t)) \quad (2)$$

$$Q(t) = \sin(w_c t + \varphi(t)) \quad (3)$$

s(t) multiplies with I(t) and Q(t) respectively for removing the carrier. After simplifying and ignoring the high frequency noise component, they can be expressed as:

$$s(t)\cdot I(t) = \sqrt{P/2}\cdot C(t)D(t)\cos[\varphi-\varphi(t)] \quad (4)$$

$$s(t)\cdot Q(t) = \sqrt{P/2}\cdot C(t)D(t)\sin[\varphi-\varphi(t)] \quad (5)$$

Where φ-φ(t) is the phase error Δφ between the local carrier frequency and IF frequency, i.e. Δφ = φ-φ (t).

The formula (4) and (5) are respectively multiplied with the PN generated by the local coder for dispreading pseudo code. They are written as follow:

$$I'(t) = \sqrt{P/2}\cdot D(t)\cos(\Delta\varphi) \quad (6)$$

$$Q'(t) = \sqrt{P/2}\cdot D(t)\sin(\Delta\varphi) \quad (7)$$

After that, integration is carried out for I'(t) and Q'(t) signal, then two components are phased by carrier phase detector and pseudo code phase detector, and then output phase error signal sends it into loop filter. Finally, it is fed back to the local NCO. Error signal can reduce the effects of noise by the loop filter.

3. Design and implementation of correlator's each module

The function of correlator is that carrier frequency error and pseudo code phase error signal is obtained by demodulating carrier which is produced through the input GPS intermediate frequency signal crossing the local carrier NCO (Voltage Controlled Oscillator) and dispreading pseudo code generated by local pseudo code generator. Error signal is extracted from the phase detector, and it filters out noise through a loop filter and controls the local NCO. Finally the navigation data can be obtained from integration signal of I path, meanwhile, Q path can obtain the noise signal by integrating [5].

The capturing and the tracking parts of the baseband processor can use the structure of correlator, so it can save design resources. The main modules of correlator are the local NCO, pseudo-code generators, mixer, integral-clear filter, phase detector and loop filter. Here each module is designed and described in detail as follow.

3.1. Local NCO

Local NCO includes carrier NCO and pseudo-code NCO. The function of the carrier NCO is generating local carrier signal, which demodulates the IF signal input, and removes the carrier signal. While the function of pseudo code NCO is providing 2.046MHz shift frequency for pseudo code generator [6]. Pseudo-code NCO has similar principles with carrier NCO. Here introduce the carrier NCO in detail.

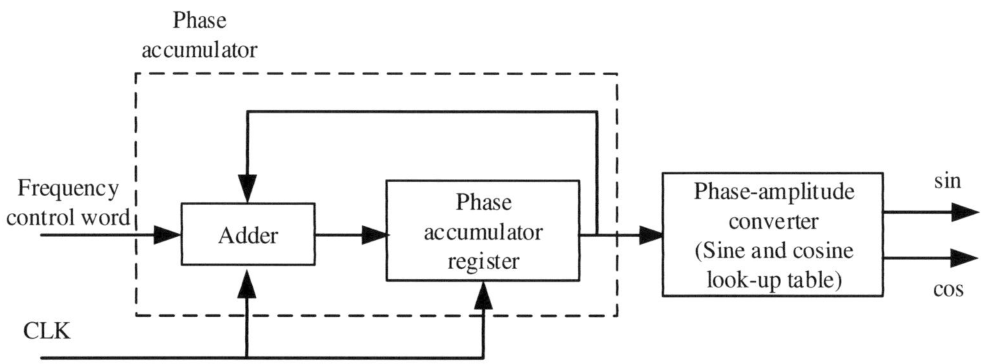

Figure 2. Block diagram of the carrier NCO

Carrier NCO is implemented by DDS (Direct Digital Synthesis) technology [7], shown in Figure 2. The frequency control word is the increasing amount of phase of the carrier within one sampling clock period. And accumulator carries out addition operation of frequency control word and the phase accumulation register phase data, then the results will be accumulated to the phase accumulator register for latching. Amplitude conversion is conducted on intercepting several highest data of phase accumulation register in term of sine and cosine look-up table, which stores the quantized values of sine and cosine signals in advance. Changing register certain high-bit data can output sine and cosine signals.

The carrier frequency of the VCO output signal f_{NCO} is:

$$f_{NCO} = \frac{f_{CLK} \times M}{2^N} \tag{8}$$

Frequency resolution is $f_{CLK}/2^N$, since the system clock frequency is equal to 50MHz, M is 29 bits (i.e., M= 29). The bit of phase accumulator register is 30 (i.e., N= 30), the carrier frequency of input signal is 3.449784MHz, then f_{NCO} = 3.449784MHz. From the formula (8), it can be obtained for the frequency control word:

$$M = \frac{f_{NCO} \times 2^N}{f_{CLK}} = \frac{3.449784 \times 10^6 \times 2^{30}}{50 \times 10^6} = 74083547\,(D) \tag{9}$$

3.2. Pseudo-code generator

The function of pseudo-code generator is producing three pseudo-code signals which are different with half a chip, including advance code, instant code and delayed code. Where the advance code and delayed code are used in pseudo-code phase detector to obtain a pseudo-code phase error. While instant code can obtain carrier frequency error in the carrier discriminator. Frequency shift of three pseudo-code signal generated by the pseudo-code generator comes from pseudo-code NCO. C/A code generation principle is shown in Figure 3.

C/A code is driven by the clock of 1.023MHz, and produced by two 10 m-sequence register (G1 and G2) by adding the two mold. The initial state of G1 and G2 are all set as "1", the output G1 is generated by the 10th storage unit of the shift register, the output G2 is obtained by selecting two memory cells of G2 shift after XOR operation.

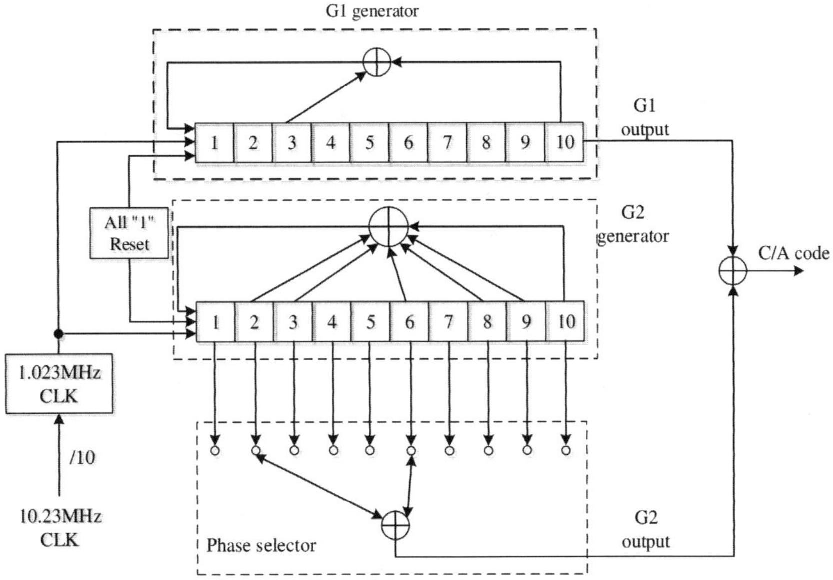

Figure 3. Schematic of C/A code generator

Different phase option of generator G2 outputs different sequences corresponding to different numbers of pseudo-satellite PRN code, and produces different delays phase to generator G2. Code phase of the GPS signal's C/A code correspondence table is shown in Table 1. In an actual circuit design, designing a single phase alone will increase the resource consumption and the difficulty of design. At present, it can take the method that directly inputs the corresponding the G2 initial phase to the satellite PRN numbers for the different satellites RPN numbers, which can reduce design resource on the basis of implementation function.

Table 1. GPS code phase signal C/A code correspondence table.

PRN Satellite number	Phase select	G2 register initial phase	The first 10 chips (octal)	PRN Satellite number	Phase select	G2 register initial phase	The first 10 chips (octal)
1	2⊕6	3EC	1440	17	1⊕4	226	1156
2	3⊕7	3D8	1620	18	2⊕5	04C	1467
3	4⊕8	3B0	1710	19	3⊕6	098	1633
4	5⊕9	360	1744	20	4⊕7	130	1715
5	1⊕9	096	1133	21	5⊕8	260	1746
6	2⊕10	12C	1455	22	6⊕9	0C0	1763
7	1⊕8	196	1131	23	1⊕3	0CE	1063
8	2⊕9	32C	1454	24	4⊕6	270	1706
9	3⊕10	258	1626	25	5⊕7	0E0	1743

10	2⊕3	374	1504	26	6⊕8	1C0	1761
11	3⊕4	2E8	1642	27	7⊕9	380	1770
12	5⊕6	3A0	1750	28	8⊕10	300	1774
13	6⊕7	340	1764	29	1⊕6	056	1172
14	7⊕8	280	1772	30	2⊕7	0AC	1453
15	8⊕9	100	1775	31	3⊕8	158	1625
16	9⊕10	200	1776	32	4⊕9	2B0	1712

3.3. Mixer

The role of the mixer is that multiplies the input IF data multiplied by the local carrier NCO signal to remove carrier. The value of IF is ± 1 and ± 3, represented by sign bit (if_sign) and data bits (if_mag) totally 2 bits; the value of the local carrier is ± 1 and ± 2, represented by sign bit (carr_sign) and data bits (carr_mag) totally 2 bits; so the mixer output range is [-6, + 6] and width is 4 bits. Due to less operation bits of the input signal, it can directly obtain an output result with bit operation. After mixing, truth value table is shown as follows in Table 2 and Table 3:

Table 2. Truth table of symbol bit.

if_sign	carrier_sign	mix_sign
0	0	1
0	1	0
1	0	0
1	1	1

Table 3. Truth table of amplitude bit.

If_mag	Carrier_mag	Mix_mag_0	Mix_mag_1	Mix_mag_2
0	0	1	0	0
0	1	0	1	0
1	0	1	1	0
1	1	0	1	1

3.4. Integrate-clear filter

Integrate-clear filter mainly implements the mixing and integrating of mixed signal and C/A code. The integration time is set to 1ms (one C/A code cycle), since the sampling rate is 16.311MHz, therefore it need to integrate 16,311 sampling points of mixed signal. The clear signal comes from dump_enable pulse of the PN code generator.

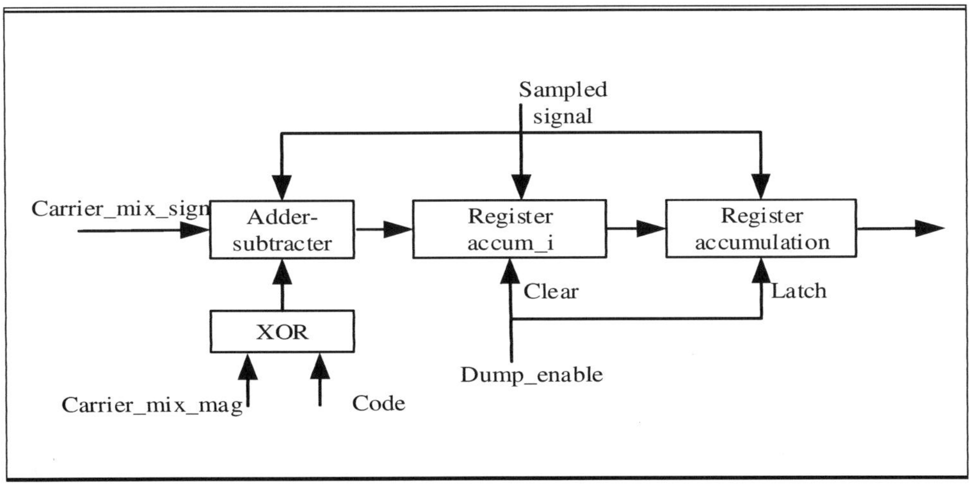

Figure 4. Block diagram of integral-clear

The block diagram of integrate-clear filter is shown in Figure 4. First, XOR operation is implemented within the sign bit of C/A code and mixed signal. If positive, it accumulates data bits of mixed signal; if negative, it minuses mixing signals data bits of mixed signal. Second, register (Accum_i) is used to store data integrated. When clear signal (dump_enable) coming, the value of register (accum_i) is latched into register (accumulation) to complete data output.

3.5. Phase detector

The carrier phase detector is mainly using the phase detector's algorithm to estimate phase error of in-phase component I and quadrature component Q. PN phase detector mainly estimates pseudo-code phase error. Both pseudo-code phase detector and carrier phase detector use CORDIC algorithm [7]. Now mainly introduce the carrier phase detector as follow.

In conventional costas loop carrier phase detector, commonly use the two-quadrant arc tangent of the phase detector (maximum likelihood estimator). Its function is atan (Qp/Ip), and its output range is [-$\pi/2$, $\pi/2$]. Meanwhile the arctan(x) operation uses the CORDIC algorithm, which is easy to implement. It can be achieved only by shift and subtraction operation, thus avoiding the complicated operation process, and being suitable for realization in the hardware circuit design.

In this design, arctan (x) function is implemented by arctan function of Xillix's CORDIC (IP core). Since the range of the CORDIC (IP core) output is [-1, +1] and the data of input and output adopt QN format [8], so the first two bits of input data represent as integer and others represent as decimal, and phase value of output is set to radians format, and its output range is [-1, +1] and its unit is rad. The output of the first three phases represent as integer, the rest represent as decimal. Both input and output of the CORDIC core use QN format, so that leading to an error result when switching negative numbers to QN format. Therefore, it is necessary to switch all input value to positive value, and then calculate the phase value. Moreover, phase symbol (I_sign ^ Q_sign) will be outputed depending on the input value (I_sign and Q_sign). If the phase symbol is 1, phase value becomes negative (negated and plus 1); phase value unchanged if the phase of the symbol is 0.

3.6. Loop filter

The noise signal is filtered out thought the loop filter to improve the signal to noise ratio. The loop of correlator uses a second order loop, which is a first-order loop filter, of which filter structure is shown in Figure 5.

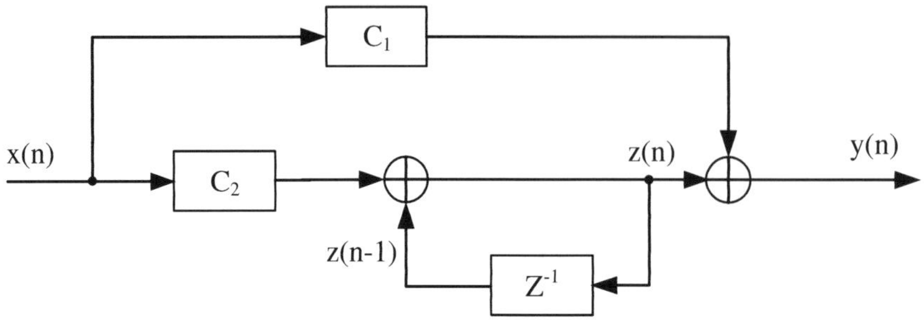

Figure 5. A block diagram of the loop filter

Fig. 5 shows:

$$y(n) = C_1 \cdot x(n) + z(n) = C_1 \cdot x(n) + C_2 \cdot x(n) + z(n-1) \tag{11}$$

Wherein parameters C1, C2 [8] are:

$$C_1 = \frac{\tau_1 T}{\tau_2} \tag{12}$$

$$C_2 = \frac{T^2}{\tau_2} \tag{13}$$

Where T is the pre-integration time, τ1 and τ2 are the coefficients of the filter transfer function. In this paper, the filter parameters for the decomposition is form of 2-n in order to reduce the computing resources and improve processing speed, which can be achieved through the register shift operation in the FPGA. The transmission coefficient employs empirical value. C1 is equal to 2-7 and C2 is equal to 2-12 in carrier tracking loop. While C1 is equal to 2-13 and C2 is equal to 2-22 in code tracking loop. Code tracking loop has the same design principles with carrier tracking loop, except different parameters.

4. Synthesis and simulation of system

After capturing successfully, Doppler frequency and PN code phase will be sent to the carrier NCO and pseudo-code NCO to obtain both initial Doppler frequency and PN code phase[9]. Then it starts tracking section. Timing simulation of correlator module is shown in correlator module timing simulation diagram[10]. It can know from Figure 6. that integration of I_prompt path obtains navigation data while integration of Q_prompt path obtains the noise signal, which accords with expectations.

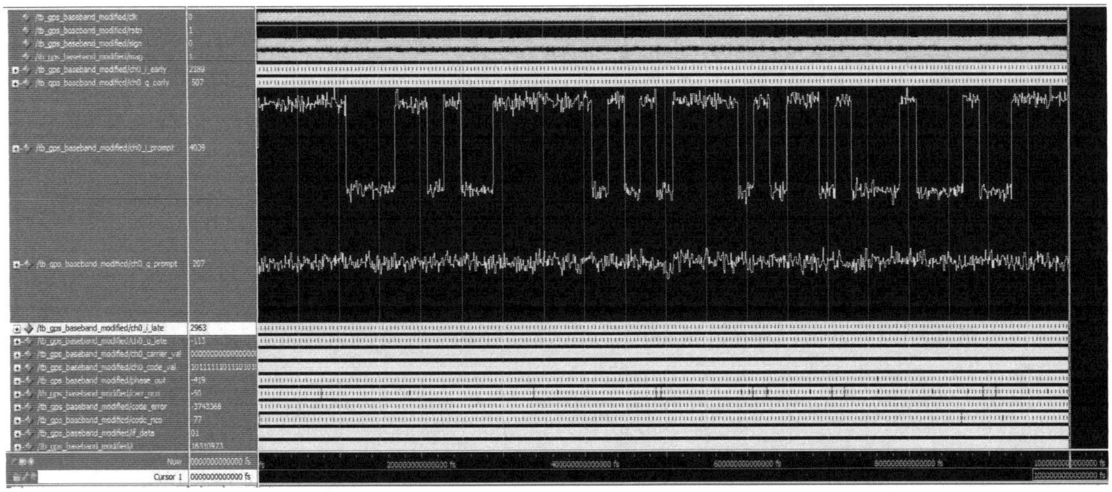

Figure 6. Correlator module timing simulation diagram.

FPGA chip resource consumption is shown in Table 4 after synthesis of correlator. The table IV shows that correlator module is less overall chip footprint, which can be easily embedded into the GPS satellite baseband signal processing chip.

Table 4. Truth table of amplitude bit.

Logic Utilization	Used	Available	Utilizati-on rate
Slice Registers	630	54576	1%
Slice LUTs	1678	27288	6%
fully used LUT-FF pairs	429	2046	21%
bonded IOBs	259	316	82%
BUFG/BUFGCTRLs	1	16	6%
DSP48A1s	4	58	7%

5. Conclusion

In this paper, the design of each module of the baseband correlator is researched and described in detail. Instead of multiplication operation, alternatively using true value operation can save design resources in the mixer design. It can omit the phase selector design when directly initializing the G2 register in pseudo-code generator design. The simulation results suggest that the correlator of this design can reduce resource consumption and stabilize output navigation data simultaneously.

Acknowledgments

This paper is supported by the 2016 Guangdong Province Science and Technology Project (2015B090910001) and the 2016 Guangzhou Science and Technology Project (201604016085).

References

[1] H. Zhang, J. Zhang and X.Z. Liang, "Development Status and Trends of Global Positioning System Technology," Instrument Standardization & Metrology, vol. 02, pp. 43-45, 2011.

[2] H. Hu, C. Yuan, C. Wu and M.H. Gao, "FPGA-based GPS Correlator Research and Implementation," Journal of System Simulation, vol. 07, pp. 1513-1517,1521, 2011.

[3] Y. Xu, Z. Zhu and Y. Yang, "Circuit Design of GPS Correlator," Chinese Journal of Electron Devices, vol. 29, pp. 722-725, 2006.

[4] C.J. Li, X. Li, "Performance of INS-aided tracking loop for GPS high dynamic receiver," Computer Science and Network Technology (ICCSNT), 2012 2nd International Conference on. IEEE, 2012, pp. 757-761.
[5] W. Hedgecock, M. Maroti and J. Sallai, "High-accuracy differential tracking of low-cost GPS receivers," Proceeding of the 11th annual international conference on Mobile systems, applications, and services. ACM, pp. 221-234, 2013.
[6] F. Moschas, S. Stiros, "PLL bandwidth and noise in 100 Hz GPS measurements," GPS Solutions, vol. 19, pp. 173-185, 2015.
[7] J.H. Chen, "Research of Tracking and Acquisition Algorithm for GPS Software Receiver Based On FPGA," Nanjing University of Posts and Telecommunications, China, 2013.
[8] J.E. Volder, "The CORDIC trigonometric computing technique," Electronic Computers, IRE Transactions on, vol. 3, pp. 330-334, 1959.
[9] Cordic Xilinx Technical Documents
[10] A.A. Zhang, Y. Du and F.J. Han, "Design and implementation of digital Costas loop on the FPGA," Electronic Engineer, vol. 01, pp. 18-20, 2006

IOP Conf. Series: Journal of Physics: Conf. Series **1026** (2018) 012017 doi:10.1088/1742-6596/1026/1/012017

Intelligent Predictive Food Traceability Cyber Physical System in Agriculture Food Supply Chain

Rui-Yang Chen

Department of Business Administration, Aletheia University, Taiwan

E-mail: a168.cloudy@msa.hinet.net

Abstract. From the emerging technologies perspective, future advanced food traceability system must consider not only intelligent system needs such as cyber physical system (CPS) from the future internet perspectives but also intelligent behavior such as smart predictive business in food supply chain. However, CPS-based food traceability system also faces many new challenges such as communication efficiency, resource integration and system architecture. This article presents the proposed approach to develop novel intelligent predictive food traceability cyber physical system by using intuitionistic-based fuzzy case-based reasoning with enterprise architecture and value stream mapping method for simulation. Successful case study and experiment demonstrated the performance of the proposed approach. The proposed approach used for traceability performance predictive behavior can contribute to identify traceable objects in the CPS-based food traceability system that are sensitive to a broader range of intelligent food traceability.

1. Introduction

Nowadays, food production in industrial manufacturing environment has to face increasing challenges to keep competence over the next decade with the advent of information system and related emerging technology such as IoT (Internet of Things) and cloud computing. These emerging technologies change over a new generation of manufacturing environment including food supply chain. In recent years, a great application about integrating IoT technology such as RFID (radio-frequency identification) and WSN (wireless sensor network) within the food supply chain has emerged [1][2]. The IoT technology is a novel paradigm attributed to the technical development of infrastructure is specified by EPCglobal (electronic product code global) architecture [3] that it defines information capture, middleware, EPCIS (EPC information system) and advanced applications, discovery service and pedigree, to integrate into this architecture. Accordingly, food production is affected by this emerging system, for example, food production traceability system. Traceability is nowadays fundamental capability to ensure the food safety and quality through information exchange for food production to tracking and tracing the status of food products operation. Pizzuti et al. [4] presents the Food Track &Trace Ontology (FTTO) framework to managing food traceability Information System. Alfredo et al. [5] proposed a novel system using the EPCglobal Architecture for the framework ‘‘RFID from Farm to Fork’’. From the above-mentioned discussion, we know these emerging technologies developed the advanced food traceability system, but these literatures review are still lack of more emerging technologies research such as cyber physical system (CPS) or intelligent system from the future internet perspectives in food supply chain. CPS plays important role in the manufacturing environment such as field device process, production machine process and manufacturing control process using IoT [6]. CPS also relied on the Industry 4.0 [7]. Some previous literature reviews proposed CPS services and applications for business performance such as new CPS software architectural [8]. The emerging CPS technology will explore the new generation intelligent system for food traceability system.

However, CPS also faces many new challenges such as communication efficiency, resource integration and system architecture from literature reviews, especially in food traceability system. From the above-mentioned saying, we know a crucial concept that food traceability cyber physical system need overcome these challenges for the emerging advanced system that is used for the intelligent generation. From this concept, this article proposes two issues. Firstly, how to present novel approach to solve the three challenges using EA and VSM method? Based on the requirements of traceability in the food chain, an intelligent object used to food physical object such as food product or machine is explored, called "traceable object". The internet of things refers to the possibility of endowing everyday objects with each other for cooperation goals [9]. Therefore, this article proposes a novel intelligent CPS-based food traceability system with intelligent object combined by EA, EPCglobal and VSM method. Second, how to show on intelligent behavior in food traceability cyber physical system. Intelligence allows operation to become more autonomous. Autonomous mechanism can help speed up food manufacturing by making it easier to make a decision proactively without human interaction, especially food traceability. From prior literature review, Amit et al. [10] proposed intelligent behavior such as composition-based predictive model to cover many practical situations using microbiological and chemical kinetic databases. Thus, food traceability efficiency is a critical aspect where such predictive behavior can have great impact. The predictive behavior can simulate a food traceability situation for food safety requirement by combining an architecture-based model of the CPS environment. Thus, this article adopts fuzzy case-based reasoning (CBR) and intuitionistic fuzzy for intelligent predictive model to simulate early food traceability efficiency. The proposed fuzzy CBR also includes a similarity and fuzzy membership function to perform predictive behavior. Aiming at traditional CBR dependence on expert or data mining, this paper presents intuitionistic-based fuzzy membership function into CBR method, namely intuitionistic-based fuzzy CBR system. The intuitionistic-based fuzzy CBR introduces predicting as a similarity algorithm of making decision based on comparisons between the new case and the previous similar one as benchmark case. Thus predictive result can help understanding this CPS-based food traceability system whether it is implemented or not in advance for self-configurable capability.

This paper is organized as follows: Section 2 introduces related work such as CBR, intuitionistic-based fuzzy. Section 3 presents the proposed approach to develop novel intelligent predictive food traceability cyber physical system. It consists of predictive algorithm for food traceability efficiency. Section 4 presents the case study and numerical example for agriculture food product. Section 5 submits evaluated experiments with three cases by utilizing sensitivity analysis among the influence of each challenge. Section 6 concludes the work.

2. Related Work

2.1.Intuitionistic Fuzzy Set

The intuitionistic fuzzy sets (IF set) as an extension of Zadeh's fuzzy sets is adapted and useful by K. Atanassov [11] to address imperfect or imprecise information. Recent, IF set theory has been successfully applied in decision making analysis by using both a membership degree and a non-membership degree [12][13]. They consider not only the degree of membership to a accept set, but also the degree of rejection such that the sum of both values is less than 1.In food traceability perspectives, precision recognition may be present in a food product on quantifying the probability of finding intended or unintended traceability information. Therefore, a novel fuzzy traceability approach to food traceability information decision making is based on intuitionistic fuzzy sets. As the food traceability information are vague, uncertain and subjective in nature, this article combines the theory of fuzzy sets with intuitionistic fuzzy approach in modelling fuzzy traceability associated with the traceable object in food supply chain and hence facilitates tracing/tracking the information up to a desired degree of relevance. Modelling food continuous processing and traditional processes by fuzzy logic is fuzzy traceability methodology to improve the handling the quality of food products [14] [15]. But a few literature reviews talk about intuitionistic fuzzy in food traceability application. Therefore, in this paper we study K. Atanassov's method using the notion of IFS theory in food traceability application as follows.

Definition1. Let a set E be fixed. An intuitionistic fuzzy set or IF set A in E is an object having the form.

$$A = \{\langle X, \mu A(X), \nu A(X)\rangle | X \in E\}, \quad (1)$$

where the functions $\mu A : E \rightarrow [0, 1]$ and $\nu A : E \rightarrow [0, 1]$ define the degree of membership and degree of non-membership respectively of the element $X \in E$ to the set A, which is a subset of E, and every $X \in E$, $0 \leq \mu A(X) + \nu A(X) \leq 1$.

Definition2. The value of

$$\pi A = 1-(\mu A(X) + \nu A(X)) \quad (2)$$

is called the degree of non-determinacy (or uncertainty) of the element $X \in E$ to the intuitionistic fuzzy set A. The amount $\pi A = 1-(\mu A(X) + \nu A(X))$ is called the uncertainty, which may cater to either membership value or non-membership value or both.

2.2. Case-based Reasoning

Case-based reasoning (CBR) is the ability of solving problem obtained from past experiences by retrieval method of similar cases. Accordingly, this article applies CBR into predictive food traceability efficiency model. In such this model, prior literature reviews had discussed related topic such as process monitoring and cloud platform as follows. Yang [16] proposed an energy-saving and case-based reasoning information agent in a cloud environment. Applying CBR into business process monitoring is successful approach [17]. This paper applies intuitionistic fuzzy-based CBR technology to the detection of the main food traceability efficiency in food supply chain.

3. Intelligent Predictive food Traceability CPS Approach

3.1. Intelligent CPS-based Food Traceability Model

This article adopts Zachman [18] EA framework to intelligent CPS-based food traceability model with two dimensional matrix which include six descriptive abstractions (data, function, network, people, time and motivation) and six player perspectives (planner, owner, designer, builder, subcontractor and enterprise). Through Zachman framework, this model proposes, food traceability value-added process is explored. The food traceability value-added process uses some tools described as follows. This CTP (Critical Traceability Point)-id identifies unique of the traceable points and further find the relationships between traceable points called CTP-relation. Furthermore, applying CTP into value stream analysis with NVA, NNVA, VA, CVA, LVA and IVA is carried out. Monden [19] presented three types such as non-value adding (NVA), necessary but non-value adding (NNVA) and value-adding (VA). To analyze external manufacturing operation across supply chain in VSM, this article presents three new concepts such as collaborative value adding (CVA), linked value adding (LVA) and interrupted value adding (IVA).In the meanwhile, this article uses four traceable identified information items consisting of production history, carbon emission consumption, inventory turnover and logistical status to act on the base level of quantities measurement. According to literatures reviews, the level of traceability can be presented four quantities such as breadth, depth, precision and access to speed the identified information as an essential food safety and quality measurement [20].

a) Intelligent Predictive Algorithm for Food Traceability Efficiency

This article proposes intelligent predictive algorithm for food traceability efficiency and is described as follows. The intelligent predictive food traceability efficiency such as sustainability, visibility and transparency are considered. According to the above-mentioned discussion, this paper applies intuitionistic fuzzy-based CBR technology to the detection of the main food traceability efficiency in food supply chain. The intuitionistic fuzzy-based CBR can intelligently afford predictive traceability behavior for traceable object and traceable point process interaction. In our proposed algorithm, the intuitionistic fuzzy-based CBR acts on intelligent predictive management mechanism. The mechanism is shown on Fig. 1. When a new traceable point takes place, the similarity is acquired through traceability base level with quantities measurement, then execute intuitionistic fuzzy-based CBR with similar case. Thus, most appropriate cases can be verified according intuitionistic fuzzy-based associate rule within CBR database for predictive management mechanism.

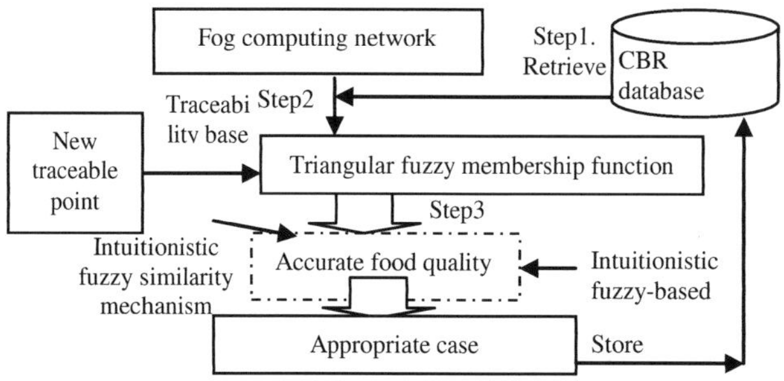

Figure 1. Intelligent predictive algorithm using intuitionistic-based fuzzy CBR.

In figure1, three steps would be useful to improve prediction performance. Step1 is to retrieve predictive capability by intuitionistic fuzzy-based associate case. Step2 is to deal with uncertainty status by triangular fuzzy membership function to verify which category for new traceable point. Step3 is to produce more accurate measure for traceability efficiency between traceable point and traceable object by intuitionistic-based fuzzy method. The intuitionistic-based fuzzy CBR database associates the correlation matrix of quantities measurement to the decision-making solution. When predictive food traceability mechanism is generated, this database applies these setting automatically. To further enhance the traceability efficiency, the intuitionistic-based case rules in the intuitionistic-based fuzzy CBR database are fully customizable to represent current correlation matrix of quantities measurement. To achieve intuitionistic-based fuzzy CBR algorithm, this article proposes intuitionistic-based fuzzy similarity algorithm as follows.

This intuitionistic-based similarity algorithm consists of three steps as follows.

Step1. Similarity algorithm

Similarity definitions could be given to intuitionistic fuzzy and intuitionistic triangular fuzzy membership function. The similarity formula of intuitionistic fuzzy CBR refers to the similar weighting algorithm in formula (3) for each traceability object:

$$\text{Total similarity} = \sum_{i=1}^{n} W_i \times Sim_i(E_i, F_i) \tag{3}$$

Where Wi refers to the weighting value of traceability base level i, and n refers to the total number of this level, and Ei refers to the value of index i of the new traceable point case, Fi refers to the value of index i of the past one case. This article uses four traceable identified information items as traceability base level consisting of production history, carbon emission consumption, inventory turnover and logistical status to act on the base of quantities measurement. The level of traceability can be presented four quantities such as breadth, depth, precision and access. Furthermore, in order to gain the Wi, we use AHP method.

Step2. $Sim_i(E_i, F_i)$

$Sim_i(E_i, F_i)$ represents the similarity degree among (0,1) between Ei and Fi from intuitionistic fuzzymembership function such as H, M and L. Its result adopts the smaller values (π_A) among H, M and L. Furthermore, the intuitionistic fuzzy CBR system predicts the traceable object that verifies quickly the accuracy food quality for a given case feature. In order to achieve the accuracy food quality, the proposed approach is stored case base that is supposed to characterize traceability base level with quantities measurement.

Step3. Traceability object ranking

To understand traceability efficiency, this predictive algorithm adopts traceability object ranking. It aims to find $\min TO_i$ ranking (π_A) efficiency degree under case k that predicts which traceability object is smaller for impacting on efficiency. The traceability object ranking algorithm consists of two matrixes such as traceable object matrix and correlation matrix. This traceable object matrix (T_i) is calculated by intuitionistic-based fuzzy CBR with (μ_A,ν_A) form as follows:

$$T_i = \lfloor TO_1, TO_2, TO_3, TO_4, TO_5, TO_6, TO_7, TO_8 \rfloor \quad i = 1\text{-}8$$

Then, the correlation matrix (P_i) is described (μ_A,ν_A) follows

$$P_i = \begin{array}{c|c} & y_1 \;\ldots\ldots\ldots\; y_n \\ \begin{array}{c} x_1 \\ . \\ . \\ x_{m-1} \\ x_m \end{array} & \begin{bmatrix} (u_{11}, v_{11}) & \ldots\ldots & (u_{1n}, v_{1n}) \\ & & \\ . & \ldots & \\ & & \\ (u_{m1}, v_{m1}) & \ldots\ldots & (u_{mn}, v_{mn}) \end{bmatrix} \end{array} \quad m, n = 8; Y_i = TO_i$$

$x_{1\sim4}$= traceability base level; $x_{5\sim8}$ = quantities measurement

Final, the traceability object ranking algorithm is described as follows.

$$\lfloor TO_1, TO_2, TO_3, TO_4, TO_5, TO_6, TO_7, TO_8 \rfloor \times \begin{array}{c|c} \begin{array}{c} x_1 \\ . \\ . \\ x_{m-1} \\ x_m \end{array} & \begin{bmatrix} (u_{11}, v_{11}) & \ldots\ldots & (u_{1n}, v_{1n}) \\ & & \\ . & \ldots & \\ & & \\ (u_{m1}, v_{m1}) & \ldots\ldots & (u_{mn}, v_{mn}) \end{bmatrix} \end{array} = \lfloor TO'_1, TO'_2, TO'_3, TO'_4, TO'_5, TO'_6, TO'_7, TO'_8 \rfloor \quad (4)$$

The Method of intuitionistic-based fuzzy CBR is characterized by the correlation between traceability object and traceability base level, then we transform these correlations to quantitative indicators to evaluate the similarities of traceability performance for predicting the future behavior. Furthermore, some cases are compared with ranking of the degree of efficiency according to repeated calculation for intuitionistic-based fuzzy CBR database. This CBR database with intuitionistic-based fuzzy rule can be used for classifying a new case by searching for the cases that have the most similar features to find similar traceability performance. It can be generated using IF-THEN statements: The statement is as followings: IF (The correlations to quantitative indicators with intuitionistic-based fuzzy membership function) THEN (similar traceability performance). According to the intuitionistic-based fuzzy rule, we can further predict traceability performance by similar traceability behavior.

4. Experiment

The experiment is evaluated with three cases by utilizing sensitivity analysis of the three degrees of intuitionistic-based fuzzy membership function such as low, medium and high among the influence of each challenge for three scenarios in the case study. Thus, table 1 shows on case1 and scenarios1 with communication efficiency. The case represents different values based on the same case with intuitionistic-based fuzzy membership function model. The experiment was conducted by simulating different scenarios with three running times based on intuitionistic-based fuzzy rule, which were defined by changing different cases in the Table 1. Table 1 and Fig. 2 show the results obtained after the simulation execution using Intuitionistic-based fuzzy rule method. It is possible to make the following conclusion according to those results using accuracy and efficiency index. The efficiency index (1-| $\Sigma\pi_{A1}$ - $\Sigma\pi_A$ |) illustrates the running times experiment result ($\Sigma\pi_{A1}$) that were achieved similarity degree of traceability performance after compared the case ($\Sigma\pi_A$) base solution. The accuracy index (1-|$(\Sigma\pi_{A1} - \Sigma\pi_A)/\Sigma\pi_A$|) illustrates the running times experiment result that were achieved accurate degree of traceability performance after compared the case base solution. If accuracy index < 0, then it = 0. Both efficiency and accuracy have a similar trend of process. The average of efficiency is 0.73 and the average of accuracy is 0.79. In general, the two index tend to improve the CPS-based traceability system becomes better (>0.5).

Table 1. Efficiency and accuracy index

No of Running times	Case1 Scenario1	Case1 Scenario2	Case1 Scenario3	Case2 Scenario1	Case2 Scenario2	Case2 Scenario3	Case3 Scenario1	Case3 Scenario2	Case3 Scenario3
Efficiency 1	0.56	0.66	0.56	0.78	0.77	0.66	0.78	0.87	0.66
2	0.69	0.91	0.71	0.59	0.81	0.67	0.79	0.81	0.67
3	0.77	0.85	0.73	0.67	0.8	0.83	0.57	0.8	0.73
Accuracy 1	0.88	0.87	0.66	0.78	0.87	0.56	0.78	0.77	0.56
2	0.89	0.81	0.77	0.79	0.91	0.77	0.89	0.91	0.77
3	0.67	0.79	0.83	0.67	0.82	0.83	0.67	0.81	0.93

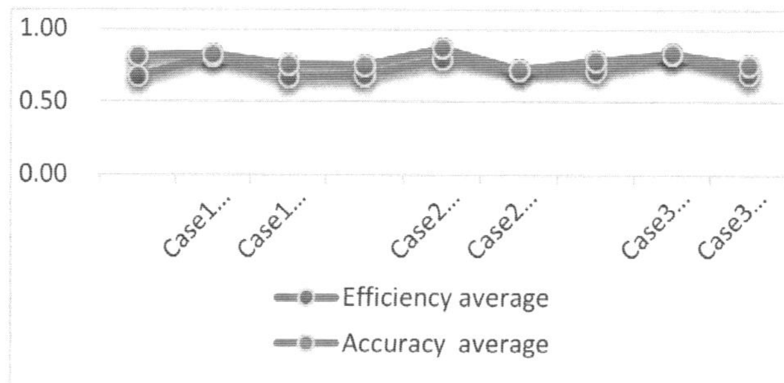

Figure 2. Efficiency and accuracy index

5. Conclusion

From the emerging technologies perspective, future advanced food traceability system must consider not only intelligent system needs such as cyber physical system (CPS) from the future internet perspectives but also intelligent behavior such as smart predictive business in food supply chain. We know a crucial concept that food traceability cyber physical system need overcome these challenges for the emerging advanced system that is used for the intelligent generation. The food traceability efficiency is a critical aspect where such predictive behavior can have great impact. Thus, this article adopts fuzzy case-based reasoning and intuitionistic fuzzy for intelligent predictive model to simulate early food traceability efficiency. In this article, the proposed approach contribution is described as follows. The Method of intuitionistic-based fuzzyCBR is characterized by the correlation between traceability object and traceability base level, then we transform these correlations to quantitative indicators to evaluate the similarities of traceability performance for predicting the future behavior. According, the proposed approach divided to two mechanisms consisting of intelligent CPS-based food traceability in food supply chain and predictive algorithm for food traceability efficiency. To understand traceability efficiency, this predictive algorithm adopts traceability object ranking method. The proposed intuitionistic fuzzy case-based reasoning of the impact of CPS-based food traceability system can be effective.

References

[1] Grabacki, S.T., Ronchetti, M., Humphrey, T., Hedgepeth, O., 2007. How it will transform packaging, distribution and handling of Alaska seafood. In: Anchorage, Alaska: International

Smoked Seafood Conference Proceedings Alaska Sea Grant College, Program, AK-SG-08-02, pp. 101–106.
[2] Kumar P., H.W. Reinitz, J Simunovic, K.P. Sandeep and P.D. Franzon, 2009. Overview of RFID technology and its applications in the food industry. Journal of Food Science 74(8), R101–R106.
[3] EPCglobal. (2015), Available in: /http://www.epcglobalinc.org/home: April.
[4] Pizzuti T., G. Mirabelli, M. Angel, S. Bobi, Fernando G.G.(2014). Food Track & Trace ontology for helping the food traceability control, Journal of Food Engineering, 120, 17–30.
[5] Alfredo P. M., Alejandro A.M., Mira T., Piero F.. (2014).Advanced traceability system in aquaculture supply chain, Journal of Food Engineering, 122, 99–109.
[6] Chong H., J. M. Jornet, E. Fadel, I. F. Akyildiz, A cross-layer communication module for the Internet of Things, Computer Networks 57 (2013) 622–633.
[7] Kagermann, H. Wahlster, W. & Helbig, J., Securing the future of German manufacturing industry: Recommendations for implementing the strategic initiative INDUSTRIE 4.0. Final report of the Industrie 4.0 Working Group, acatech, April 2013: 78.
[8] Bhave, A., Garlan, D., Krogh, B., Rajhans, A., Schmerl, B. Augmenting Software Architectures with Physical Components. In: Proceedings of the Embedded Real Time Software and Systems Conference ERTS2th (2010).
[9] Giusto D., A. Iera, G. Morabito, L. Atzori (Eds.), The Internet of Things, Springer, 2010. ISBN: 78-1-4419-1673-0.
[10] Amit H., A. Dhall, A. K. Datta, D. Glenn Black, P.M. Davidson, J. Li, S. Zivanovic, A user-friendly general-purpose predictive software package for food safety, Journal of Food Engineering, Volume 104, Issue 2, May 2011, Pages 173-185.
[11] Atanassov K., Intuitionistic fuzzy sets, Fuzzy Sets and Systems 20 (1986) 87-96.
[12] Zhou L., W.Z. Wu, On generalized intuitionistic fuzzy rough approximation operators, Inform. Sci. 178 (2008) 2448–2465.
[13] Huang B., D.K. Wei, H.X. Li, Y.L. Zhuang, Using a rough set model to extract rules in dominance-based interval-valued intuitionistic fuzzy information systems, Inform. Sci. 221 (2013) 215–229.
[14] Skoglund T., P. Dejmek, Fuzzy Traceability: A Process Simulation Derived Extension of the Traceability Concept in Continuous Food Processing Food and Bioproducts Processing, Volume 85, Issue 4, December 2007, Pages 354-359.
[15] Allais I., N. Perrot, C. Curt, G. Trystram, Modelling the operator know-how to control sensory quality in traditional processes, Journal of Food Engineering, Volume 83, Issue 2, November 2007, Pages 156-166.
[16] Yang S. Y., 2013. Developing an energy-saving and case-based reasoning information agent with Web service and ontology techniques, Expert Systems with Applications, 40 3351–3369
[17] Montani, S., Leonardi, G., & Bramer, M. (2010). A case-based approach to business process monitoring. In Intelligence in theory and practice III. Boston: Springer.
[18] Zachman, J.: The framework for information systems architecture. IBM Syst. J. 26 (3), 276–292 (1987).
[19] Monden, Y., Toyota Production System: An Integrated Approach to Just-in-Time, 2nd ed., Industrial Engineering and Management Press, Norcross, GA, 1993.
[20] McEntire, J. C., Arens, S., Bernstein, M., Bugusu, B., Busta, F. F., Cole, M., et al. (2010). Traceability (product tracing) in food systems: an IFT report submitted to the FDA, volume 1: technical aspects and recommendations. ComprehensiveReviews in Food Science and Food Safety, 9, 92-158.

IOP Conf. Series: Journal of Physics: Conf. Series **1026** (2018) 012018 doi:10.1088/1742-6596/1026/1/012018

Optimizing and Modeling for Plastic Injection Molding Process using Taguchi Method

T Kiatcharoenpol[1], T Vichiraprasert[1]

[1] Department of Industrial Engineering, Faculty of Engineering, King Mongkut's Institute of Technology Ladkrabang, Bangkok, Thailand.

Abstract. The objective of this work is to study process parameters of Plastic injection molding. Taguchi method, an unconventional design of experiment, is used as an approach to optimize the process parameters to improve quality characteristic of work-piece. The two responses of experiment are volume shrinkage and total displacement. Orthogonal array (L16) is used to conduct experiments. The result shows that there are three statistically significant factors out of seven factors or process parameters. Such three factors are Melt temperature, Packing time and Cooling time. Linear equations derived from experimental results are constructed and suitable conditions can be obtained from a computer based response optimizer.

1. Introduction

The Taguchi method is a useful tool to optimize designs for high performance and quality with limited experiment. As this method is based on statistics, it is prevalent in engineering, especially in the field of manufacturing process. This work, Taguchi method is used to find the suitable condition to produce the best quality of work-piece in a plastic molding process. The initial process parameters or factors studied are Filling time, Melt temperature, Mold temperature, Maximum injection pressure, Packing time, Maximum packing pressure, and Cooling time. And, uncontrollable factors studied are Air temperature and Eject temperature. The important factors were determined through analysis of variance technique (ANOVA). The linear equations were developed according to significant factors. The quality characteristics of work-piece as responses of experiment are volume shrinkage and total displacement. Then, the primary objective of this work is to develop the mathematic modeling in order to optimize and find the suitable condition that can produce the work-piece with minimum volume shrinkage and total displacement. The experimental plan based on Taguchi approach and experimental conditions are presented in later sections. The result of analysis and conclusion are provided in the last section.

2. Literature review

Taguchi method contains system design, parameters design and tolerance design to achieve a robust design in order to enhance practical product quality [1-2]. Taguchi design provides a powerful and efficient method for designing processes that operate consistently and optimally over a variety of conditions [3]. The use of Taguchi Method involves with following step [4] ; a) Identify the main function and it side effects, b) Identify the noise factors, testing conditions, and quality characteristics, c) Identify the objective function to be optimized, d) Identify the control factors and their levels, e) Select the orthogonal array matrix experiment, f) Conduct the matrix experiment, g) Analyze the data, predict the optimum levels and performance, and h) Perform the verification experiment and plan the future action. The application of Taguchi approach is widely used in various industries [5-7].

3. Experiment plan for a plastic injection molding process

Case study conducted in this place is Injection Molding Process by using injection molding simulation software. Usually, injection molding processes in Figure 1 are as follow: Screw will rotate and retract in melting process in order to achieve setting volume of injection molding as defined. The liquid plastic is injected into the molding. Packing will be done to maintain pressure in order to compensate for shrinkage of the plastic. A cooling process of work piece is started until temperature of work piece reach to eject temperature and work piece is ejected out of the molding [8].

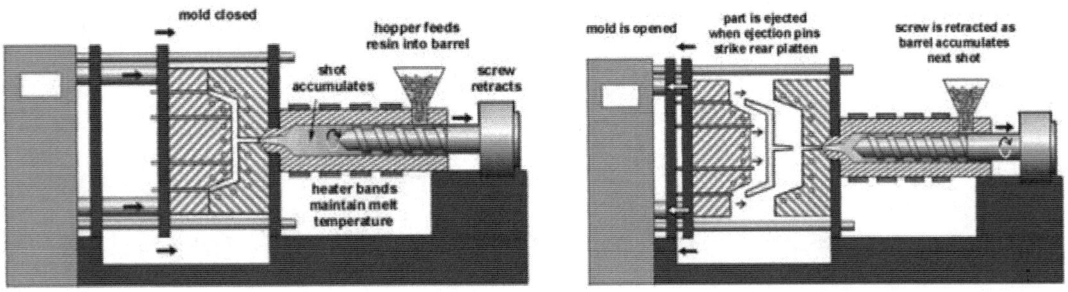

Figure 1. Plastic injection mold process.

The purpose of the experiment is to study factors that have most effect on Volume shrinkage and Total displacement of work-piece. There are seven controllable factors of essential process parameters and two uncontrollable factors as noise parameters. The factors and their levels are presented in Table 1 below.

Table 1. List of controllable factors and theirs levels for the experiment.

Factors	Level		Units
	Low (-)	High (+)	
A. Filling time (A)	2	3	sec.
B. Melt temperature (B)	215	230	°C
C. Mold temperature	35	60	°C
D. Maximum injection pressure	50	80	%
E. Packing time	4	6	sec.
F. Maximum packing pressure profile value	30	50	%
G. Cooling time	12	14	sec.
Uncontrollable Factors			
H. Air temperature	15	40	°C
J. Eject temperature	50	99	°C

The Volume shrinkage is measured in percentage (%) and the Total displacement is measured in millimeter (mm.) are major responses of this experiment. Plastic injection computer programs are used to analyze for injection molding process to find the best value of work- piece design and measure two responses as shown in Figure 2 and 3.

IOP Conf. Series: Journal of Physics: Conf. Series **1026** (2018) 012018 doi:10.1088/1742-6596/1026/1/012018

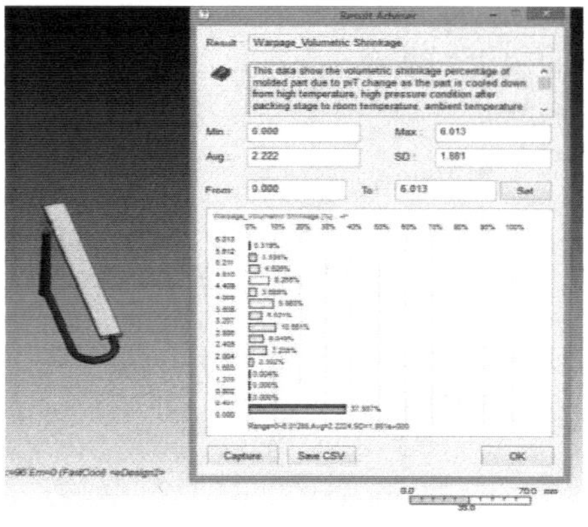

Figure 2. Measure of volume shrinkage.

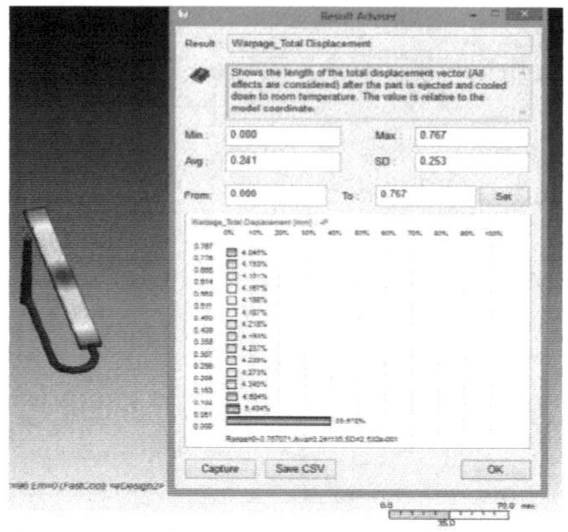

Figure 3. Measure of total displacement.

4. Data analysis and result

4.1. Experiment data of orthogonal array table

In this work, the Orthogonal Arrays (L16) are used to study the seven factors and theirs interaction. The 16 different experimental conditions are set as shown in Table 2. Four replicates were performed in each condition and the Volume shrinkage (%) and Total displacement (mm.) are measured as responses. The overall experiments are run on 64 experiments.

Table 2. Design of experiment data.

Runs	A	B	C	D	E	F	G	Volume Shrinkage (%) (four replicates)	Total Replacement (mm.) (four replicates)
1	-	-	-	-	-	-	-	12.146, 12.142, 10.652, 10.652	1.299, 1.299, 0.960, 0.960
2	-	-	-	+	-	+	+	08.528, 09.092, 07.540, 07.540	1.107, 1.193, 0.852, 0.852
3	-	-	+	-	+	-	+	07.769, 07.769, 06.192, 06.192	1.116, 1.116, 0.774, 0.774
4	-	-	+	+	+	+	-	09.093, 09.093, 07.541, 07.541	1.197, 1.197, 0.856, 0.856
5	-	+	-	-	+	-	-	10.213, 10.213, 08.683, 08.683	1.220, 1.220, 0.880, 0.880
6	-	+	-	+	+	+	+	08.436, 08.436, 06.872, 06.872	1.135, 1.135, 0793, 0.793
7	-	+	+	-	-	-	+	10.135, 10.135, 08.602, 08.602	1.215, 1.215, 0.875, 0.875
8	-	+	+	+	-	+	-	14.837, 14.649, 13.199, 13.199	1.361, 1.359, 1.021, 1.021
9	+	-	-	-	+	-	+	07.545, 07.545, 05.964, 05.964	1.120, 1.120, 0.776, 0.776
10	+	-	-	+	+	+	-	08.924, 08.924, 07.369, 07.369	1.181, 1.181, 0.840, 0.840
11	+	-	+	-	-	-	-	11.072, 11.072, 09.557, 09.557	1.279, 1.279, 0.939, 0.939
12	+	-	+	+	-	+	+	08.900, 08.900, 07.344, 07.344	1.177, 1.177, 0.835, 0.853
13	+	+	-	-	-	-	+	09.862, 09.862, 08.324, 08.342	1.201, 1.201, 0.860, 0.860
14	+	+	-	+	-	+	-	13.170, 13170, 11.695, 11.695	1.328, 1.328, 0.989, 0.989
15	+	+	+	-	+	-	-	10.105, 10.105, 08.591, 08.591	1.210, 1.210, 0.870, 0.870
16	+	+	+	+	+	+	+	08.031, 08.031, 06.459, 06.459	1.123, 1.123, 0.781, 0.781

The ANOVA test, a statistical technique, is used to identify significant factors affecting those two responses. The results are shown in Figure 4 and 5. The confidential level is set at 95% (α 0.05) and the p-valve below 0.05 is statistically significant.

Analysis of Variance for Means

Source	DF	Seq SS	Adj SS	Adj MS	F	P
B	1	6.6967	6.6967	6.6967	29.32	0.000
E	1	22.5192	22.5192	22.5192	98.60	0.000
G	1	24.9595	24.9595	24.9595	109.28	0.000
B*E	1	0.7042	0.7042	0.7042	3.08	0.113
B*G	1	0.8523	0.8523	0.8523	3.73	0.085
E*G	1	2.8344	2.8344	2.8344	12.41	0.006
Residual Err	9	2.0556	2.0556	0.2284		
Total	15	60.6219				

Figure 4. ANOVA result of volume shrinkage.

Analysis of Variance for Means

Source	DF	Seq SS	Adj SS	Adj MS	F	P
B	1	0.004047	0.004047	0.004047	32.40	0.000
E	1	0.033721	0.033721	0.033721	269.96	0.000
G	1	0.042388	0.042388	0.042388	339.34	0.000
B*E	1	0.000648	0.000648	0.000648	5.19	0.049
B*G	1	0.000314	0.000314	0.000314	2.51	0.148
E*G	1	0.002213	0.002213	0.002213	17.72	0.002
Residual Err	9	0.001124	0.001124	0.000125		
Total	15	0.084454				

Figure 5. ANOVA result of total displacement.

The significant factors to Volume shrinkage are Melt temperature (B), packing Time (E), Cooling time (G) and interaction of E and G (EG). For Total displacement, the significant factors are Melt temperature (B), packing Time (E), Cooling time (G), interaction of B and E (BE), and interaction of E and G (EG).

4.2 Process modeling and optimization

The results of significant factors are used to construct a linier modeling. The equation for predicting Volume shrinkage and Total displacement are shown in equation (1) and (2), respectively.

$$Y_1 = 9.1731 - 0.6470\,B + 1.1864\,E + 1.2490\,G + 0.4209\,E*G \quad (1)$$

$$Y_2 = 1.03795 - 0.01590\,B + 0.04591\,E + 0.05147\,G - 0.00636\,B*E + 0.01176\,E*G \quad (2)$$

Y_1 = Volume shrinkage (%)
Y_2 = Total displacement (mm.)
B = Melt temperature
E = Packing time
G = Cooling time

In order to optimize those two responses simultaneously, A response optimizer is used to facilitate the procedure. All equations are performed in computer software to select the suitable input condition to meet the target outputs as shown in Figure 6.

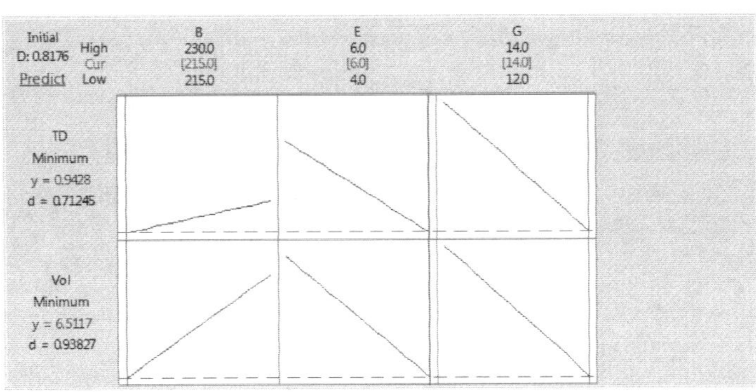

Figure 6. Response optimizer for minimizing volume shrinkage and total displacement.

It is found that the minimum of volume shrinkage and total displacement can be achieved with the optimization tool. The optimal condition are low level of Melt temperature (B) and high level of Packing time (E) and Cooling time (G) ,which provide the 6.5% of Volume shrinkage and 0.94 mm. of Total displacement.

5. Conclusion
The application of the Taguchi method in optimization of Plastic injection molding process is introduced. The limited experiments of Orthogonal array (L16) as advantage is set to be experimental plan. The 64 experimental runs are carried out and volume shrinkage and total displacement as responses are measured. Based on ANOVA test, three factors and their interactions are found statistically significant. Such factors are Melt temperature (B), Packing time (E), Cooling time (G), interaction of B and E (BE), and interaction of E and G (EG). The designed quality of an example work-piece is set to be the minimum of volume shrinkage and total displacement as target of optimizing procedure. The solution yields Melt temperature of 215 'C, Packing time of 6 sec., and Cooling time of 14 sec. in association with volume shrinkage of 6.5% and total displacement of 0.94 mm. Then, the parameter design of Taguchi method provides a simple, systematic and efficient methodology for modeling and optimizing a case study process.

6. References
[1] Taguchi G and Konishi S 1987 *Taguchi Methods, Orthogonal Arrays and Linear Graphs, Tools for Quality Engineer* American Supplier Institute p 8-35

[2] Foster W T 2000 Basic Taguchi Design of Experiments, *Proc. National Association of Industrial Technology Conference* Pittsburgh USA

[3] Athreyal S and Venkatesh Y D 2012 Application of Taguchi Method for Optimization of Process Parameters in Improving the Surface Roughness of Lathe Facing Operation *International Refereed Journal of Engineering and Science (IRJES)* Volume 1 Issue 3 p 13-19

[4] Karna S K and Sahai R 2012 An Overview on Taguchi Method *International Journal of Engineering and Mathematical Sciences* Volume 1 p11-18

[5] Prashar A 2016 A Conceptual Hybrid Framework for Industrial Process Improvement : Integrating Taguchi Methods, Shainin System and Six Sigma *Production Planning and Control* Volume 27 issue 16 p1389-1404

[6] Goyal T, Walia R S and Sidhu T S 2017 Taguchi and Utility Based Concept for Determining Optimal Process Parameters of Cold Sprayed Coatings for Multiple Responses *International Journal on Interactive Design and Manufacturing* Volume 11 Issue 4 p 761-769

[7] Xu Y, Lu H, Zhou D, Zheng J and Zhang J 2017 Image Matching Optimization Based on Taguchi Method and Adaptive Spatial Clustering with SIFT *International Journal of Pattern Recognition and Artificial Intelligence* Volume 31 Issue 11

[8] Malloy A R 1994 *Plastic Part Design for Injection Molding* Carl Hanser Verlag New York p 20

A High-efficiency FPGA-based Accelerator for Convolutional Neural Networks using Winograd Algorithm

Y Huang[1,2], J Shen[1,2], Z Wang[1], M Wen[1,2] and C Zhang[1,2]

[1] Science and Technology on Parallel and Distributed Laboratory, National University of Defence Technology, Changsha 410073, China
[2] College of Computer, National University of Defence Technology, Changsha 410073, China

Abstract. Convolutional neural networks (CNNs) are widely used in many computer vision applications. Previous FPGA implementations of CNNs are mainly based on the conventional convolutional algorithm. However, the high arithmetic complexity of conventional convolution algorithm for CNNs restricts the performance of accelerators and significantly increases the challenges of design. It has been proved that the Winograd algorithm for CNNs can effectively reduce the computational complexity. Although a few FPGA approaches based on the Winograd algorithm have been implemented, their works are lake of evaluation on the performance for different tile sizes of the Winograd algorithm. In this work, we focus on exploring the possibility of using the Winograd algorithm to accelerate CNNs on FPGA. First, we propose an accelerator architecture applying to both convolutional layers and fully connected layers. Second, we use high level synthesis tool to expediently implement our design. Finally, we evaluate our accelerator with different tile sizes in terms of resource utilization, performance and efficiency. On VUS440 platform, we achieve an average 943 GOPS for overall VGG16 under low resource utilization, which reaches higher efficiency than the state-of-the-art works on FPGAs.

1. Introduction

Convolutional neural networks (CNNs) have achieved great success in many computer vision applications [1], such as image recognition, speech recognition. Recently, the computational complexity of CNN has significantly increased with the continuous improvement of its recognition accuracy. Since CPUs can hardly provide the massive parallelism required by CNNs, many hardware accelerators, such as GPU, FPGA [2-12], and ASIC [13], have been developed to improve the performance of CNNs. Of these designs, FPGA-based accelerators have become a particularly option due to their high energy efficiency and reconfigurability.

The existing FPGA implementations of CNNs are mainly based on conventional convolution algorithm [4, 7]. However, due to its high arithmetic complexity, more intensive design efforts are required to design an efficient accelerator for 2D CNNs on FPGA. As the current trend of CNN topologies is towards deeper, the increasing computational complexity of CNNs makes it difficult to design an efficient accelerator. Therefore, it is necessary to reduce arithmetic complexity of convolution operation by using fast convolution algorithm. In this work, we propose a FPGA accelerator design based on the Winograd algorithm. It is evident that using the Winograd algorithm can effectively reduce the arithmetic complexity [14], thereby achieving better acceleration performance on the FPGA platform compared to the conventional convolution algorithm.

A few prior works [2, 11] have focused on implementing an accelerator with the fixed tile sizes of the Winograd algorithm, which are lack of evaluation on the performance of different tile sizes. However, the Winograd algorithm contains different tile sizes and the arithmetic complexity reduction of convolution operation for each tile size is different. In this work, we aim to evaluate our accelerator with different tile sizes.

In this paper, we make the following contributions:

- An accelerator design using the Winograd algorithm on FPGA is proposed.
- An analytical model is developed to explore the design space.
- As a case study, we implement the accelerator for state-of-the-art CNNs: VGG16. Our accelerator is evaluated with different tile sizes. Moreover, the implementation on VUS440 achieves a peak performance of 1412 GOPS and an average performance of 943 GOPS.

The rest of the paper is organized as follows: Section 2 introduces the background of CNNs and the Winograd algorithm. Section 3 presents the details of accelerator design. Section 4 describes the design space exploration. Section 5 shows our experimental results. Section 6 discusses related works and Section 7 makes conclusion.

2. Background

2.1. Convolutional Neural Networks

CNNs extended from artificial neural network has been used to solve many complex problems [2]. CNNs are composed of series of layers, including convolution layer, ReLU layer, pooling layer and fully connected layer. Of these layers, convolution layer makes major contributions to CNNs, where the convolution operations occupy most of computation time [15]. Therefore, in this work our design focus on accelerating convolutional layers. In addition, our design can be reused in fully connected layers but in a suboptimal performance. A typically convolution layer contains input feature maps, filters and output feature maps. Each layer outputs M feature maps by the convolution of a set of N input feature maps of size $H \times W$ and M sets of $N \times K \times K$ filters. For each of M sets, N filters of size $K \times K$ slide across the corresponding N input feature maps with a stride of S. At each location, N filters multiply-accumulate with the overlapping elements of input feature maps to generate one output value. Each of output feature maps has the size of $R \times C$, where $R = \frac{H-K}{s} + 1$, $C = \frac{W-K}{s} + 1$. The computation pattern of convolution layers can be expressed as follows:

$$Out[m][r][c] = \sum_{n=0}^{N} \sum_{i=0}^{K} \sum_{j=0}^{K} In[n][S \times r + i][S \times r + j] \times W[m][n][i][j] \quad (1)$$

Where In and Out denotes the input and output feature maps of 3-dimensional arrays and W indicates the filters.

2.2. Winograd Convolution Algorithm

Winograd Convolution is a new class of fast algorithm for CNN which derived from Winograd's minimal filtering algorithms [16]. The Winograd algorithm reduces the number of multiplications through a series of transformations on input tile and filters. To simply discussion, we describe computing m outputs with r-tap FIR filter as $F(m,r)$. For $F(2,3)$, We define the input vector $\boldsymbol{x} = (x_0, x_1, x_2, x_3)$, filter $\boldsymbol{w} = (w_0, w_1, w_2)$ and output vector $\boldsymbol{y} = (y_0, y_1)$.

Firstly, by the Winograd algorithm, the input vector and filter are transformed into the following vectors:

$$\boldsymbol{x}' = \left(x_0 - x_2, x_1 + x_2, x_2 - x_1, x_1 - x_3\right), \boldsymbol{w}' = \left(w_0, \frac{w_0 + w_1 + w_2}{2}, \frac{w_0 - w_1 + w_2}{2}, w_2\right)$$

Secondly, by calculating the dot product of $\boldsymbol{x}'$ and $\boldsymbol{w}'$, we can get the following results:

$$f_1 = \left(x_0 - x_2\right)w_0\ ,\ f_2 = \left(x_1 + x_2\right)\frac{w_0 + w_1 + w_2}{2}$$

$$f_4 = \left(x_1 - x_3\right)w_2,\ f_3 = \left(x_2 - x_1\right)\frac{w_0 - w_1 + w_2}{2}$$

Finally, the Winograd algorithm for $F(2,3)$ give the following equation:

$$\boldsymbol{y} = \begin{bmatrix} y_0 \\ y_1 \end{bmatrix} = \begin{bmatrix} x_0 & x_1 & x_2 \\ x_1 & x_2 & x_3 \end{bmatrix} \begin{bmatrix} w_0 \\ w_1 \\ w_2 \end{bmatrix} = \begin{bmatrix} f_1 + f_2 + f_3 \\ f_2 - f_3 - f_4 \end{bmatrix} \quad (2)$$

It can be seen from Equation (2) that the Winograd algorithm only requires 4 multiplications to compute the output y, while the standard convolution algorithm uses $2 \times 3 = 6$ multiplications. Winograd algorithm can also be written in matrix form:

$$y = A^T[(Gw) \text{e } (B^T x)] \quad (3)$$

Where e denotes element-wise multiplication. A, G and B are constant matrices, which are determined by the sizes of x and w. For $F(2,3)$, the matrices are

$$B^T = \begin{bmatrix} 1 & 0 & -1 & 0 \\ 0 & 1 & 1 & 0 \\ 0 & -1 & 1 & 0 \\ 0 & 1 & 0 & -1 \end{bmatrix}, G = \begin{bmatrix} 1 & 0 & 0 \\ \frac{1}{2} & \frac{1}{2} & \frac{1}{2} \\ \frac{1}{2} & -\frac{1}{2} & \frac{1}{2} \\ 0 & 0 & 1 \end{bmatrix}, A^T = \begin{bmatrix} 1 & 1 & 1 & 0 \\ 0 & 1 & -1 & -1 \end{bmatrix} \quad (4)$$

Similarly, 2D Winograd algorithm for $F(m \times m, r \times r)$ can be expressed as follows:

$$y = A^T[(GwG^T) \text{e } (B^T xB)]A \quad (5)$$

Where now w is an $r \times r$ filter and x is an $n \times n$ $(n = m + r - 1)$ input tile. In general, the Winograd algorithm for $F(m \times m, r \times r)$ requires $(m + r - 1) \times (m + r - 1)$ multiplications [17], and the standard algorithm uses $m \times m \times r \times r$ multiplications. The 2D Winograd algorithm for $F(m \times m, r \times r)$ can be used to compute convolution layers in CNNs, which can effectively reduce the arithmetic complexity. Table 1 shows the arithmetic complexity reduction for different tile sizes. It can be seen from table 1 that different tile sizes lead to different arithmetic complexity reduction.

Table 1. The arithmetic complexity reduction for different tile sizes.

$F(m \times m, r \times r)$	Multiplications		
	$m = 2, r = 3$	$m = 4, r = 3$	$m = 6, r = 3$
Standard algorithm	36	144	324
Winograd algorithm	16	36	64
Arithmetic complexity reduction	2.25x	4x	5.06x

3. Accelerator design

In this section, we bring forward a FPGA accelerator design based on 2D Winograd algorithm.

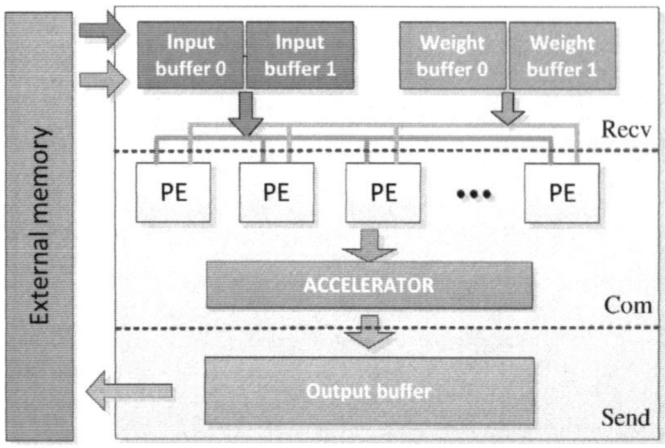

Figure 1. Overview of accelerator design

3.1. Accelerator Overview

Figure 1 shows an overview of our accelerator architecture. It can be divided into three principal modules, which are the receiving module, the computation module and the sending module. The receiving module is responsible for fetching the input feature maps from the off-chip memory. The computation module contains several processing elements and the accumulator. The Winograd PEs fetch the input data from the on-chip buffers and send the processed data to the accumulator buffer. Finally, the cached output results are written back to off-chip memory through the sending module.

3.2. PE

Our PE is designed based on Winograd convolution algorithm. As shown in figure 2, the Winograd PE is separated into three parts. PE1 is the transformation of input tile; PE2 is the transformation of filter; PE3 contains both dot-product units and transformation unit. More clearly, for computing $F(m\times m, r\times r)$, each time we need fetch $n\times n$ input tile from input feature maps, where current input tile is obtained by sliding the last input tile by m stride. In section 2, we mentioned the regular expression of Winograd convolution algorithm. As shown in equation (5), *A*, *G* and *B* are constant matrix already given. In fact, matrix multiplication in equation (5) can be designed as processing the rows and columns of matrix with multiplication and addition operation. Owing to different transformation for different tile size, we design several PEs for each of tile size.

As we can see from the figure 2, PE1 generate x_1 by calculating input tile column by column, and then processing x_1 row by row to acquire U. The transformation in PE2 and PE3 are the similar method. Two matrices U and V with the same scale ($n\times n$) are fed to PE3. PE3 multiply U and V by dot-product operation and transform M to Out. The overall structure of PEs is better for dataflow.

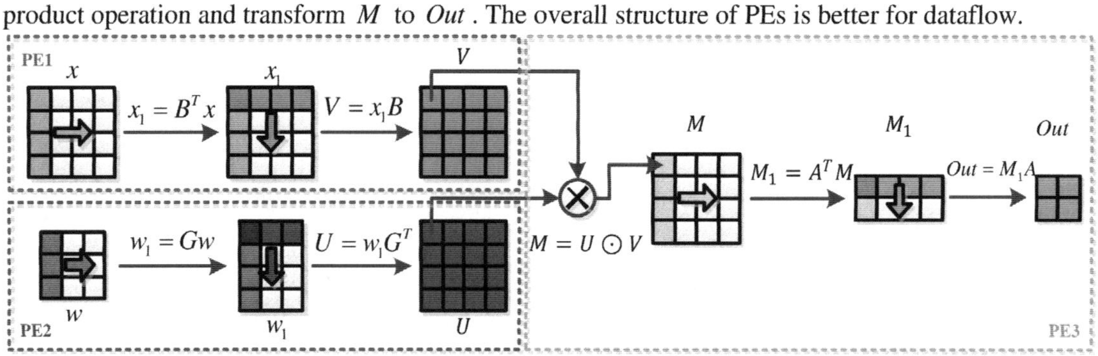

Figure 2. Winograd PE

3.3. Parallel Design of PE

The parallel design of PE is the key point to improve throughput of computation module. As we can see from the figure 2, PE1 and PE2 are two separate processes which can be executed in parallel. The code of computation module designed by HLS is shown in figure 3. Both Loop3 and Loop4 are unrolled, where parallelism degree of PE1 is T_N and parallelism degree of PE2 and PE3 is $T_N \times T_M$. In fact, PE1 and PE2 executed in parallel consume lots of DSP and LUT, which causes Loop3 and Loop4 cannot be completely unrolled. Considering the short latency of the transformation of filters, we put PE2 out of the whole loop and let it to be executed first. The transformed filters are stored in the caches before executing PE3.

Figure 4 shows the completely parallel structure of computation module. T_N input tiles are transformed by PE1 in parallel. Then T_N transformed input tiles and T_M sets of T_N transformed filters fetched from on-chip buffers are fed to PE3 in parallel. PE3 generates T_M sets of T_N output tiles in a cycle. For each group, we accumulate T_N output tiles and store the result in output buffers. It is notable that the current results are generated from T_N input feature maps while we have N input feature maps. Therefore, the computation module should be executed $\frac{N}{T_N}$ times. The complete output results are given by accumulating these $\frac{N}{T_N}$ sets of results.

```
Loop1:for(trr = 0; trr < TR; trr += m){
Loop2:   for(tcc = 0; tcc < TC; tcc += m){
#pragma HLS PIPELINE
Loop3:       for(tn = 0; tn < TN; nn++){
#pragma HLS UNROLL
                 Load input_tile(trr, tcc, tn)
                 Winograd_PE1(input_tile, input_tile1);
Loop4:           for(tm = 0; tm < TM; tm++){
#pragma HLS UNROLL
                     Load filter(tm, tn)
                     Winograd_PE2(filter, filter1);
                     Winograd_PE3(input_tile1, filter1,tile_temp);
                     Output_temp(trr, tcc, tm)+= tile_temp; }}}}
```

Figure 3. Pseudo code of computation module

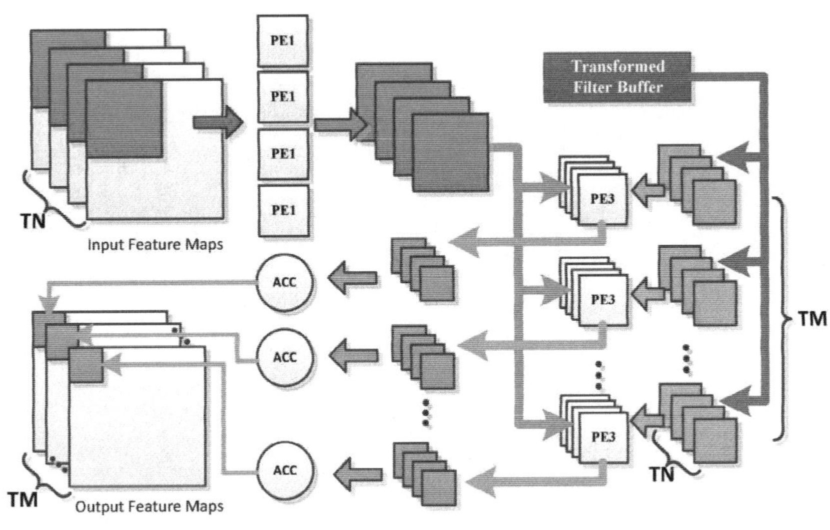

Figure 4. Computation engine

3.4. Memory Access

Communication is the main constraint in the optimization of accelerator design [2]. All the input feature maps and filters are stored in external memory. Owing to the large amount of input data of CNNs and limited on-chip resource, it is difficult for the accelerator to fetch all the data from the external memory at the same time. In this work, the input data are divided into several groups and the accelerator processing the data group by group. More clearly, for a set of N input feature maps of size $H \times W$ convoluted with M sets of $N \times K \times K$ filters, each time accessing external memory the accelerator fetch a set of T_N input feature maps of size $T_W \times T_H$ and T_M sets of $T_N \times K \times K$ filters. However, due to the limited memory bandwidth, there is a memory access delay when writing the input feature maps and filters to the on-chip buffers. As shown in figure 1, all the data processed the by computation module comes from on-chip buffers. Therefore, the computation module could not be executed until the receiving module finish writing the input data to on-chip buffers. Obviously, the memory access delay has an impact on the optimization of the whole system. To overlap data communication and computation, double buffers should be applied in the receiving module. Double buffers used by ping-pong operation contains four sets: two buffers for input feature maps and two buffers for filters (here collectively referred to as *input buffer 0* and *input buffer 1*). Computation module fetch the data from *input buffer 0* while receiving module writes the off-chip data to *input buffer 1*, and the next time computation module fetch the data from *input buffer 1* while receiving module writes the data to *input buffer 0*.

3.5. Fully Connected Layers

The designed PEs can also be used to execute fully connected layers. In the fully connected layers, each node of the layer is connected to all nodes of the upper layer, as shown in figure 5(a). In this work, we use the dot-product units to achieve the computation of fully connected layers. Figure 5(b) describes the implementation of fully connected layers. The batch size is determined by the size of dot-product array. Moreover, for $F(m \times m, r \times r)$, the batch size is $n \times n$. T_N input tiles and T_N weights are fed to T_N dot-product arrays in a cycle. The input tile consists of $n \times n$ elements fetched from $n \times n$ different input features. For each dot-product array, $n \times n$ elements of input tile are multiplied by the same weight. Dot-product arrays generates T_N sets of $n \times n$ outputs which are sent to accumulator. These T_N sets of $n \times n$ outputs will be added point-to-point to acquire the final $n \times n$ outputs.

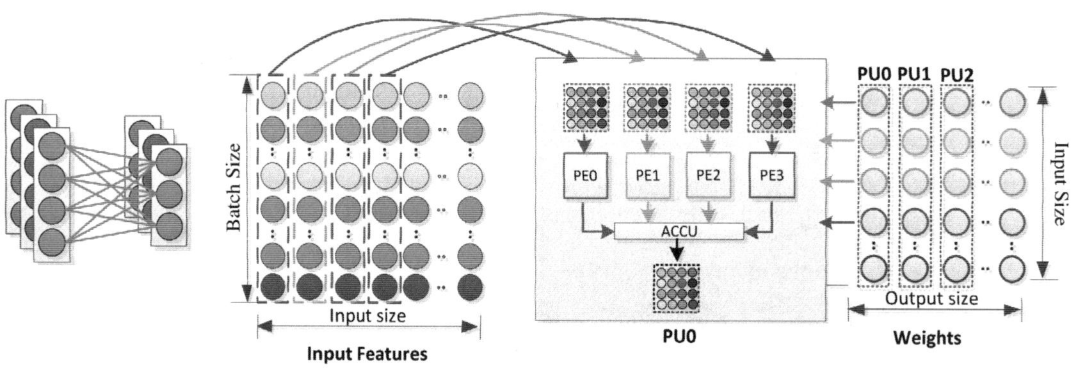

(a) FC layer (b) The process of FC layer mapping

Figure 5. FC layer implementation

4. Design Space Exploration

In this section, we build an analytical model to theoretically estimate the performance of our accelerator. For $F(m \times m, r \times r)$, the design of accelerator adopts several parameters including input output parameters $\langle N, M, W, H, R, C, K, S \rangle$ and tile parameters $\langle T_N, T_M, T_R, T_C \rangle$. Among these, N is the number of input feature maps, M is the number of output feature maps, W and H are the width and height of input feature maps, R and C are the width and height of output feature maps, $K \times K$ (short for $Ksize$) is the size of filters and S is the stride of convolution. T_N, T_M are tiling factors for N and M, and T_R, T_C are tiling factors for R and C.

In the following, we model the execution time of a convolution layer. For convenient to analysis, we first model the individual execution time of receiving module, computation module and sending module. Execution time of these three modules are calculated by the following equations.

$$T_{recv} = \frac{Q_{recv} \times DW}{BW} \tag{6}$$

$$Q_{recv} = Q_{in} + Q_w, Q_w = T_M \times T_N \times Ksize \tag{7}$$

$$Q_{in} = T_N \times (S \times T_R + K - S) \times (S \times T_C + K - S) \tag{8}$$

$$T_{send} = \frac{Q_{send} \times DW}{BW} \tag{9}$$

$$Q_{send} = T_M \times T_R \times T_C \tag{10}$$

$$T_{com} = \left(\frac{T_R \times T_C \times II}{m \times m} + I_{depth} \right) \times \frac{1}{Freq} \approx \frac{T_R \times T_C \times II}{m \times m \times Freq} \tag{11}$$

Where DW is the data width of input feature maps, weights and output feature maps which are 16-bit fixed point precision. BW is the bandwidth of the off-chip memory. II denotes the iteration interval of the pipeline of computation. I_{depth} is the depth of pipeline, which can be ignored due to its little impact on total execution cycle. Freq is the clock frequency of FPGA. Considering the ping-pang operation (section 3) between two modules, we can model the total execution time as:

$$\begin{aligned} T_{total} &= \frac{R}{T_R} \times \frac{C}{T_C} \times \left(\left\lceil \frac{M}{T_M} \right\rceil \times \left\lceil \frac{N}{T_N} \right\rceil \times \max \langle T_{recv}, T_{com} \rangle + T_{send} \right) \\ &\approx \frac{R}{T_R} \times \frac{C}{T_C} \times \left\lceil \frac{M}{T_M} \right\rceil \times \left\lceil \frac{N}{T_N} \right\rceil \times \max \langle T_{recv}, T_{com} \rangle \end{aligned} \tag{12}$$

T_{send} can be ignored, because it has little impact on the total time. Due to limited bandwidth, both communication and computation are probably the constraint of optimization. To make the accelerator compute-bound [3], there should be a $T_{recv} < T_{com}$. Thus, we can get a bandwidth expression with lower bound:

$$BW_{req} \geq \frac{Q_{recv} \times DW \times m \times m \times Freq}{T_R \times T_C \times II} \tag{13}$$

If the actual bandwidth is out of this range, the accelerator would be memory-bound.

The total operations of a convolution layer are:

$$OP_{total} = 2 \times Ksize \times N \times R \times C \times M \tag{14}$$

Assuming the memory bandwidth meets the minimum requirement of compute-bound, the calculation of the accelerator performance can be defined as:

$$Per = \frac{OP_{total}}{T_{total}} = \frac{2 \times Ksize \times N \times M \times m \times m \times Freq}{\left\lceil \frac{N}{T_N} \right\rceil \times \left\lceil \frac{M}{T_M} \right\rceil \times II} \tag{15}$$

From the above expression of performance, we can modify the parameters $\langle T_N, T_M \rangle$ to maximize the performance. However, it is necessary to consider the on-chip resource when improving performance. Next, we will focus on modelling the utilization of DSP. Since we design the PEs based on Winograd algorithm, the dot-product operation of PE3 mostly consume DSP. Moreover, when the input tile $n > 4$, the transformation in PEs consumes DSP due to its high complexity. It is notable that a multiplication of 16-bit fixed point only consumes one DSP. Assuming the number of DSP required by the transformation is DSP_{trans}, we can calculate the consumption of DSP by the following equation.

$$DSP = T_N \times T_M \times \alpha(II) \times (n \times n + DSP_{trans}) \tag{16}$$

As shown in the above equation, parameter $\langle T_N, T_M \rangle$ directly affects the consumption of DSP. Since the actual iteration interval of the pipeline of computation is more than one cycle, the parallelism degree of PEs cannot reach $T_N \times T_M$. Therefore, we add a parameter α to the equation, and α is related to the iteration interval ($\alpha = \frac{1}{II}$).

5. Experimental Results

5.1. Experimental Setup

In this work, we implement our accelerator design by using Vivado HLS 2016.4, which can turn C program into hardware description language required by FPGA. Two FPGA platforms used to evaluate our design are Xilinx VC709 and VUS440. Xilinx VC709 integrates a Virtex-7 690t FPGA and two 4GB DDR3 DRAMS. The working frequency of this work is 150MHz. VUS440 consists of a Xilinx VCU440 FPGA and an 8GB DDR4, which evaluate our accelerator with 200MHz working frequency. The CNN model used for case studies is VGG16. All the convolution layers of VGG16 use 3×3 filters, which is suitable for Winograd algorithm.

5.2. Result Analysis and Comparison

In our experiment, we evaluate the Winograd algorithm for both $F(2 \times 2, 3 \times 3)$ and $F(4 \times 4, 3 \times 3)$. Our designed accelerator contains several tile parameters $\langle T_N, T_M, T_R, T_C \rangle$ for CNNs. The uniform tile parameters and off-chip bandwidth for all layers of VGG16 are shown in table 2. The other tile parameters are assigned according to the size of each layer. It is worth noting that T_N in *Conv1* should be changed to 3 due to its only 3 channels of input features.

Table 2. Uniform cross-layer parameters.

Parameter	T_N	T_M	BW_{off}
Design Value	4	64	256bit

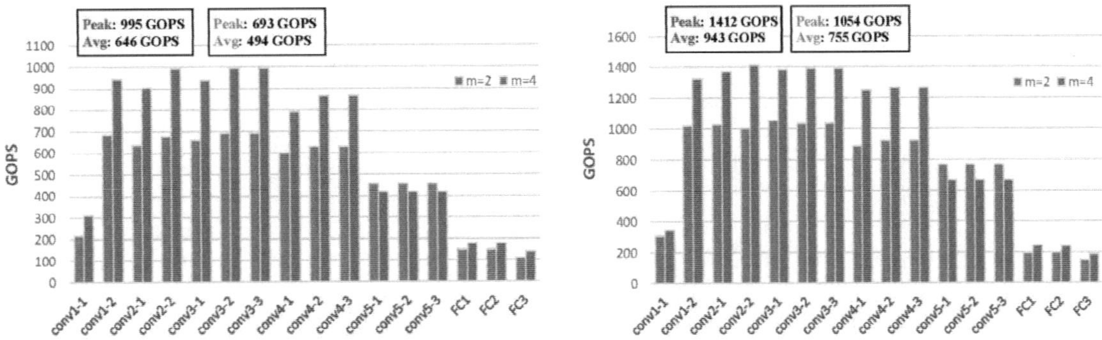

(a) VC709 platform (b) VUS440 platform

Figure 6. Results on multiple FPGA platform for different tile sizes

Figure 6 shows the results of all convolution layers on different FPGA platform for two sizes of accelerator. First, on both two platforms, Winograd algorithm for $F(4\times4,3\times3)$ achieve higher peak performance (995 GOPs for VC709 and 1412 GOPs for VUS440) and higher overall performance (646 GOPs for VC709 and 943GOPs for VUS440) than $F(2\times2,3\times3)$. Moreover, the performance of $F(4\times4,3\times3)$ is higher than $F(2\times2,3\times3)$ in each layer of VGG16 except Conv5. As shown in table1, the Winograd algorithm for $F(4\times4,3\times3)$ achieves a nearly 2x arithmetic complexity reduction over $F(2\times2,3\times3)$. Therefore, it is reasonable that $F(4\times4,3\times3)$ achieves better performance. Second, under the same size, the results on VUS440 are better than VC709's. However, it can be seen from figure 6 that layer1-1 performs the worst result in convolutional layers for both two sizes. It is because layer1-1 has only 3 input channels, which even cannot perform ping-pang operation. Third, as the convolutional layer goes deeper, there is a downward trend on performance for both $F(4\times4,3\times3)$ and $F(2\times2,3\times3)$. The main reason is that the size of input feature maps are getting smaller from conv1 to conv5, which causes inefficient memory access.

Table 3 gives the resource utilization of designed accelerator with two sizes on different FPGA platform. Thanks to the Winograd algorithm, DSP are no longer the mainly consumption of on-chip resource. Instead, the utilization of LUT increases a lot. On the VC709 platform, $F(2\times2,3\times3)$ consumes about 39% of DSP while $F(4\times4,3\times3)$ uses 21% of DSP. For further analysis, in the case of the same T_N and T_M, the consumption of DSP is not only related to input tile size but also associated to the iteration interval of the pipeline of computation according to equation (16) In our implementation, the actual iteration interval for $F(2\times2,3\times3)$ and $F(4\times4,3\times3)$ are 3 and 11. Thus the theoretical results given by equation (16) are mostly same with the actual results. As we can see from the table 3, VUS440 contains less DSP but much more LUT (5.8X) than VC709. In general, VUS440 platform present the less overall resource utilization, which can be allowed to work at higher clock frequency.

Comparison with prior FPGA works are shown in table 4. We mainly compare DSP, overall performance and DSP efficiency. As we can see, DSP efficiency of our work is 0.86 on VC709 and 1.2 on VUS440, which significantly outperform other prior works. In this work, 16-bit fixed point data are used to evaluate the designed accelerator. The advantage of 16-bit fixed point is that addition operation does not require DSP and multiplication operation consumes only one DSP, while the Winograd algorithm reduce the number of multiplications and requires more adders to deal with the additional transformation. That's why our implementation can reach such a high DSP efficiency. For the overall performance, our implementation achieves better results than other works except [2]. However, the work in [2] uses Intel FPGA SDK for OpenCL while we use Vivado HLS which can hardly reach a high frequency with over 90% utilization of DSP. Therefore, there is no comparability between our works and [2].

Table 3. FPGA resource utilization.

Device		DSP	BRAM	LUT	FF
VC709	Available	3600	2940	433K	866K
	Utilization(m=2)	1402(39%)	1352(46%)	199K(46%)	225K(26%)
	Utilization(m=4)	756(21%)	1376(47%)	326K(75%)	438K(51%)
VUS440	Available	2880	5040	2518K	5037K
	Utilization(m=2)	1402(49%)	1352(27%)	199K(8%)	225K(4%)
	Utilization(m=4)	756(26%)	1376(27%)	326K(13%)	438K(9%)

Table 4. Comparison with prior FPGA implementations.

	[2]	[7]	[8]	[4]		Ours	
FPGA	Arria10 GX1150	Zynq XC7Z045	Stratix-V GSD8	Ultrascale KU060	Virtex 690T	Virtex690T	VCU440
CNN	AlexNet	VGG	VGG	VGG	VGG	VGG	VGG
Freq(MHz)	303	150	120	200	150	150	200
Precision	16-bit float	16-bit fixed	16-bit fixed	16-bit fixed	16-bit fixed	16-bit fixed	16-bit fixed
DSP Utilization	1476	780	1963	1058	2833	756	756
Throughput (GOPs)	1382	137	117	266	354	646	943
Efficiency (GOPs/DSPs)	0.93	0.18	0.06	0.25	0.12	0.86	1.2

5.3. Comparison with CPU and GPU

In this section, we compare our FPGA design with CPU and GPU solution. Table 5 shows the comparison between our accelerator for $F(4\times 4, 3\times 3)$ and CPU and GPU. CPU and GPU solution are optimized by using OpenBLAS and CuDNN libraries. On VC709 platform, our accelerator achieves an average performance of 11x over CPU solution and 5.7x energy efficiency over GPU solution. VUS440 platform presents better results, where our implementation achieves a 16x speedup over CPU and 8x energy efficiency over GPU.

Table 5. Comparison with CPU/GPU platform.

Platform	CPU	GPU	FPGA	
Device	E5-2680	K40	VC709	VUS440
Technology	22nm	28nm	28nm	20nm
Freq	2.8GHz	1GHz	150MHz	200MHz
Power(W)	115	250	25	25
VGG average(Gops)	58.7	1174.0	646	943
Speed up	1x	20.0x	11.0x	16.0x
Energy efficiency (Gops/W)	1x	9.2x	51.0x	74.5x

6. Related Works

Recently, there have been several FPGA-based accelerators for CNN. Among these, many accelerators are implemented with conventional convolution algorithm. Communication and computation are the main constraint of accelerator design. The work in [3] build a roofline model to maximize the compute resource. Many works focus on implementing CNNs layer by layer, which lead to bottleneck of memory access. The work in [9] propose a design of fused-layer CNN accelerator to minimize the off-chip data transfer. In [7], the author proposes a dynamic-precision data quantization to improve the bandwidth and resource utilization. Some works also use fast algorithm to reduce the computational complexity. The work in [10] designs an accelerator based on FFT. The work in [2] and [11] use fixed tile size of Winograd algorithm to design the accelerators. In our work, we use Vivado HLS to implement our accelerator design based on Winograd algorithm and evaluate the performance of different sizes.

7. Conclusion

In this work, we present a FPGA-based accelerator design based on Winograd algorithm. Moreover, our designed architecture applies to both convolutional layers and fully connected layers. We also build an analytical model to estimate the performance and resource utilization. In our experiment, we use VGG16 to evaluate our accelerator with different tile size. Synthesizing the overall performance and resource utilization, the design for $F(4\times 4, 3\times 3)$ achieves better performance.

8. Acknowledgment

The authors gratefully acknowledge supports from National Key Research and Development program under No.2016YFB1000401 and No.2016YFB1000403; National Nature Science Foundation of China under NSFC No.61502509 and 61402504.

9. References

[1] Yang K, Dou Y, Lv S, Zhang F and Lv Q 2016 Relative distance features for gait recognition with Kinect *J. Vis. Commun. Image. R* **39** p 209-217

[2] Aydonat U, O'Connell S, Capalija D, Ling AC and Chiu GR 2017 An OpenCL(TM) Deep Learning Accelerator on Arria 10 *Field-Programmable Gate Arrays*

[3] Zhang C, Li P, Sun G, Guan Y, Xiao B and Cong J 2015 Optimizing fpga-based accelerator design for deep convolutional neural networks *Proc. of the 2015 ACM/SIGDA Int. Symp. on Field-Programmable Gate Arrays* (New York: ACM) p 161-170

[4] Zhang C, Fang Z, Zhou P, Pan P and Cong J 2016 Caffeine: Towards uniformed representation and acceleration for deep convolutional neural networks *Int. Conf. on Computer-Aided Design* (Austin) p 1-8

[5] Farabet C, Poulet C, Han J Y and LeCun Y 2009 Cnp: An fpga-based processor for convolutional networks *Int. Conf. on Field Programmable Logic and Applications* (Prague: IEEE) p 32-37

[6] Ma Y, Cao Y, Vrudhula S and Seo J S 2017 Optimizing Loop Operation and Dataflow in FPGA Acceleration of Deep Convolutional Neural Networks *Proc. of the 2017 ACM/SIGDA Int. Symp. on Field-Programmable Gate Arrays* (New York: ACM) p 45-54

[7] Qiu J, et al 2016 Going deeper with embedded fpga platform for convolutional neural network *Proc. of the 2016 ACM/SIGDA Int. Symp. on Field-Programmable Gate Arrays* (New York: ACM) p 26-35

[8] Suda N, Chandra V, Dasika G, Mohanty A, Ma Y, Vrudhula S, Seo J and Cao Y 2016 Throughput-optimized OpenCL-based FPGA accelerator for large-scale convolutional neural networks *Proc. of the 2016 ACM/SIGDA Int. Symp. on Field-Programmable Gate Arrays* (New York: ACM) p 16-25

[9] Alwani M, Chen H, Ferdman M and Milder P 2016 Fused-layer CNN accelerators *49th Annual IEEE/ACM Int. Symp. on Microarchitecture* (Taipei) p 1-12

[10] Zhang C and Prasanna V K 2017 Frequency Domain Acceleration of Convolutional Neural Networks on CPU-FPGA Shared Memory System *FPGA* p 35-44

[11] DiCecco R, Lacey G, Vasiljevic J, Chow P, Taylor G and Areibi S 2016 Caffeinated FPGAs: FPGA Framework For Convolutional Neural Networks *Int. Conf. on Field-Programmable* Technology (Xi'an) p 265-268

[12] Liu Z, Dou Y, Jiang J, Xu J, Li S, Zhou Y and Xu Y 2017 Throughput-Optimized FPGA Accelerator for Deep Convolutional Neural Networks *ACM. T. Reconfig. Techn* **10** p 17

[13] Chen T, Du Z, Sun N, Wang J, Wu C, Chen Y and Temam O 2014 Diannao: A small-footprint high-throughput accelerator for ubiquitous machine-learning *Proc. of the 19th Int. conf. on Architectural support for programming languages and operating systems* (New York: ACM) p 269-284

[14] Lavin A and Gray S 2016 Fast algorithms for convolutional neural networks *Proc. of the IEEE Conf. on Computer Vision and Pattern Recognition* p 4013-4021

[15] Cong J and Xiao B 2014 Minimizing computation in convolutional neural networks *Int. Conf. on artificial neural networks* (Switzerland) p 281-290

[16] Winograd S 1980 *Arithmetic complexity of computations* vol 33, ed J W Arrowsmith L and Bristol E (Philadelphia: Siam)

[17] Winograd S 1980 On multiplication of polynomials modulo a polynomial *SIAM J. Comput*, (Philadelphia: Siam) p 225-229

Contributions from the Drone Delivery System in Thailand to Environmental Pollution

J Koiwanit

Faculty of Engineering, King Mongkut's Institute of Technology Ladkrabang, Bangkok 10520, Thailand

Abstract. In recent years, the great development of science and technology together with social movements and commerce has shaped the design and use of commercial drones. This particular emerging technology has boosted the speed and ability of online shopping systems to provide customers ever faster delivery. In addition, Global Warming Potential (GMP) is the most serious impact to people and environment. Carbon dioxide (CO_2) from transportation is one of the main contributors contributing to GHG emissions which result in higher number of GWP. Drones should have a reduced GWP impact and other environmental impacts in which these help reduce the number of delivery trucks or other transportation options traveling on roads. In addition, as road infrastructure challenges especially in rural areas, drone delivery systems are able to transport freight across remote supply chain. The use of drones appears to be an important approach because they are not only inexpensive and easily available but can also travel on individual direct routes. In this study by establishing a life cycle assessment (LCA) framework, all emissions from a drone delivery system in rural areas in Chiang Mai, Thailand will be evaluated. The results show that the drone's part production is the main contributor of all impact categories.

1. Introduction

With the pace of urbanization and population growth, e-commerce volumes grow is one of the megatrends of this 21st century. Thailand posted an impressive annual growth rate of 17.95 for its e-commerce market in 2016. This amounted to \$3010 million in revenues and this is estimated to be reaching \$5826.9 million by 2020 [1]. This market is extended across the country through not only in urban but also in rural areas. However, negative implications of this market especially in rural areas include different road and terrains, long destination, and pollution. Drones tend to have energy consumption and carbon dioxide emissions advantages over trucks [2,3]. As a result, drones are introduced to deliver parcels directly to customers at their doorsteps. Given the potential benefits, many delivery providers have been testing the use of drones for commercial package delivery [4]. Drones have also been used in different applications such as military purposes, parcel delivery, emergency deliveries of medication, etc. [4,5]. Regarding the LCA papers published on commercial drone production, most of them focused on battery and their use phase [6–8], while none of them have discussed different types of about drones' full life stage. The analysis of the state of the art underline that the battery production and use phase were frequently evaluated, while a lack attention was paid to all parts production together with their use stage. Therefore, the aim of this study is to follow a "from cradle to grave" approach from an LCA perspective.

2. Methodology: LCA study of drone delivery services

2.1 Goal and scope definition

This study is to evaluate environmental impacts over a full life of a drone delivery system in rural areas in Chiang Mai, Thailand. The case study was chosen since the city is the largest city in northern Thailand providing a significantly increase in its economic growth but still has a rural lifestyle. The study takes a cradle-to-grave approach in which the environmental effects associated with coal mining, electrical generating station operations, drone raw materials, drone productions, drone use phase, and disposal will be taken into accounted.

2.1.1. Functional unit. The most common way to compare the life cycle impact of one vehicle to another is to measure emissions per kilometer driven over their lifetimes. Therefore, the functional unit selected in this comparative study is 1 km travelled. Even though running fuel is computed differently during on and off-peak times, this study assumed that the average electricity consumption rate is constant at any period of time.

2.1.2. System boundaries. The system boundary considers all unit processes included in the system under the study. This study covers the full life cycle of a drone delivery system or a cradle-to-grave study by considering the processes of resource extraction to produce drone, fuels, electricity, and resource depletion through the drone use phase. The system boundaries of the drone is shown in Figure 1.

Figure 1. System boundaries for drone delivery system.

2.1.2.1. Temporal and geographical boundaries. Since there is no available design lifespan for drone delivery systems specific to the Thai context, the design lifespan for the purpose of this study is assumed that drones would be constructed and operated for 500 hours using the experts' estimates. The 500 hours is mainly from drones' operation. However, it is possible that the vehicles will be able to operate more than their design lives.

2.1.2.2. Technological boundaries. This study evaluates the modern transportation delivery technology and its emission impacts. There is one systems analyzed, which is a drone delivery for an online shopping system. This study evaluates the drone delivery services to compare it with other transportation options for an online shopping system in the future.

2.1.3. Life Cycle Impact Assessment (LCIA) method. The CML 2001 method has been accepted and widely used by the Thailand Environment Institute (TEI) and Thai government [9,10]. CML 2001 is used in this study together with GaBi 7, an LCA software product of PE International, Germany.

3. Model description

Since drones are designed for use in different environments and they vary in size depending on the number of packages and their weight. These result in different energy use and distance to be traveled. The model description of the drone delivery system is presented as follows.

- Drones that can autonomously operate during flights that can last up to 30 minutes [11]
- Drone with its capacity of carrying products up to 5 kgs
- The order must be smaller than 12 inches long, 12 inches width, and 10 inches high to fit in the cargo box.
- The delivery location must be within a 10 mile radius of the order fulfillment center [11]

IOP Conf. Series: Journal of Physics: Conf. Series **1026** (2018) 012020 doi:10.1088/1742-6596/1026/1/012020

4. Results

4.1 Global Warming Potential (GWP)

This section presents only the Global Warming Potential (GWP) result of the full life cycle of the drone for home delivery system. The environmental impact in GWP impact category is shown in Figure 2. Global warming is the increase of average temperature in the atmosphere, which causes adverse effects on human healths and the environment. The primary GHG that contribute to global warming include not only CO_2, but also methane (CH_4), nitrous oxide (N_2O), hydrofluorocarbons (HFCs), perfluorocarbons (PFCs), and sulphur hexafluoride (SF_6). In this study, the global warming impact is 0.079 kg CO_2-Equiv. in the drone delivery system. The impact was primarily due parts production, which accounted for 99.2% of the impact. The main substances that contributed to global warming include CO_2, CH_4, and N_2O, which accounted for 20.47%, 17.67, 15.72%, and 15.37%, respectively.

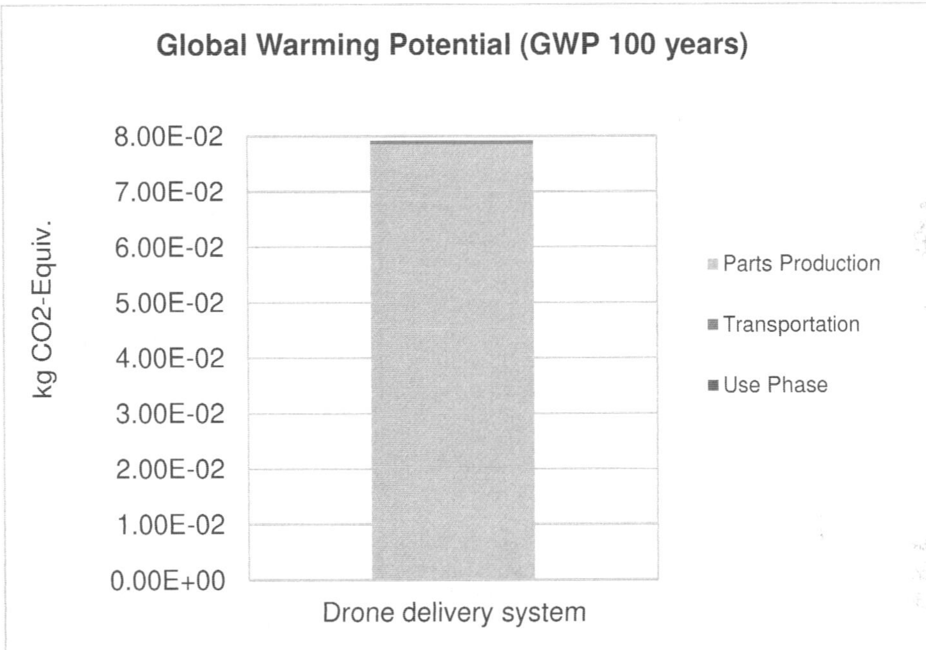

Figure 2. Global warming potential from part productions, transportation, and drone use phase.

5. Discussion

The impact in all impact categories was mainly due to parts production ranging from 93 to almost 100% of the impact. The results from the GHG emissions of the drone delivery system shows that the dominant contributor to GWP impact category is the parts operation. Parts operation consists of (1) coal mining in which parts operation requires coal as an input for electrical generating station operation, (2) electrical generating station operation, and (3) parts production. The drone delivery system is consequently environmentally friendly especially when they are under operation. This will achieve substantial energy savings during the drone operation.

6. Sensitivity analysis

As carbon fibers production for the cargo box and Li-ion production for the battery showed highest contribution to all impact categories, carbon fibers production was selected as a case study to illustrate the significance of operational factors in this study.

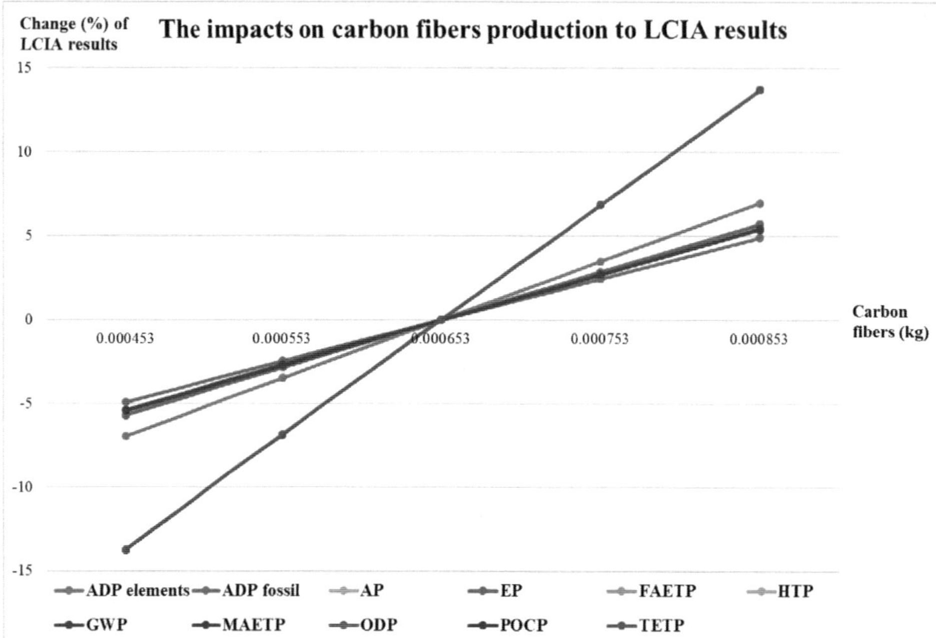

Figure 3. The impact on carbon fibers production to LCIA results.

In Figure 3, the X axis demonstrates the carbon fibers consumed, while the Y axis shows the percentage changes of the LCIA results. If more carbon fibers are consumed, more emissions are emitted and consequently, increases in all impact categories will occur. Carbon fibers production is the main contributor to the impact category of Human Toxicity Potential (HTP), Freshwater Aquatic Ecotoxicity Potential (FAEP), Marine Aquatic Ecotoxicity Potential (MAEP), and Terrestric Ecotoxicity Potential (TETP). This is because of the amount of heavy metals emitted to the environment.

7. Conclusions

A drone delivery system has potentially becoming one of the key delivery options to operate such deliveries to customers in remote areas. This is because of their key advantages, which include 1) speed, 2) the ability to traverse difficult terrain or fly over water, and 3) their reduced environmental impact. Even though drones are widely known as a delivery tool to enable rural people's participation in the global economy, there is a lack of research on their environmental impact. As its advantages in reducing numbers of environmental impacts, this study evaluated its emissions over its full life stage using an LCA. LCA result from this study indicate that the parts production, consisting of carbon fibers and Li-ion productions for cargo box and battery, respectively, dominate the most to GWP impact category. The main recommendation would be to further improve the material used for cargo box and increase efficiencies of battery. This will lead to decrease in emissions and wastes that contribute to each impact category. A full sensitivity analysis will be conducted in order to identify all important parameters that affect the most of the LCA results. In addition, governments can affect the adoption of the drone delivery system. These will allow for drones to be widely used in rural areas. To achieve filling these gaps, drones would be one of the best delivery options to offer home delivery service even though this widespread usage still needs time.

8. Acknowledgements

I am grateful for the financial support from Research Seed Grant for New Lecturer from KMITL Research Fund.

9. References

[1] Bangkok Entrepreneurs 2016 E-commerce is now the hot cake!!! | TeCS 2016. Retrieved from http://www.bangkok-entrepreneurs.com/2016/06/14/e-commerce-is-now-the-hot-cake/ (Accessed 1 October 2017)

[2] Langston J 2017 Drone vs. truck deliveries: Which create less carbon pollution?. Retrieved from https://phys.org/news/2017-05-drone-truck-deliveries-carbon-pollution.html (Accessed 7 December 2017)

[3] Gulden T 2017 *The energy implications of drones for package delivery: A geographic information system comparison.* No. RR-1718/1-RC (RAND Corporation)

[4] Lee H L, Chen Y, Gillai B and Rammohan S 2016 *Technological disruption and innovation in last-mile delivery* (U.S.A: Graduate School of Stanford Business)

[5] Heutger M and Kückelhaus M 2014 *Unmanned aerial vehicles in logistics a DHL perspective on implications and use cases for the logistics industry* (Germany: DHL Customer Solutions & Innovation)

[6] Weideli D 2013 *Environmental analysis of US online shopping*. Master Thesis for MIT Center for Transportation & Logistics, Cambridge MA, U.S.A.

[7] Miller P 2015 *Johns.Matthey Technol. Rev.* **59**(1) 4

[8] Belmonte N, Luetto C, Staulo S, Rizzi P and Baricco M *Challenges* **8**(1) 9

[9] Lohsomboon P and Jirajariyavech A 2004 Life cycle inventory for cement product and steel making towards sustainable development. (Thailand: Thailand Environment Institute (TEI))

[10] Wankanapon P, Chindapol S and Tantasavasdi C 2013 *Built* **2** 43

[11] Pandit D D V and Poojari A 2014 *IOSR-JBM* **16**(11) 6

Novel Vulnerability Metrics for Interdependent System based on System Controllability

L Faramondi, G Oliva and V Piemonte

University Campus Bio-Medico di Roma, Faculty of Engineering, via Álvaro del Portillo 21, 00128, Rome, Italy.

Abstract. In this paper we develop novel vulnerability metrics for interdependent critical infrastructures or economic sectors based on the concept of controllability. Specifically, we consider the Input-Output Inoperability Model, that represents the dynamics of the dependencies and interdependencies among infrastructures or sectors during an adverse event or malicious attack, and we argue that the more the system is *controllable* via an external perturbation that represents the fault or attack, the more the system is vulnerable. Therefore, we analyze the vulnerability of the system in terms of the magnitude of the associated controllability matrix. Moreover, based on the proposed index, we develop a simple defensive strategy to reduce the effect of an attack. A validation of the approach with respect to real data concludes the paper.

1. Introduction

Due to the diffusion of emerging technologies (e.g., cloud services, IoT systems and mobile computing), modern infrastructures or economic sectors are now part of the same macro-system; since their functionalities are essential to the health, safety, security or economic well-being of a nation or community, such an increased degree of mutual dependency is both a blessing and a curse. In fact, from one side the increasing degree of cooperation among sectors or infrastructures is improving the quality of our everyday life; however, from another side, the growing dependency of such systems on each other is posing a serious threat to our society, because of the possibility of cascading effects in the event of a natural disaster, human error or terroristic attack.

In this view, control theory represents a valuable framework for analyzing such systems, since such a discipline provides useful tools and methodologies to deal with dynamical systems, disturbances, model uncertainties, feedback and interdependency. Among other ideas, the concept of *controllability* appears particularly valuable to quantify the effect of the propagation of faults and cascading events [1, 2, 3].

Consider an attacker aiming at disrupting the services and the functioning of a group of densely interrelated critical infrastructures or sectors; in this perspective, the ability of the attacker to control the system can be regarded as highly disruptive.

We point out that, in the literature, the relation between control and attacks has been widely investigated. In the context of Cyber Physical Systems, in [4] the authors face the problem of identifying a set of *driver nodes*, i.e., nodes in a network that guarantee the structural controllability of a system. The relation between attack vulnerability and controllability has been also studied in [5], where the authors analyze the effect of node deletion from a graph representing the structure of an interconnected system, following several attack strategies and considering different network topologies. A similar topological approach has been adopted in [6] to study the robustness of power grids to node or edge disruptionor in [7, 8, 9, 10], where vulnerable nodes are identified via optimization techniques. In [11] random attacks or faults have been regarded as a major cause of damage to interconnected networks. An interesting study about the relation between the attacks and controllability of networks is presented by Liu et al. [12]; in this work, the authors show that network controllability can be conveniently studied in the framework of percolation theory.Finally, in [13] possible countermeasures are investigated considering a network of networks scenario.

In this paper we develop a new metric able to highlight the fragility of a subset of interdependent systems, services, or sectors, based on the controllability of the system, with particular reference to the Input-Output Inoperability Model [14,15,16]. In order to show the effectiveness of the proposed index, we compare it with well established descriptors of the vulnerability of such a model. Moreover, we develop a strategy to guide the defense of the different sectors in order to achieve an arbitrarily small value of the proposed index.

The outline of the paper is as follows: In Section 2 we provide some preliminary definitions; in Section 3 we provide a short description about the input output models with a focus on the Inoperability Input Output model; the definition of the new metric and the novel defensive strategy are discussed in Section 4; results and discussions are collected in Section 5 with the aim to validate the proposed index; finally, some conclusive remarks are collected in Section 6.

2. Preliminaries

2.1. General preliminaries

In the following, we denote by $card(X)$ the cardinality of a set X. Moreover, we denote matrices and vectors via cursive letters, and we use k_m to indicate a vector in $\mathbb{R}^m$ whose components are all equal to k. We represent by $||A||_1$ the *entrywise 1-norm* of an $n \times n$ matrix A, i.e.,

$$||A||_1 = \sum_{i=1}^{n} \sum_{i=1}^{n} |a_{ij}|. \tag{1}$$

2.2. LTI dynamic system controllability

Let a continuous-time linear time-invariant (LTI) system having the form

$$x(t) = Ax(t) + Bu(t), \tag{2}$$

where $x \in \mathbb{R}^n$ represents the state variable vector, $u \in \mathbb{R}^p$ is the input vector, A is the $n \times n$*dynamic matrix* having constant entries a_{ij}, and B is the $n \times p$*input matrix*, having entries b_{ij}. The above system is*controllable* in for any finite time window $[t_0, T]$ we can find an input signal $u(t)$ for $t \in [t_0, T]$ able to steer the state $x(\cdot)$ from any initial value to any final value.

According to Kalman [17], System (2) is controllable if and only if the*controllability matrix*

$$R = [B \quad AB \quad \dots \quad A^{n-1}B] \tag{3}$$

is full row-rank, i.e., if$rank\ (R) = n$. Notice that, if$rank\ (R) = q < n$, then the state space $\mathbb{R}^n$ can be decomposed in a*controllable subspace* of dimension q and a*noncontrollable subspace* of dimension $n - q$.

3. Inoperability input output model

The traditional *Input-Output Model* is a popular static equilibrium model to represent macroeconomic interaction among the economic sectors of a country or region. Such a tool is often used to describe the growth of an economy due to sector expansion or technology adoption [18, 19]. More in detail, let us consider the model defined by Equation 4 for$n$ interconnected industry sectors (e.g., transportations, manufacturing, retail trade, etc.); vector $x \in \mathbb{R}^n$ represents the set of production outputs of each sector, while vector $c \in \mathbb{R}^n$ represents the set of consumer demand for goods, services, and usage in each sector; finally, the $n \times n$ matrix A is the so-called*Leontief matrix* whose elements a_{ij} (Leontief technical coefficients) represent the ratio of the output produced by the i-th infrastructure that is required by the j-th infrastructure in order to generate one unit of production. In a nutshell, the

Input-Output Model reads as follows:

$$x = Ax + c. \tag{4}$$

Note that Equation (4) captures the equilibrium between demand and production at the different sectors and describes how industry output (x) is a function of intermediate demand (Ax) among sectors and final demand (c).

Following the economic equilibrium theory of Leontief [18, 19], a static *Demand-Reduction Model* for n industrial sectors is given by:

$$\Delta x = A\Delta x + \Delta c, \tag{5}$$

where Δx represents the difference between the planned (x_0) and the degraded production (x_d), Δc is the difference between the planned final demand (c_0) and the degraded final demand (c_d).

Based on the latter model, in [14] a model able to capture the dependency betyween sectors or infrastructures is developed, by introducing the notion of *inoperability* on an infrastructure/sector, i.e., its inability (as a percentage) to operate properly due to the degraded production.

Let $q \in \mathbb{R}^n$ be the vector whose entries (q_i) represent the inoperability of the i-th sector. With the aim to describe how*inoperability* propagates, [14] introduces an $n \times n$ transformation matrix

$$P = [diag\{x_0\}]^{-1}. \tag{6}$$

The inoperability is then computed by applying the transformation to Equation (5), i.e.,

$$q = P\Delta x. \tag{7}$$

The outcome of such a transformation is the *Input Output Inoperability Model* (IIM) [15, 16]

$$q = A^*q + c^* \Rightarrow q = (I - A^*)^{-1}c^* = D^*c^*, \tag{8}$$

where $A^* = PAP^{-1}$, and $c^* = P\Delta c$. In this view, vector c^* describes the perturbation in demand resulting from a disruptive event, and is the driver of inoperability in the demand-reduction, while A^* is the*interdependency matrix*, which can be derived from economic data as in [20], or using other sources as in [21].

In [14] the static IIM defined in Equation 8 is further extended by introducing a dynamics, i.e.,

$$\dot{q}(t) = K(A^* - I)q(t) + Kc^*(t), \tag{9}$$

where matrix K is the *industry resilience* coefficient matrix (it has a relation with the convergence rate towards an equilibrium), and each element k_{ii} measures the resilience of the $i-th$ infrastructure. The k_{ii} coefficient can also be seen as the recovery rate with respect to adverse or malicious events. In the absence of concrete information, it is common practice to set $K = I$.

3.1. Key sectors and criticality measures for IIM

Analogously to*component importance measures* (CIMs) in the reliability engineering literature, key sector analysis is used in the economic input-output literature to identify the criticality that certain industry sectors have on the economy, and the productivity of a country or region. In the literature, two indices for expressing the criticality of an infrastructure within a given scenario have been introduced in [22, 23], based on their economic input-output counterpart [24, 25]. These two indices are referred to here as the *dependency index*, and the*influence gain*.

The *dependency index* is defined as the sum of the coefficients of matrix A^* along a row, shown in (10).

$$\delta_i = \sum_{j=1}^{n} a_{ij}^* \tag{10}$$

The dependency index measures the robustness of the ith infrastructure with respect to the inoperability of the other infrastructures, representing a measure of the residual operational capability of the ith infrastructure when every other infrastructure is fully inoperable. The lower the value, the greater the ability of the ith infrastructure to maintain functionality despite the inoperability of the other sectors upon which it relies. On the other side, $\delta_i \geq 1$ means that the operability of the ith infrastructure might be nullified even if the other infrastructures still have some residual operational capability.

Another index, the *influence gain*, measures the influence that one infrastructure exerts over the other infrastructures. It is defined as the sum of the coefficients of matrix A^* along a column, as in

(11).

$$\gamma_i = \sum_{j=1}^{n} a_{ji}^* \tag{11}$$

A large influence gain suggests that the inoperability of the i-th infrastructure induces significant degradations on the entire system. When $\gamma_i > 1$, the adverse effects in terms of inoperability induced by cascading phenomena on the other infrastructures are amplified.

Ordering the sectors in decreasing order of δ_i provides a ranking of sectors that are most impacted by inoperability in other sectors, and likewise, a decreasing order of γ_i results in a ranking of sectors that are most impactful on other sectors. In a more absolute measure, Hirschmann [26] suggests that*key sectors* are those sectors for which the normalized version of indices (10) and (11) are greater than 1.

4. IIM vulnerability/robustness metrics based on controllability

In this section we develop novel vulnerability/robustness metrics of the IIM model that are based on the concept of controllability. In fact, in this section we argue that the more an IIM system is controllable by means of the input $c^*(t)$, the more an adverse event or a terroristic attack may have severe consequences on the overall system.

Let us consider a dynamic IIM model as din Equation (9), and for the sake of simplicity suppose that it holds

$$c^*(t) = Bf(t), \tag{12}$$

where $f(t)$ is a scalar input that represents an attack signal, and B is a column vector whose entries b_i model the vulnerability of the i-th infrastructure to an attack or failure affecting it directly. In more details, we define the entries b_i in the range $[0 \ldots 1]$, where $b_i = 0$ represents an infrastructure that can not be directly affected by $f(t)$ and $b_i = 1$ represents an infrastructure vulnerable to the direct application of $f(t)$.

The dynamic IIM model can be reformulated as follows

$$\dot{q}(t) = K(A^* - I)q(t) + KBf(t). \tag{13}$$

In the following remark, we argue that being controllable is a source of vulnerability for the IIM system in Equation (13).

Remark 1 *If the pair $(K(A^* - I), KB)$ is controllable, then all the infrastructures can be made completely inoperable in finite time, i.e., for any $T > 0$ there is a choice of $f(t)$ for all $t \in [0, T]$ such that $q(T) = 1$.*

A consequence of the above Remark is that, for a given choice of B representing one or more concurrent failures or attacks, the fact that the pair $(K(A^* - I), KB)$ is not controllable implies that some infrastructures/sectors (actually, a subspace of the state of the overall state space) can not be made inoperable no matter the magnitude of such combined failures. Hence, we point out that lack of controllability is strongly related with the resilience of the IIM model.

Notice that, when the system is controllable, there are, in general, no hints on the exact shape of $c^*(\cdot)$ that yields complete inoperability. There is, however, an although theoretical way for an attacker to obtain the complete inoperability of all the infrastructures/sectors in "zero time", as outlined in the next proposition.

Proposition 1 *Let $\delta(t)$ be a Dirac pulse and let $\delta^{(m)}(t)$ be the m-th derivative of $\delta(t)$. Let an IIM system having the form of Equation (13) and suppose that B is a column, i.e., that there is just one input. Moreover, let R be the controllability matrix of the pair $(K(A^* - I), KB)$ and suppose that the system in Equation (13) is controllable. Then, the initial condition of the system can be forced to 1_n (complete inoperability) by choosing*

$$f(t) = g_1\delta(t) + g_2\delta^{(1)}(t) + \cdots + g_n\delta^{(n-1)}(t), \tag{14}$$

where

$$[g_1, \cdots, g_n]^T = g = R^{-1} 1_n. \quad (15)$$

Proof 1 *It is well known (see for instance [27]) that, with this choice of $f(t)$, the initial condition is given by $q(0) = Rg$. Since the system is controllable, R is invertible and the statement is proved.*

The above result implies that an attack on a single infrastructure can force the inoperability of all the infrastructures to 1 in "zero time"; clearly, such an abstraction is not feasible in practice, but represents an intuitive justification of the following vulnerability metric.

Remark 2 *Based on Equation (15) we argue that the effort necessary to generate the signal attack $f(t)$ depends on the magnitude of g. Therefore, in the following, we model the magnitude of the failure/attack by $||g||_1$. In this view, considering Equation (15) we observe that it holds*

$$||g||_1 = ||R^{-1}g||_1 \leq ||R^{-1}||_1 ||1||_1 = \sqrt{n}||R^{-1}||_1, \quad (16)$$

meaning that $||R^{-1}||_1$ can be regarded as a measure of the attack effort.

Note further that, if R^{-1} exists, $||R^{-1}||_1$ gives a rough measure of the fragility of the system; indeed when $||R^{-1}||_1 \gg 1$ a large effort is required to cause large damages, while $||R^{-1}||_1 \ll 1$ implies that the complete inoperability can be reached with a small attack effort and the system is highly vulnerable.

The above remark is the foundation for the definition of a novel vulnerability index. Note that in the following definition we adopt $||R||_1$ as a measure that is inversely proportional to the attack effort. Indeed, we point out that $||R^{-1}||_1$ is a well defined index only if there is just one input and the system is controllable, while by considering R we can extend the index to the case of multiple inputs and not completely controllable systems.

Let us consider a scenario where a subset of infrastructures $\Phi \subseteq \{1, \ldots, n\}$ is involved in the attack. Considering Equation (13), we express matrix$B$ as a vector having entries $b_i \in (0,1]$ if the i-th infrastructure is in Φ, and such that $b_i = 0$, otherwise. Notice that the exact value of the entry b_i depends on the degree of robustness of the infrastructure i with respect to the direct application of the attack signal $f(t)$.

Definition 1 (Fragility Index) *Let us define the Fragility Index ν of an infrastructures subset $\Phi \subseteq \{1, \ldots, n\}$ as follows:*

$$\nu(\Phi) = ||R||_1, \quad (17)$$

where R is the controllability matrix of the pair $(K(A^* - I), KB)$ and B reflects the impact of $f(t)$ on the infrastructure subset Φ.

The fragility index ν, being inversely proportional to the attack/damage effort, can be regarded as a measure of the overall vulnerability of an IIM system with respect to an adverse event or malicious attack affecting a specific infrastructures subset. Indeed, if $\nu(\Phi) \gg 1$ the subset Φ is a vulnerable subset, instead, If $\nu(\Phi) \ll 1$ the subset Φ represent a resilient subset of infrastructures.

4.1. Defensive strategy

We now show that the fragility index can be the basis to guide the decision-maker in the protection of the different sectors. In fact, if the sectors are protected in a way such that the coefficients b_i are replaced by new coefficients $\tilde{b}_i$, which are inversely proportional to the sector fragility index, then we can impose arbitrarily small fragility index to the resulting system.

Theorem 1 *Let $\nu(\{k\})$ the fragility index of the sector k. Such an index can be reduced to an arbitrarily small value ρ by setting $\tilde{b}_k = \rho\nu(\{k\})^{-1}$.*

Proof 1 *For the sake of simplicity let $\hat{A} = A^* - I$ and $K = I$. Let e_i be the i-th vector in the canonical basis of $\mathbb{R}^n$; i.e., e_i has all entries equal to 0, except the i-th entry that is equal to 1.*

IOP Conf. Series: Journal of Physics: Conf. Series **1026** (2018) 012021 doi:10.1088/1742-6596/1026/1/012021

Moreover, let us denote by $\hat{A}^p_{i,j}$ *the entries of the matrix* $\hat{A}^p$. *Let* $\Phi = \{k\}$, *and consider the associated fragility index* $\nu(\{k\})$, *defined in Equation (17). Since* $\nu(\{k\}) = \| R \|_1$, *it holds*

$$\nu(\{k\}) = \left\| [e_K \hat{A} e_K \;\cdots\; \hat{A}^{n-1} e_K] \right\|_1 = 1 + \sum_{i=1}^{n} |\hat{A}_{i,K}| + \cdots + \sum_{i=1}^{n} |\hat{A}^{n-1}_{i,K}|. \quad (18)$$

Let us now adopt a defensive strategy by choosing:

$$\tilde{b}_K = \rho \nu(\{k\})^{-1};$$

the new fragility index $\tilde{\nu}(K)$ *is given by*

$$\begin{aligned}\tilde{\nu}(K) &= \left\| [\rho \nu(\{k\})^{-1} e_K \rho \nu(\{k\})^{-1} \hat{A} e_K \;\cdots\; \rho \nu(\{k\})^{-1} \hat{A}^{n-1} e_K] \right\|_1 \\ &= \rho \nu(\{k\})^{-1} (1 + \sum_{i=1}^{n} |\hat{A}_{i,K}| + \cdots + \sum_{i=1}^{n} |\hat{A}^{n-1}_{i,K}|).\end{aligned} \quad (19)$$

At this point, plugging Equation (18) in the above Equation we have that it holds

$$\tilde{\nu}(K) = \rho \nu(\{k\})^{-1} \nu(\{k\}) = \rho. \quad (20)$$

This completes our proof. ■

The following result is a rather obvious consequence of the above theorem.

Corollary 1 *Let* $r = [r_1, \dots, r_n]^T \in \mathbb{R}^n$ *the column vector whose entries are the new desired fragility indices. We can obtain the desired values by applying the following defensive strategy:*

$$\tilde{b} = [r_1 \nu(1)^{-1} \;\; \dots \;\; r_n \nu(n)^{-1}]^T. \quad (21)$$

5. Case study

With the aim to validate the effectiveness of the proposed Fragility Index and its use as a base for the implementation of a defensive strategy, in this section, we propose an experimental analysis with respect to a real world case study.

5.1. Data set

We consider the 15×15 sector case study in [20, 28, 29] from the 2005 BEA annual IO tables [30]. The 15 considered sectors represent a relevant subset of the industry and infrastructure sectors in the US, i.e.,: 1 - Agriculture, 2 - Mining, 3 - Utilities, 4 - Construction, 5 - Manufacturing, 6 - Wholesale trade, 7 - Retail trade, 8 - Transportation, 9 - Information, 10 - Finance, 11 - Business services, 12 - Educational service and health care, 13 - Arts and entertainment, 14 - Other services, 15 - Government. The A^* matrix for the 15-sector example is given in [28].

5.2. Analysis of the proposed index

The upper plot in Figure 1 shows the value of the fragility index ν_i by considering infrastructure sets Φ with cardinality one, i.e., by inspecting the effect of an attack/damage directly affecting one infrastructure at the time. Note that, for the sake of simplicity, we assume $K = I$ and $b_i = 1$ if the i-th infrastructure is directly involved in the attack. We observe that, in this case, we obtain $\nu(\Phi) > 1$ for all infrastructures, and the proposed index ranges from around $\nu(\{1\}) \approx 1.62$ for Agriculture, to about $\nu(\{3\}) \approx 6.46$ for the Utilities sector. This suggests that a perturbation having the structure of the one discussed in Proposition 1 is amplified from 1.62 to 6.46 times, depending on the affected sector.

The lower plot in Figure 1 shows a comparison among the indices $\nu(\Phi)$ (blue bars), the dependency indices δ_i (yellow bars) and the influence gains γ_i (grey bars) for each infrastructure, normalized with respect to the average. According to the plot, sectors 1 and 2 correspond to a dependency index δ_i which is significantly above the average, while sectors $\{3,5,8,9,11\}$ are slightly above the average. As for the influence gain γ_i, we have that sector 5 has a remarkably high value with respect to the average, while sectors $\{10,11,15\}$ are slightly above the average.

According to Hirschmann [26], therefore, sectors 5 and 11 can be regarded as critical. The results for the controllability index show that sectors $\{3,4,5,7,12,13,14,15\}$ are above the average, and such a ratio ranges from about 1.03 times the average (sectors 13) to about 1.83 times (sector 3). Such results highlight that the proposed controllability index ν_i is able to capture different, possibly complementary, aspects with respect to previous indices. Indeed, while δ_i and γ_i model the direct effect of the inoperability of the infrastructures in the short term, the proposed index models the overall effect of the attack/failure, with a larger time scale.

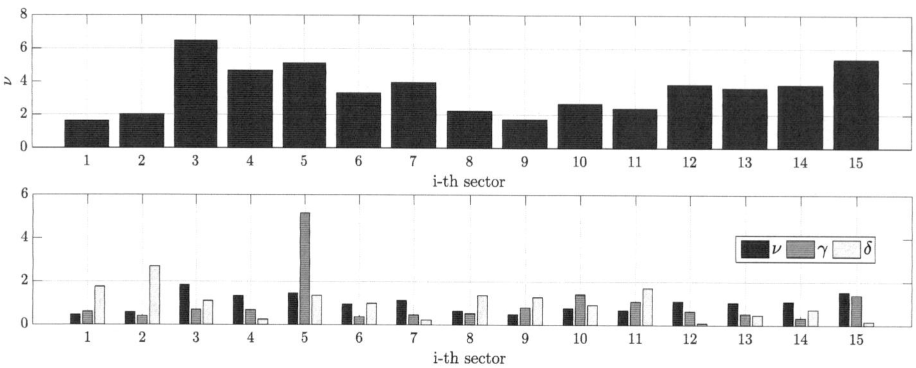

Figure 1. Vulnerability Metrics for 2005 US Infrastructure Sectors case study.
On top: Distribution of Fragility Indices ν for each sector
On bottom: Comparison of the indices ν (dark blue bars), dependency indices δ (yellow bars), and influence gains γ (grey bars). Each index is normalized with respect to the average of its distribution

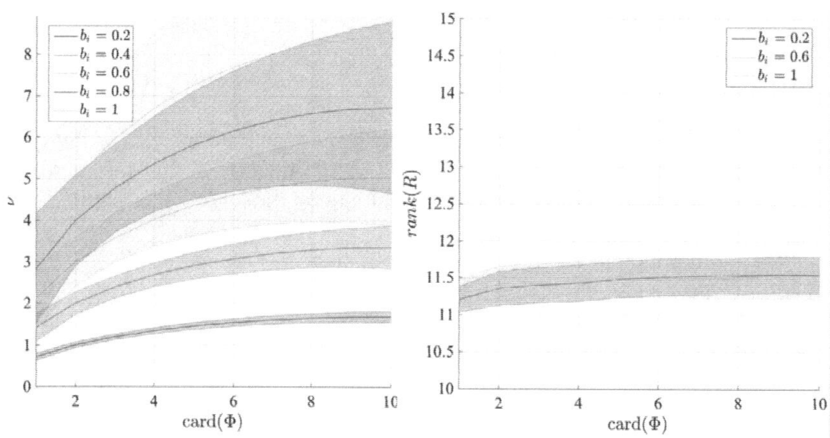

Figure 2. Fragility indices and controllability matrix rank for multiple attack and resilience coefficients combinations.
On left: Average value and variance of the Fragility Index ν considering multiple attacks ($card(\Phi) > 1$) and multiple resilience values b_i
On right: Average value and variance of matrix R considering multiple attacks ($card(\Phi) > 1$) and multiple resilience values b_i

Figure 2 provides additional insights on the proposed metric. Specifically, in Figure 2(left) we report the fragility index $\nu(\Phi)$ for attacks affecting multiple sectors. In the figure, the average value and the variance of $\nu(\Phi)$ are shown considering multiple attack plans Φ with different cardinality. For each fixed cardinality in the range [1, ..., 10], the results in terms of mean value and variance are computed considering all the existing combination of attacks against the considered sectors. Moreover the same results are shown also considering different values of $b_i \in \{0.2,0.4,0.6,0.8,1\}$ in order to model the effect of the attack depending on the robustness/resilience of the sector or infrastructure being attacked. As an example, for $\Phi = \{2,4\}$ and $b_i = 0.2$ the associated B matrix is defined as $B = [0,0.2,0,0.2,0,\cdots,0]^T$.

IOP Conf. Series: Journal of Physics: Conf. Series **1026** (2018) 012021 doi:10.1088/1742-6596/1026/1/012021

The results in Figure 2 (left) highlight that, for each choice of b_i, the fragility index increases with increasing number of sectors involved in the attack. Moreover, the effectiveness of the proposed metric in describing the effect of the attack is supported by the increment of fragility index for high values of b_i; in fact, high values of b_i represent the fact that the i-th sector or infrastructure is prone to the attack. In the same plot when an high resilience level is considered (i.e. $b_i = 0.2$) the fragility index is reduced to low levels. In Figure 2 (right) the rank of the controllability matrix is reported in terms of average value and variance by varying the number of sectors involved in the attack $|\Phi|$ and their resilience b_i. The results shown that for each pair of parameters Φ and b_i the rank of the controllability matrix is always in the range [11 ... 12].

5.3. Defense strategy based on Fragility Index

The following simulation campaign is analysed in order to prove the utility of the defensive strategy based on the fragility indices as explained by Theorem 1. In more details, starting from the same scenario analysed in Figure 1, we adopt a defensive strategy by setting the entries of the new matrix $\tilde{b}$ selecting

$$\tilde{b} = 0.5[\nu(1)^{-1} \quad \dots \quad \nu(n)^{-1}], \tag{22}$$

where $\nu(1), \dots, \nu(n)$ are the fragility indices collected in the bar plot in Figure 1(top). As highlighted in Figure 3 the new fragility indices $\tilde{\nu}$ are all equals to 0.5. On the other side, the utility of this kind of defensive strategy is not highlighted by the two other metrics (dependency index, influence gain) because they are based only on the entries of the matrix A^* without considering the resilience coefficients b_i.

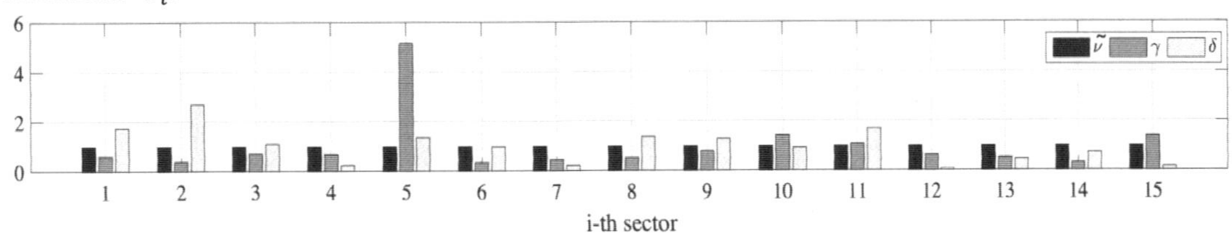

Figure 3. Comparison of the indices $\tilde{\nu}$ (dark blue bars), dependency indices δ (yellow bars), and influence gains γ (green bars) in case of defensive strategy. Each index is normalized with respect to the average of its distribution.

6. Conclusions

In this paper we provide a novel metric to evaluate the fragility of a subset of interconnected systems (infrastructures or economic sectors). With respect to the state of the art the proposed fragility index takes into account both the interdependencies and the reactions of each component to an external attack. Hence, it is able to catch the effects of cascade faults or attack by providing an accurate estimation about the fragility of the system. Moreover, we prove the effectiveness of the proposed index as a basis for a defensive strategy against external unpredicted malicious input. Future works will aim at comparing the proposed index with other structural vulnerability metrics, and at analyzing the relation between the fragility index and sector specific vulnerabilities metrics.

7. References

[1] Pasqualetti F, Zampieri S, and Bullo F 2014 Controllability metrics, limitations and algorithms for complex networks *IEEE Transactions on Control of Network Systems* vol. 1, no. 1, pp. 40–52.

[2] Wu-Yan E, Betzel R F, Tang E, Gu S, Pasqualetti F and Bassett D S 2017 Benchmarking measures of network controllability on canonical graph models *arXiv preprint* arXiv:1706.05117.

[3] Summers T H, Cortesi F L and Lygeros J 2016 On submodularity and controllability in complex dynamical networks *IEEE Transactions on Control of Network Systems* vol. 3, no.1, pp. 91-101.

[4] Cam H 2014 Controllability and Observability of Risk and Resilience in Cyber-Physical Cloud Systems *Secure Cloud Computing* pp. 325–343, Springer New York.

[5] Lu Z and Li X 2016 Attack Vulnerability of Network Controllability *PloS one* vol. 11, no. 9 e0162289.
[6] Solé R V, Rosas-Casals M, Corominas-Murtra B and Valverde S 2008 Robustness of the European power grids under intentional attack *Physical Review E* vol. 77, no. 2, 026102.
[7] Faramondi L, Oliva G, Pascucci F, Panzieri S and Setola R 2016 Critical node detection based on attacker preferences *Proceedings of the 24th Mediterranean Conference on Control and Automation (MED)* pp. 773-778.
[8] Faramondi L, Oliva G, Setola R, Pascucci F, Amideo A E and Scaparra M P 2017 Performance Analysis of Single and Multi-objective Approaches for the Critical Node Detection Problem *Proceedings of the International Conference on Optimization and Decision Science* pp. 315-324.
[9] Faramondi L, Setola R, Panzieri S, Pascucci F and Oliva G Finding Critical Elements in Infrastructure Networks *International Journal of Critical Infrastructure Protection.*
[10] Faramondi L, Oliva G, Panzieri S, Pascucci F, Schlueter M, Munetomo M and Setola R Submitted Network Structural Vulnerability: A Multi-Objective Attacker Perspective *IEEE Transactions on Systems, Man, and Cybernetics: Systems*.
[11] Holme P, Kim B J, Yoon C N and Han S K 2002 Attack vulnerability of complex networks *Physical review E* vol. 65, no. 5, 056109.
[12] Liu Y Y, Slotine J J and Barabási A L 2011 Controllability of complex networks *Nature* vol. 473(7346), pp. 167-173.
[13] Liu X, Peng H and Gao J 2015 Vulnerability and controllability of networks of networks *Chaos, Solitons & Fractals* vol. 80, pp. 125-138.
[14] Haimes Y, Horowitz B, Lambert J, Santos J, Lian C and Crowther K 2005 Inoperability input-output model for interdependent infrastructure sectors I: Theory and methodology *Journal of Infrastructure Systems* vol. 11, n. 2, pp. 67–79.
[15] Haimes Y and Jiang P 2001 Leontief-based model of risk in complex interconnected infrastructures *Journal of Infrastructure Systems* vol. 7, n. 1, pp. 1–12.
[16] Santos J R and Haimes Y Y 2004 Modeling the demand reduction input-output (I-O) inoperability due to terrorism of interconnected infrastructures *Risk Analysis* vol. 24, pp. 1437–1451.
[17] Kalman R E 1963 Mathematical description of linear dynamical systems *Journal of the Society for Industrial & Applied Mathematics, Series A: Control* vol. 1, no. 2, pp. 152–192.
[18] Leontief W W 1966 *Input-Output Economics*, Oxford University Press.
[19] Miller E M and Blair P D 2009 *Input-Output Analysis: Foundations and Extensions* 2nd ed., Cambridge University Press.
[20] Barker K and Rocco C M 2011 Evaluating uncertainty in risk-based interdependency modeling with interval arithmetic *Economic Systems Research* vol. 23, n. 2, pp. 213–232.
[21] Oliva G, Panzieri S and Setola R 2011 Fuzzy dynamic input-output inoperability model *International Journal on Critical Infrastructure Protection* vol. 4, n. 3–4, pp. 165–175.
[22] Setola R 2010 How to measure the degree of interdependencies among critical infrastructures *Int. J. of System of Systems Engineering* vol. 2, n. 1, pp. 38–59.
[23] Setola R, De Porcellinis S, and Sforna M 2009 Critical infrastructure dependency assessment using the input-output inoperability model *International Journal of Critical Infrastructure Protection* vol. 2, n. 4, pp. 170–178.
[24] Chenery H B and Watanabe T 1958 International comparisons of the structure of production *Econometrica* vol. 26, n. 4, pp. 98–139.
[25] Rasmussen P 1956 *Studies in Intersectorial Relations*, North Holland, 1956.
[26] Hirschmann A O 1958 *The Strategy of Economic Development*, Yale University Press.
[27] Kailath T 1980 *Linear systems. vol. 1* Englewood Cliffs, NJ: Prentice-Hall.

[28] Barker K and Haimes Y Y 2009 Uncertainty analysis of interdependencies in dynamic infrastructure recovery: Applications in risk-based decisionmaking *Journal of Infrastructure Systems* vol. 15, n. 4, pp. 394–405.
[29] Oliva G, Setola R and Barker K 2014 Fuzzy Importance Measures for Ranking Key Interdependent Sectors Under Uncertainty *IEEE Transactions on Reliability* vol. 63, n. 1, p. 42–57.
[30] Bureau of Economic Analysis 2007 *Input-Output Accounts Data* http://www.bea.gov.

A Sitting Posture Surveillance System Based on Kinect

Si He

College of Information Engineering, Wuhan University of Technology, Wuhan, China

Abstract. Modern life has made more and more people become sedentary, and long periods of unhealthy posture can lead to myopia, cervical spine and lumbar disease. Therefore, this paper presents a sitting posture surveillance system based on Microsoft Kinect. By getting joints' three-dimensional position information from Kinect's real-time skeleton tracking technology, and input the data into computer. After processing the joints' location information with posture recognition algorithm, computer recognizes the user's sitting posture, and warns the incorrect postures.

1. Introduction

In daily life, whether it is work, study or relaxation, sitting is the most commonly used posture, but the long-term incorrect sitting posture can cause great harm to the health of the human body. Unhealthy sitting posture can lead to the distance too close to the eye, and if things continue in this way, the ciliary of the eyes may lose their ability to regulate themselves and cause myopia[1]. And long periods of abnormal physiological curvature of the spine are likely to cause spinal deformity and various cervical diseases.

The occurrence and even getting into a habit of bad sitting posture is often involuntary, timely remind or correction from outside can effectively help keeping good sitting posture to habit. Traditional posture correction methods, besides human supervision and reminding, mostly are wearable physical orthotics or simple electronic orthotics based on acceleration sensors or level sensors. The effect is immediate but the precision is insufficient, and wearing for a long time can also hamper the body movement, bringing extra burden to users[2]. By contrast, using visual capture technology, through video or static images, judging posture by processing visual data is more accurate and flexible[3].

Gesture recognition is a hotspot in the field of machine vision and is widely used in human-computer interaction, behavior analysis, multimedia application and sports science[4].

2. Kinect

With the rapid development of camera technology in recent years, the cameras with 3D camera has gradually entered our horizon. TOF cameras emerged in the 1990 s, the principle of which is to calculate the flight time of light and obtain the depth information of objects. Compared with the traditional 2D cameras, TOF cameras can easily separate foreground and background in the scene, thus the TOF cameras have the advantage in the aspect of undetermined target tracking and recognition. But the cost and relatively low resolution of the TOF cameras affected its development.

For the first time in November 2010, Microsoft launched a motion sensing device—Kinect. Kinect has functions of body and skeleton real-time tracking, motion capture and the microphone input, and can identify a series of actions. People can stand in front of the device and interact with computers

IOP Conf. Series: Journal of Physics: Conf. Series **1026** (2018) 012022 doi:10.1088/1742-6596/1026/1/012022

through a variety of gestures. Kinect greatly reduce operating costs of human-computer interaction. Therefore, the research on the natural human-computer interaction based on Kinect has become a hotspot[5].

Different from normal cameras, Kinect senses the outside world with a CMOS infrared sensor. The sensor apperceives outside environment by black and white spectrum: pure white represents an infinite close, pure black represents infinity far, according to the physical distance to the sensor, objects are represented by corresponding grey level. It can sensitively perceive the space within the device's field of vision and generate a depth image represents the surrounding environment. The sensor generates a depth image stream at the frequency of 30 frames per second, making a real-time 3D reconstruction of the surrounding environment. Kinect detects 3D depth images and transfers them to skeletal tracking systems. The system can track the operator's 20 bone nodes, including the torso, limbs and fingers. In order to understand the user's action, Microsoft used machine learning technology and established a large image database, realized intelligent recognition that can understand the meaning of the user's movement as far as possible[6].

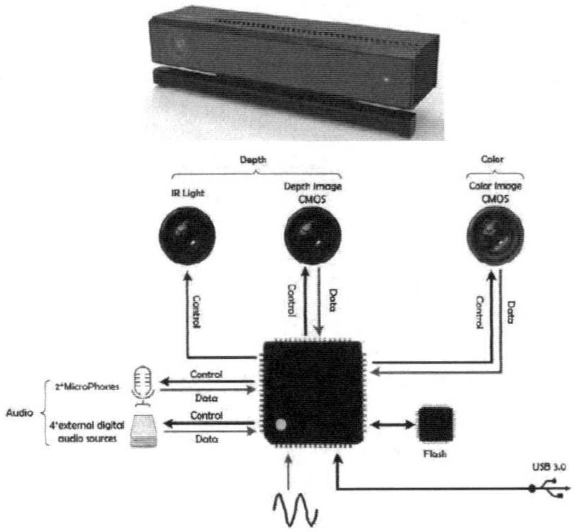

Figure 1. Physical and schematic diagram of Kinect.

In June 2014, Microsoft launched the upgraded Kinect, Kinect2.0, with its physical and schematic diagram as shown in figure 1. Compared with the first-generation Kinect product, Kinect2.0 that we use in this paper enhanced the depth sensing capability, which can distinguish smaller objects through higher depth fidelity and dramatically improved noise base. Video resolution is upgraded from VGA level to 1080p level, providing full high definition video and smooth scene experience, the data processing amount can be up to 2GB/s; With USB3.0 interface, the delay reduced to 2 frames/s; Microsoft has also used the TO method to replace the existing "light measurement" method, which makes Kinect2.0 faster and more accurate than the first generation. The vision of Kinect2.0 increased 60%, support 6 operators at the same time, and can track each person's 25 joints information at the same time, optimize the tracking algorithm, and make the tracking posture more accurate and stable. Improved the detection performance of infrared sensor that can even work in the dark. In addition, Kinect2.0 also provides thumb tracking, fingertip tracking, gesture recognition, so that more detail elements can be added to future motion sensing applications.

3. Joint location

3.1 Posture recognition

Posture is a relationship of static position the between body parts, a posture can be uniquely determined by the relative position of joints, and the complexity of the postures directly affects the complexity of the recognition algorithm.

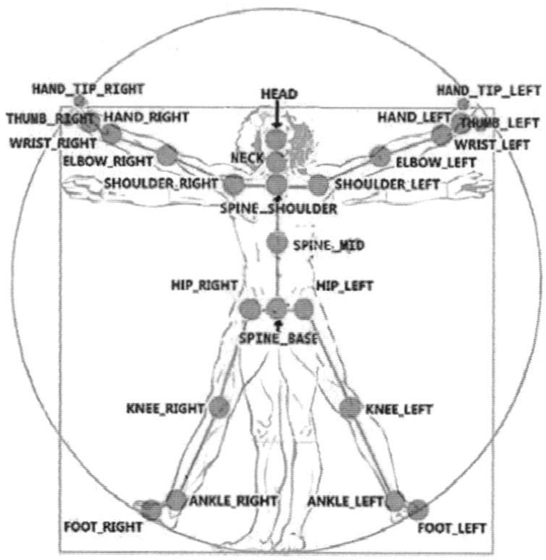

Figure 2. Distribution of joints.

3.2 Skeleton capture

Because the amount of information of the depth image that the Kinect generates is limited, the Kinect will need extra information beyond depth data to develop applications with real value. That's why the skeleton tracking technology is presented. The characteristics of skeleton recognition are the integration of anthropometry and gait identification, identify skeleton by the physical characteristics and behavioral characteristics of the human body. Specifically, the skeleton recognition system accesses the human body static information from Kinect's skeleton tracking function, then processes the depth image with methods such like matrix transformation and machine learning to get the coordinates of the joints[7]. Figure 2 shows the 25 joints that Kinect2.0 captures.

4. Posture identification

4.1 Judgement the credibility of joint's information

Because of occlusion and other causes in the process of body movement, Kinect might lose some joints or recognize the wrong information of the joints, which can cause a serious impact on the recognition of the sitting posture. It is necessary to judge the credibility of the joints and recover the information.

4.1.1 Preliminarily judge according to the length of each part

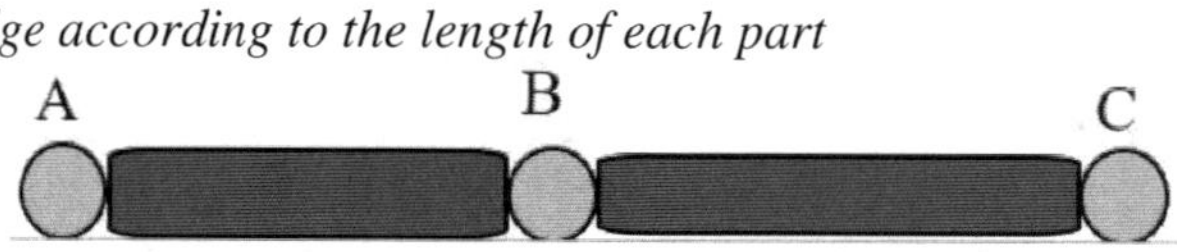

Figure 3. Schematic of arm joints.

Take the arm as an example, figure 3 shows the schematic of the human arm joints, A is the shoulder joint, B is the elbow joint, and C is the wrist joint. Suppose the length of the upper arm is L_{AB}, lower arm's length is L_{BC}, joints' information of the Kth frame are $A(x_a, y_a, z_a)$, $B(x_b, y_b, z_b)$, $C(x_c, y_c, z_c)$, and assumes that the information of A is reliable, to determine the credibility of joint B. L_{ABK} is the length of the upper arm of the Kth frame, and set a threshold τ. When $L_{ABK} \in (L_{AB} - \tau, L_{AB} + \tau)$, it is considered that the joint B is initially reliable, otherwise it is considered untrustworthy.

4.1.2 Further judge according to the continuity of movement

The movement of the joints is continuous, within a certain amount of time the velocity of the joints is limited, if the velocity beyond a certain range that the joint's information is not credible.

First assumes that the joint B moves in uniform motion in each frame, set the time interval between the two frames as t, the coordinates of joint B at the (K-1)th frame and the Kth frame are $B(x_{b,K-1}, y_{b,K-1}, z_{b,K-1})$ and $B(x_{b,K}, y_{b,K}, z_{b,K})$. Therefore, the velocity of joint B at the Kth frame is:

$$\begin{cases} V_{Bx,K} = \frac{x_{b,K}-x_{b,K-1}}{t} \\ V_{By,K} = \frac{y_{b,K}-y_{b,K-1}}{t} \\ V_{Bz,K} = \frac{z_{b,K}-z_{b,K-1}}{t} \end{cases} \quad (1)$$

Similarly, according to the formula can evaluate the velocity of joint B in each direction at the (K-1)th frame, set the threshold of velocity in each direction as (v_x, v_y, v_z). According to a few frames before, estimate the velocity of B at the Kth frame is $(V'_{Bx,K}, V'_{By,K}, V'_{Bz,K})$. And the real-time velocity of joint B at the Kth frame of should meet the formula 2, otherwise it is considered unreliable.

$$\begin{cases} V_{Bx,K} \in (V'_{Bx,K} - V_x, V'_{Bx,K} + V_x) \\ V_{By,K} \in (V'_{By,K} - V_y, V'_{By,K} + V_y) \\ V_{Bz,K} \in (V'_{Bz,K} - V_z, V'_{Bz,K} + V_z) \end{cases} \quad (2)$$

4.2 Recovery of the joint's information

4.2.1 Recovery of intermediate joint

If joint A and C is known and reliable, B is not. The arm model shows that the joint B should be on the surface of the sphere Q_A which centered at the joint with a radius of L_{AB} and the sphere Q_C which centered at the joint with a radius of L_{BC} at the same time, and the two spheres intersect at the circle S_b as shown in figure 4.

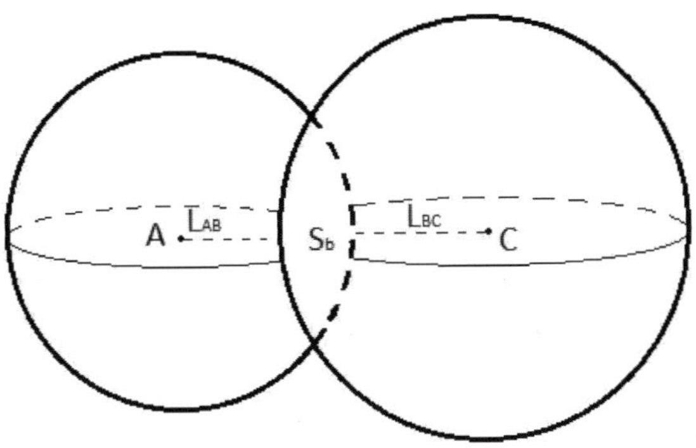

Figure 4. Two spheres intersect at S_b.

Suppose the coordinate of the joint B at the Kth frame is $P_{b,K}(x_{b,K}, y_{b,K}, z_{b,K})$. Therefore:

$$\begin{cases} L_{AB}^2 = (x_{b,K} - x_{a,K})^2 + (y_{b,K} - y_{a,K})^2 + (z_{b,K} - z_{a,K})^2 \\ L_{BC}^2 = (x_{b,K} - x_{c,K})^2 + (y_{b,K} - y_{c,K})^2 + (z_{b,K} - z_{c,K})^2 \\ (x_{b,K} - x_{b,K-1})^2 + (y_{b,K} - y_{b,K-1})^2 + (z_{b,K} - z_{b,K-1})^2 = V_{Bx,K-1}^2 + V_{By,K-1}^2 + V_{Bz,K-1}^2 \end{cases} \quad (3)$$

There may be one intersection or two intersections situations. There is only one joint B, which can exclude two intersections situation. There might be no solution because of the errors in the length measurement and human tracking.

4.2.2 Recovery of terminal joint

For the recovery of terminal joint, can be performed according to the length of L_{BC} and the velocity of the joint C. We know the average velocity of the joint C, and the speed keeps constant in a very short time, assuming that $V_{Cx} < V_{Cy} < V_{Cz}$. Selects the two smaller velocity components of the joints, and then there is:

$$\begin{cases} x_{c,K} - x_{c,K-i} = x_{c,K-1} - x_{c,K-i-1} \\ y_{c,K} - y_{c,K-i} = y_{c,K-1} - y_{c,K-i-1} \\ L_{BC}^2 = (x_{c,K} - x_{b,K})^2 + (y_{c,K} - y_{b,K})^2 + (z_{c,K} - z_{b,K})^2 \end{cases} \quad (4)$$

It is knowable from above that the angular velocity of joint C of i frames before is valued in the range of $(\alpha - \delta, \alpha + \delta)$. It is credible if the angular velocity of the recovered joint C valued within the range, if not, then select the other two velocity components V_{Cy} and V_{Cz}, if is still not credible, select V_{Cx} and V_{Cz}. As all three groups is beyond the range of threshold value, then take the group which is the most close to α.

4.3 Coordinate transformation

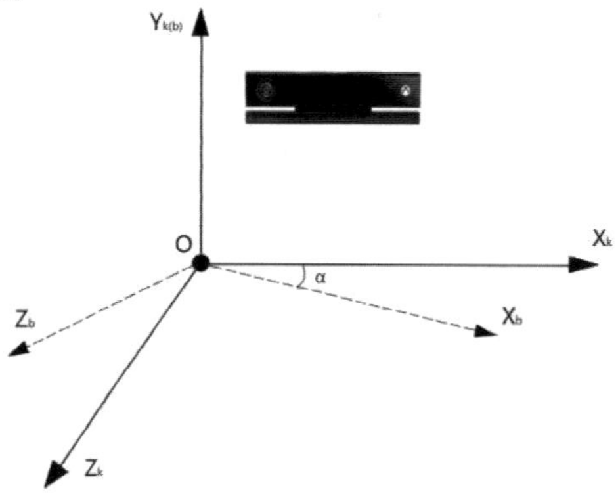

Figure 5. Coordinate systems of body and Kinect.

Because of the need for work, users can't always directly face the sitting posture surveillance system. When the user's body rotate a small angle(AG_α) around the axis that is perpendicular to the horizontal plane, as shown in figure 5, the coordinate system with subscript k represents the coordinate system of Kinect, and the coordinate system with subscript b represents the coordinate system of the user's body. When a forward leaning motion occurs, the coordinates of shoulders and neck would lean to the negative direction of Z_b axis relative to the coordinates of spine or hip, the projection would not only on the negative direction of Z_k axis as when user is facing the device, but also create a component in the X_k direction, and would be judged as leaning to the right. Similarly, the left leaning motion would create a forward leaning component, and the right leaning motion would create a backward leaning component, so directly use coordinate system of Kinect can't recognize the postures accurately. Therefore, coordinate transformation is required before posture recognition, transform the joints' location data from Kinect coordinate system to the body coordinate system.

In most cases, the shoulders maintain a same altitude in the sitting posture. Due to the relative stable location relationship between the user's shoulders. Calculate the angle AG_α between the connection line of the Shoulder_Left joint and the Shoulder_Right joint and the line that is parallel to the X_k direction. As AG_α is known, the coordinate transformation is to transform the coordinates from Kinect coordinate system to the body coordinate system by anticlockwise rotating the coordinate system X_kOY_k around the Y_k axis by AG_α degrees, and the coordinate transformation formula is shown in formula 5.

$$\begin{bmatrix} x_b \\ y_b \\ z_b \end{bmatrix} = \begin{bmatrix} cos\alpha & 0 & -sin\alpha \\ 0 & 1 & 0 \\ sin\alpha & 0 & cos\alpha \end{bmatrix} \begin{bmatrix} x_k \\ y_k \\ z_k \end{bmatrix} \quad (5)$$

4.4 Evaluation of tilt angle

In correct sitting posture, the cervical spine and lumbar vertebrae should be straight and perpendicular to the horizontal plane. When the cervical spine and lumbar vertebra are not in the same line or not perpendicular to the horizontal plane, the user is leaning to the left, right, forward or doing other bad posture. Therefore, the core of judging the sitting posture is to evaluate the tilt angle of the cervical vertebra and lumbar vertebra. Since the system is generally placed in a relatively close area and is mainly used to monitor the posture and movement in sitting, it is generally only need to detect the 17 joints include Spine_Base joint and above. For easy detection, the nonlinearity of the curvature of the spinal is ignored.

Using the law of cosines, we use the coordinates of the Head joint, Neck joint and Spine_Shoulder joint to calculate the tilt angle of cervical vertebra (AG_{CV}), and calculate the tilt angle of lumbar vertebra (AG_{LV}) with the coordinates of the Spine_Shoulder joint, Spine_Mid joint and Spine_Base joint. Because we only need to evaluate the side tilt angle and the front tilt angle, the joints' coordinates are projected onto the X_bOY_b plane and the Y_bOZ_b plane.

To evaluate the front tilt angle of cervical vertebra (AG_{fCV}) in the plane of Y_bOZ_b, we calculate the distance from Head joint to Neck joint, Neck joint to Spine_Shoulder joint, and Head joint to Spine_Shoulder joint first, as the formula 6. Then evaluate the AG_{fCV}.

$$\begin{cases} D_{fHN} = \sqrt{(y_H - y_N)^2 + (Z_H - Z_N)^2} \\ D_{fNSpSh} = \sqrt{\left(y_N - y_{SpSh}\right)^2 + \left(Z_N - Z_{SpSh}\right)^2} \\ D_{fHSpSh} = \sqrt{\left(y_H - y_{SpSh}\right)^2 + \left(Z_H - Z_{SpSh}\right)^2} \end{cases} \quad (6)$$

$$AG_{fCV} = \frac{\arccos(D_{fHN}^2 + D_{fNSpSh}^2 - D_{fHSpSh}^2)}{2D_{fHN}D_{fNSpSh}}$$

Similarly, the front tilt angle of lumbar vertebra (AG_{fLV}), the side tilt angle of cervical vertebra (AG_{sCV}) and the side tilt angle of lumbar vertebra (AG_{sLV}) can also be evaluated.

Because the side tilt is divided into left-leaning and right-leaning, and the cosine law used in this system is scalar calculation, it can't get the direction of the side tilt angle directly. Therefore, it is necessary to add a direction judgment module, which adopts the most direct method of judgment here. If the cervical spine and spine are on the left side of the vertical line, it is left-leaning, whereas the other side is right-leaning.

5. Implement of the system

The sitting posture surveillance system is based on BodyBasics-D2D of Kinect for Windows SDK, using C++ language in the Visual Studio development environment. Acquisition and processing the 17 upper body joints data, identifying the user's sitting posture in real time according to the above-mentioned evaluation method of the tilt angle of the cervical vertebra and lumbar vertebra, and report the bad posture in real time. The interface of the system as shown in figure 6, with the user's skeleton information shown on the left side and the evaluation value of tilt angle of the cervical vertebra and lumbar vertebra be shown on the right in real time. When the user is sitting in the right posture, the sitting posture identification displays "Perfect sitting posture!" When the body is too forward, that is, any AG_{fCV} or AG_{fLV} is too large, use 15 degrees as the determining threshold here, sitting posture identification displays "You are leaning forward, please adjust sitting posture in time!" Similarly, when the body is too left or right, the sitting posture identification displays "You are leaning to the left, please adjust sitting posture in time!" and "You are leaning to the right, please adjust sitting posture in time!"

Due to contingency of the occurrence of bad sitting posture, such as the user is adjusting sitting position or get up to get items, or error occurs in joint data, the system would give the wrong posture remind and record wrong statistics. And it can seriously disturb the proper work of the user if the warning of incorrect sitting posture is too often, so on the basis of the above sitting posture surveillance system, add the time statistics module for each sitting posture. With the module, the system reports the error posture if and only if a certain posture's duration exceeds a certain limit (here set to 5 seconds), if the posture is corrected during the time, the timer will be reset. When and only

IOP Conf. Series: Journal of Physics: Conf. Series **1026** (2018) 012022 doi:10.1088/1742-6596/1026/1/012022

when next error sitting posture occurs, the timer starts again. And in the meantime, there will be no any prompt information which greatly enhance the practicability of the system.

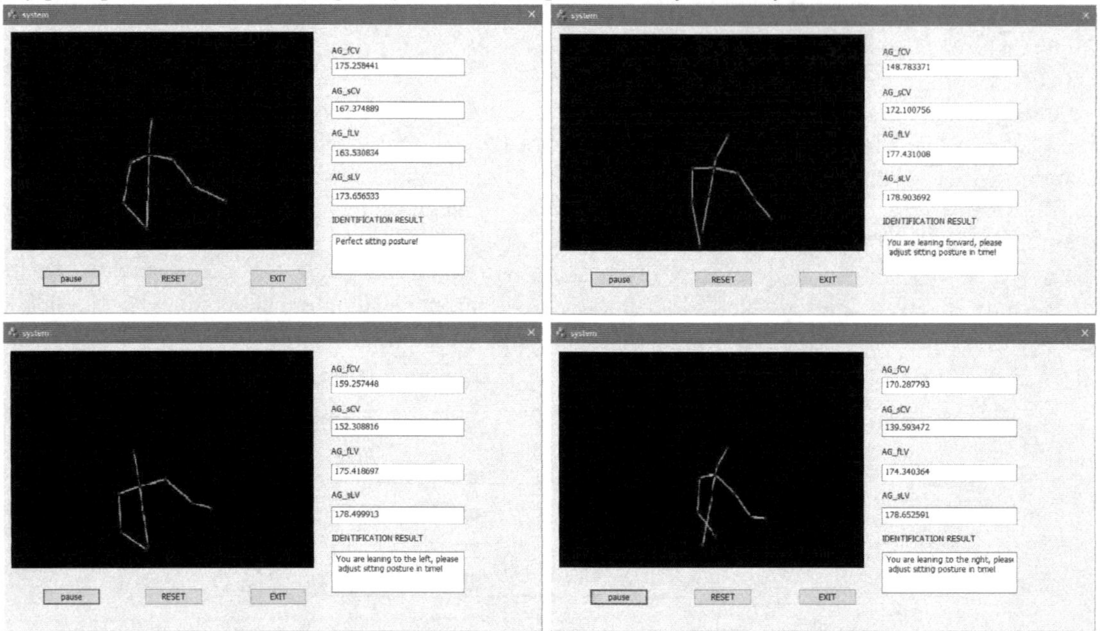

Figure 6. Interface of the system.

6. Conclusion

In this paper, the sitting posture surveillance system based on Microsoft Kinect motion-sensing development has successfully implement the real-time surveillance of the sitting posture of the user, with high accuracy of recognition. The system is only a combination of a Kinect device and a computer, with the advantages of low cost and high practicability. Although in most cases, Kinect has high recognition rate, and can get the joints' coordinates accurately. For the joints covered by others parts of the body, the result of recognition is not satisfactory. The research in restoring the coordinates of the covered joints can further improve the accuracy rate of body recognition.

7. References

[1] Jones-Jordan L A, Mitchell G L, Cotter S A, etc. 2011 Visual activity before and after the onset of juvenile myopia *Investigative Ophthalmology & Visual Science* **52(3)** 1841-1850.

[2] Jaimes A 2005 Sit straight (and tell me what I did today): a human posture alarm and activity summarization system In *Proceedings of the 2nd ACM workshop on Continuous archival and retrieval of personal experiences* 23-34.

[3] Wu L, Li K, Wu C 2010 A sitting posture surveillance system based on image and signal processing technology in *2010 2nd International Conference on Computer Engineering and Technology (ICCET)*, IEEE, 692-695.

[4] Kar A, Amitabha A, Guha M P 2010 Skeletal Tracking using Microsoft Kinect *Methodology* 1-11.

[5] Han J, Shao L, Xu D, et al. 2013 Enhanced computer vision with Microsoft Kinect sensor: a review *IEEE Transactions on Cybernetics* **43(5)** 13-18.

[6] Boulos M N, Blanchard B J, Walker C, et al. 2011 Web GIS in practice X: a Microsoft Kinect natural user interface for Google Earth navigation *International Journal of Health Geographics*, **10(1)** 45.

[7] Webb J, Ashley J 2012 Beginning Kinect Programming with the Microsoft Kinect SDK: Apress.

IOP Conf. Series: Journal of Physics: Conf. Series **1026** (2018) 012023 doi:10.1088/1742-6596/1026/1/012023

A Chinese Chessboard Calibration Method in Chess-Playing Robot by Machine Vision Sensing

Jian Jiao[1, 2], Changcun Huang[1, 2],Hongliang Liu [1, 2], Guo Zhang [1, 2]

[1]Chongqing Creation Vocational College, No. 28 west section of Yong Chuan Chang Zhou road, Chongqing, China
[2]Chongqing Engineering Research Center of Industrial Robots Application Service, Chongqing, China

Abstract. In view of Chessboard calibration in Chess-Playing Robot by machine vision sensing, a new Chinese chessboard calibration method based on external aided planar pattern is given. Introducing the advantages of the international chessboard calibration technology, which can overcome the difficulties of lens distortion imaging and non-orthographic projective imaging in which cases the former calibration methods can't be used. This method aligns the planar calibration pattern to the chessboard in such a way that we can get the chessboard grid points' Image-Coordinate-System coordinates by detecting the planar pattern's corners and obtain the mapping of a group of points' s World-Coordinate-System coordinates and Image-Coordinate-System coordinates, which is easy to operate and robust to noises. Workpiece location by machine vision methods in industrial application also faces the problems of lens distortion imaging and non-orthographic projective imaging, so it would be beneficial to introduce this method's idea into industrial areas.

1. Introduction

Chess-playing robot (CPR) based on machine vision [1] uses camera to obtain the image of chessboard and chess piece, furthermore, the game algorithm of artificial intelligence is further analyzed to the position of situation on the chess board, and determine the position of piece placement. The machine vision target positioning, robot control and other core technologies can be widely used in industry, information and other industries.

Machine vision as a link in CPR which has two core technologies: chessboard calibration and chess pieces position. Chess pieces position is to identify the position of the pieces in the image by machine vision algorithm. Chessboard calibration refers to finding the match between the actual position of the chessboard and the image position. With this correspondence, after finding the position of the chess pieces in the image, the position of the chess pieces in the actual chessboard can be known.

Chinese chessboard includes go chessboard, Chinese chessboard, gobang chessboard, et al., all of which are vertical and horizontal crossing, middle hollow square lattice structure. The Chinese chess board consists of 9*8 square lattices, except for the diagonals of individual squares, the rest are hollow. The go chessboard is composed of 18*18 hollow lattices. In order to calibrate it, the mark point interpolation algorithm and the line detection intersection algorithm are usually used. The mark point interpolation algorithm is easy to implement, but it's difficult to overcome the camera imaging distortion and the positioning offset caused by the non orthographic projection of the image. It's also difficult to solve the problem of camera distortion by finding intersection point by line detection. Distortion correction has been reported in the literature, but the optical principle of distortion should be modeling compensation, at the same time, the compensation precision needs to be considered according to the computational complexity and the compensation parameters are nonlinear variables in the calibration model, this's difficult to obtain the global optimal solution. If correction of nonorthographic projection is used, the relationship of conversion relation between camera coordinate system and world coordinate system should be calibrated in advance, moreover, if the lens distortion is considered in the calibration process, more complex nonlinear optimization is needed.

This paper focuses on a new calibration method for Chinese chessboard. By mechanical positioning for the target, which avoids the computational complexity and compensation approximation of optical distortion modeling compensation, it also avoids the nonlinear optimization problem in the calibration process with lens distortion. Provided that the mechanical positioning reaches a certain precision, the problem of lens distortion and non orthographic projection can be solved, it's easy to implement and easy to operate.

2. Chessboard Calibration

In machine vision chess robot system, the purpose of chessboard calibration is to find out the correspondence between the actual position and the image position of the chessboard, that is to find the pixel coordinates of each position of the chessboard in the image. Suppose that the world coordinate system of the spatial location where the chessboard is located is (x, y, z) ,the world coordinate of the chessboard points are (x_{ij}, y_{ij}, z_{ij}); Suppose that the image coordinate system is (u, v) after Chessboard imaging, the coordinates of the checkerboard points in the image are (u_{ij}, v_{ij}) , $i = 1,2,3,\cdots M$, $j = 1,2,3,\cdots N$. The task of chessboard calibration can be transformed into the mapping relation between the world coordinate system and the image coordinate system of chessboard grid points, that is:

$$f : (x_{ij}, y_{ij}, z_{ij}) \rightarrow (u_{ij}, v_{ij}) \tag{1}$$

As long as the world coordinate of the grid points and the corresponding image coordinates are obtained, and the mapping relation of formula (1) is obtained, the chessboard calibration task is completed. (x_{ij}, y_{ij}, z_{ij}) can be obtained by locating the mechanical origin and calculating the offset after the establishment of the world coordinate system, the corresponding image coordinates (u_{ij}, v_{ij}) needs to be extracted by machine vision algorithm. Because the chessboard grid size is fixed, so as long as access to the world coordinate chessboard corner point, according to the two-dimensional array subscript position chessboard lattice point, we can calculate the coordinates of each lattice point of the world coordinate. This paper mainly discusses the acquisition method of grid point image coordinate (u_{ij}, v_{ij}) based on machine vision.

3. Calibration Analysis

3.1. Calibration Method

Point interpolation algorithm is in the four corners of the board set markers, and detect the image coordinates of the mark points, once the detected image coordinates of the four corners of the Chess board, if the imaging process is a positive projection, the grid points inside the chessboard can be calculated by the uniform linear interpolation algorithm. (u_{ij}, v_{ij}) is represented by I_{ij} here, the image coordinates of four corners (u_{11}, v_{11}) , (u_{1N}, v_{1N}) , (u_{M1}, v_{M1}) , (u_{MN}, v_{MN}) are represented by C_1 , C_2 , C_3 , C_4 Separately. Corresponding to four red star markers in clockwise direction, the grid point image

IOP Conf. Series: Journal of Physics: Conf. Series **1026** (2018) 012023 doi:10.1088/1742-6596/1026/1/012023

coordinate which lie the first column on the left is evenly interpolated as:

$$I_{i1} = C_1 + \alpha_1(i-1) \tag{2}$$

$$\alpha_1 = \frac{C_3 - C_1}{M}$$

The grid point image coordinate which lie the last column on the right is evenly interpolated as:

$$I_{iN} = C_2 + \alpha_N(i-1) \tag{3}$$

$$\alpha_N = \frac{C_4 - C_2}{M}$$

By the evenly interpolated of the first column lattice coordinate I_{i1} and the last column lattice coordinates I_{iN}, the coordinates of the grid points for each row can be obtained as:

$$I_{ij} = I_{i1} + \beta_i(j-1) \tag{4}$$

$$\beta_1 = \frac{I_{iN} - I_{i1}}{N}$$

Formula(2) to (4) represent the whole process of uniform linear interpolation algorithm for marker points.

However, lens distortion is a common phenomenon in the imaging process. Especially for barrel distortion of wide-angle lens[2], If it isn't corrected, it will not be used in precision measurement or camera calibration, which also affects the chessboard calibration technique based on marker points.

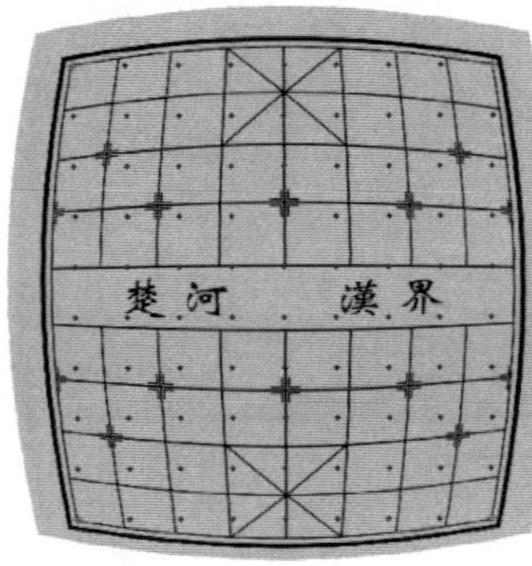

Figure 1. Interpolation effect of marker points in barrel distortion image.

As shown in figure 1, a Chinese chessboard with a barrel distortion is formed, and the straight line becomes a curved curve. The red points on the four corners are marked points, the linear interpolation points are blue dots, and the blue dots are not matched with the grid points (the intersection points of the straight lines).

If the imaging process isn't a positive projection, that means the optical axis of the camera isn't perpendicular to the chessboard plane, it will also cause the projective mapping of the change of the length ratio of the line segment. At this point, the chessboard of the image is no longer evenly divided into the line segments, and the mark point method can not work, as shown in figure 2.

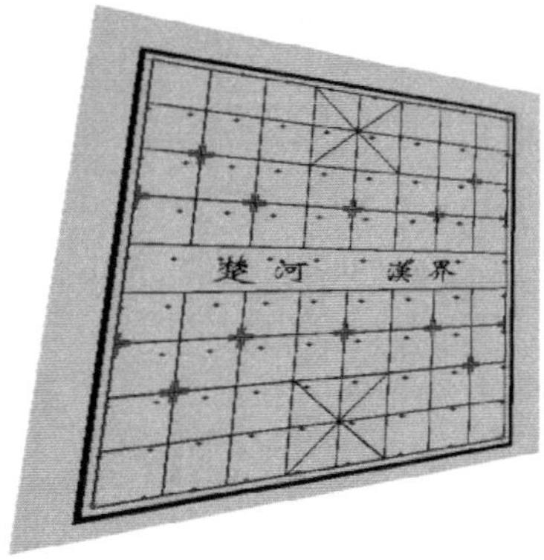

Figure 2. Interpolation effect of non orthographic image marker points.

Another Chinese chessboard calibration technology is based on line detection algorithm for intersection point, it can't be applied in the case of barrel distortion similar to Figure 1.

As shown in Figure 3, there is an image without distortion which requires to detect the boundary line region of middle green, but the boundary line becomes a curve in the case of barrel distortion. Figure4 is a straight line detected by Hough transform; it can be seen that the detected straight line has seriously deviated from the image that should have a "straight line" position, it's impossible to complete the chessboard calibration task in this case.

Figure 3. Edge of image without distortion.

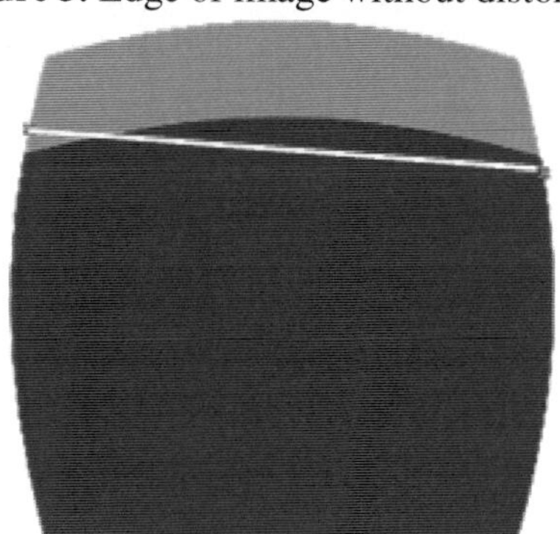

Figure 4. Straight line detection result of distorted edge.

Although the international chessboard is made up of 8*8 squares, but the squares are filled with solid black and white or other colors.

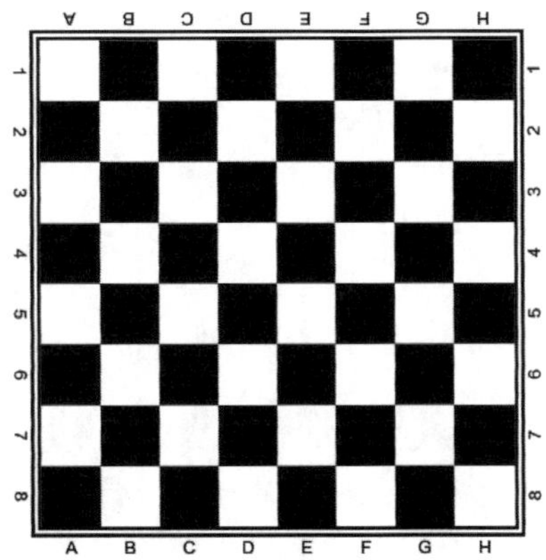

Figure 5. International chessboard.

As shown in Figure 5, because the square lattice is solid, the corner feature is obvious, therefore, the calibration method based on corner detection is usually used.

Corner detection such as the most classic Harris method, the algorithm based on local region gray information is used to detect the property of the change surface of the gray value of the pixel in a local region; if the point corresponds to the minimum value of the change surface of gray value, the movement of any direction will lead to the increase of the gray change value, and then the point is a corner. If (u,v) is the coordinate position of one pixel in the local region of an image, $I(u,v)$ is the pixel gray value at the (u,v), the tiny movement of each coordinate position changes is $\Delta x, \Delta y$, when the position of each position in the local area is moved slightly, the sum of the square of the gray value of the pixel is equal to the sum of the square of the change of the gray value of the pixel:

$$E(\Delta x,\Delta y)=\sum_u\sum_v w(u,v)(I(u+\Delta x,v+\Delta y)-I(u,v))^2 \tag{5}$$

Where $w(u,v)$ is weighted average coefficient of (u,v). $I(u+\Delta x,v+\Delta y)$ is expanded by first order Taylor approximation:

$$I(u+\Delta x,v+\Delta y)\approx I(u,v)+I_x(u,v)\Delta x+I_y(u,v)\Delta y \tag{6}$$

Where $I_x(u,v)$ and $I_y(u,v)$ are the first derivative of x direction and y direction separately, then put them into formula (5):

$$E(\Delta x,\Delta y)\approx\sum_u\sum_v w(u,v)(I_x(u,v)\Delta x+I_y(u,v)\Delta y)^2 \tag{7}$$

Express in the form of a matrix:

$$E(\Delta x,\Delta y)\approx(\Delta x,\Delta y)A\begin{pmatrix}\Delta x\\ \Delta y\end{pmatrix} \tag{8}$$

$$A=\sum_u\sum_v w(u,v)\begin{bmatrix}I_x^2 & I_xI_y\\ I_xI_y & I_y^2\end{bmatrix}=\begin{bmatrix}\langle I_x^2\rangle & \langle I_xI_y\rangle\\ \langle I_xI_y\rangle & \langle I_y^2\rangle\end{bmatrix} \tag{9}$$

Where the symbol "$\langle\rangle$" stands for "mean". If the two eigenvalue of the matrix A is a larger positive number, the change of each direction will lead to the increase of gray value in this local

IOP Conf. Series: Journal of Physics: Conf. Series **1026** (2018) 012023 doi:10.1088/1742-6596/1026/1/012023

region, there is a corner point in the local region.

Therefore, corner detection only change information associated with the local gray value of the image, the image distortion will not affect local properties, so the corner detection algorithm has the capability of anti lens distortion. Because the international chessboard has black and white gray change obviously, the corner feature information rich features, robustness and corner detection can guarantee, therefore, the plane calibration board is also the style of international chessboard; (Zhang, Z., 2000); according to the obvious feature of corner feature of the chessboard pattern, the corner points are extracted to calculate the internal and external parameters of the camera.

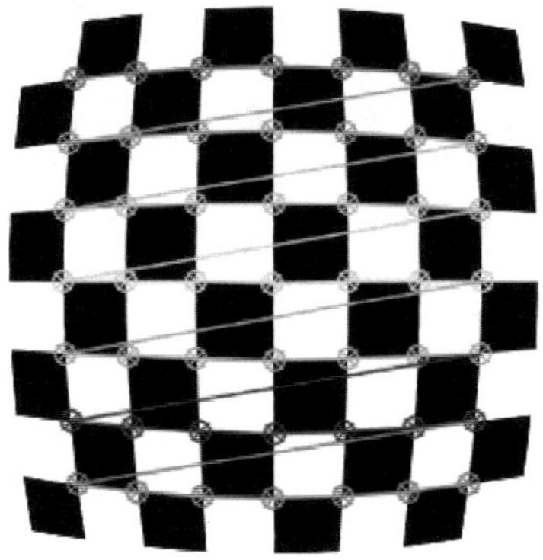

Figure 6. Bucket distortion chessboard corner extraction.

In figure 6, each corner of the chessboard has a small ball of color. The center of the circle of the small ball represents the position of the extracted corner; although the bucket distortion is serious, corner extraction algorithm still works.

3.2. New Calibration Method

A method is proposed here, the international chessboard calibration technology is applied to Chinese chessboard calibration, which will have remarkable effect to overcome the drawbacks of the existing methods.

To Chinese chess board as an example, The position of the chess pieces on the chessboard is locked by 9*8 square lattices. Make a black and white plate with 10*9 squares as the calibration board. The size of the grid is equal to the size of the squares on the chessboard, as shown in figure 7.

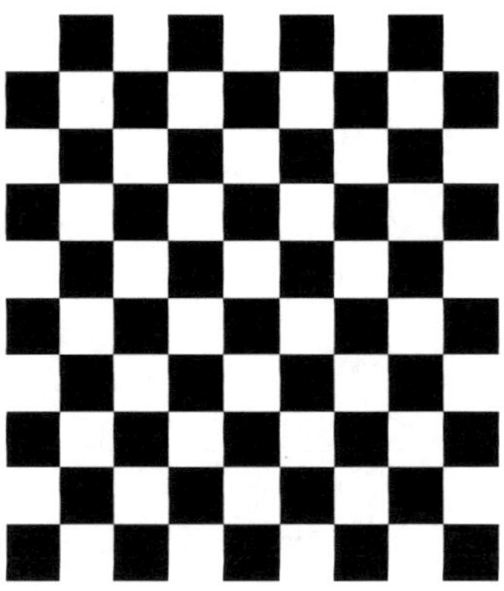

Figure 7. Chinese chess board calibration board.

Obviously, the calibration plate removes the outermost circle square lattice, which is consistent with the overall size of the chessboard size and the size of the inner square lattice is equal. Through mechanical positioning, the calibration plate removes the outermost part of the plate and aligns with the chessboard; the calibration plate removes each inner square of the outermost circle and is aligned with the square corresponding to the chessboard. After alignment, the world coordinates of each corner point of the calibration board are the world coordinates of each grid point of the chessboard:

$$(x_{ij}, y_{ij}, z_{ij}) = (x'_{ij}, y'_{ij}, z'_{ij}) \tag{10}$$

Where $(x'_{ij}, y'_{ij}, z'_{ij})$ is the world coordinates of corner calibration board. Therefore, the chessboard calibration can be achieved by getting the correspondence between the corner coordinate of the calibration board and the image coordinates.

$$f:(x'_{ij}, y'_{ij}, z'_{ij}) \rightarrow (u_{ij}, v_{ij}) \tag{11}$$

The world coordinate $(x'_{ij}, y'_{ij}, z'_{ij})$ of the corners of the calibration plate is the world coordinate (x_{ij}, y_{ij}, z_{ij}) of the chessboard grid points, which can be obtained by the calculation of the origin mechanical positioning and the two-dimensional array subscript offset calculation, because the feature of the corner image coordinate (u_{ij}, v_{ij}) of the calibration board is obvious, it can be acquired robustly. After calibration, take the index board down, you can machine vision robot chess.

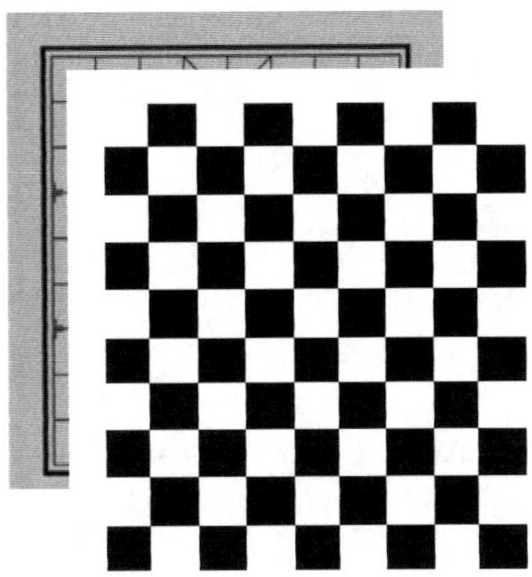

Figure 8. Alignment diagram of calibration board and chessboard.

Similarly, the chess board (18*18 square number) the plane calibration board just to create a 19*19 squares the number of black and white as the calibration plate, square lattice equal square lattice size and the chess board, you can follow the above method of the same.

The calibration board is one more square than the actual chessboard, it provides the feature information that can detect the corner points for the local area of the grid on the outermost part of the chessboard, in order to make the outermost point of the checkerboard image coordinates can also be obtained.

4. Simulation

Taking the go chessboard as an example, the calibration is carried out according to the method described above. The calibration of orthographic projection is carried out first.

Figure 9. Corner extraction effect of calibration plate.

Figure 9 is a corresponding, aligned calibration board, corner extraction using corner detection library function in OpenCV: the calibration of corner plate of each internal department has a color of small ball, corner position of small ball center representative to extract; visible extracted to 19*19 corner points successfully. The calibration board is removed to obtain the image of the chessboard, the extracted 19*19 corners are drawn into the chessboard image. If the corners are aligned with the chessboard grid points, the calibration is successful. The orthographic projection template data are compared with the calibration data, as shown in table 1. As shown in Figure 10 of the red dot, i.e. 19*19corner points plotted on the chessboard image in the corresponding and checkerboard points coincide well.

Table 1. Comparison of original projection template data and calibration data

Raw template data (pixels)		Calibration data (pixels)	
u_{ij}	v_{ij}	u_{ij}	v_{ij}
128	35	126.786	34.71629
149.444	35.0556	148.534	35.00293
170.889	35.1111	170.2682	35.48389
192.333	35.1667	191.7842	35.87283
......	......	......	......
493.389	439	492.4495	438.4564
515	439	514.306	438.8669
average error：1.3381，mean square error：0.4482			

The chessboard distortion and non orthographic chessboard image of go chessboard are calibrated respectively by using the same method.

Figure 11 is the corresponding calibration board image and corner extraction effect, all corner points can be extracted, the blue point is the effect of the uniform linear interpolation of the marker points which deviate from the checkerboard point obviously in figure 12; the red point is the result obtained by new calibration method in this paper, which is significantly better than the mark point method.

In the projection mapping of non orthographic projection, the nature of the chessboard's physical lattice length is no longer existing in the image.

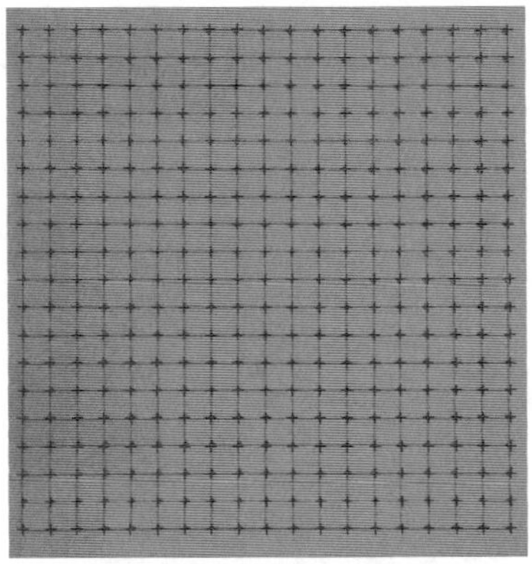

Figure 10. Calibration effect.

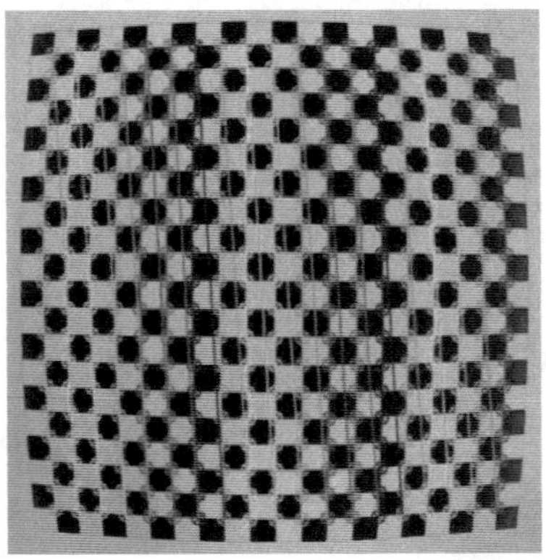

Figure 11. Corner extraction effect chart of barrel distortion calibration plate.

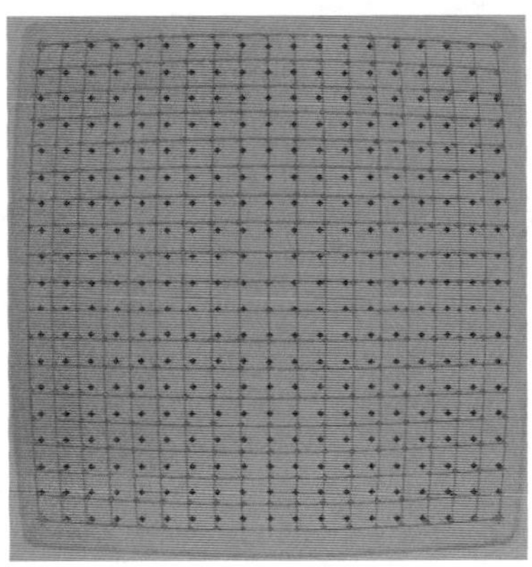

Figure 12. Comparison with mark point method.

Figure 13 is the corner extraction effect of the non orthogonal projection subscript plate, and it is not affected by the non orthographic projection. Figure 14 shows that new calibration method is better than the mark point method in the non orthographic imaging condition.

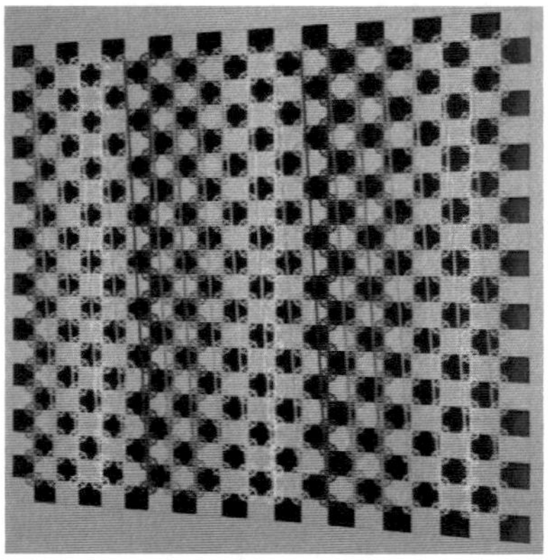

Figure 13. Corner extraction effect of non orthographic projection calibration plate.

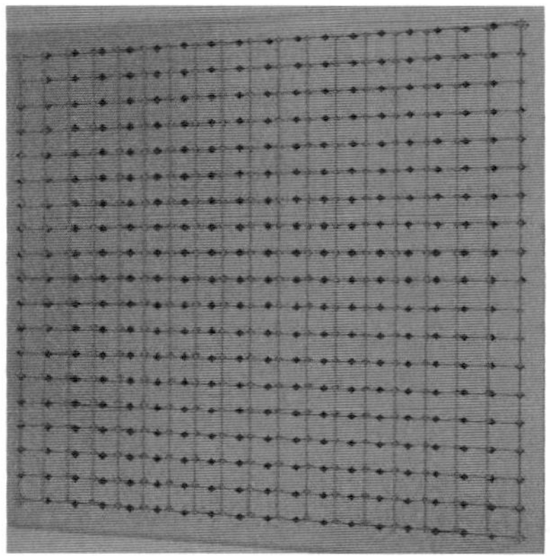

Figure 14. Compares with mark point method.

5. Conclusions

Image distortion will make straight line into curve, but non orthographic projection will change the isometric character of chessboard square, which will lead to the failure of existing Chinese chessboard calibration technology. The corner detection method is usually used in international chessboard calibration, however, the Chinese chessboard can not be directly applied to the Chinese chessboard because there are only straight lines intersecting vertically and horizontally, and there are not enough obvious corner features. In this paper, a new calibration method is proposed, which is based on the classical method of camera calibration, and relies on the calibration object, the calibration technique of the international chessboard is introduced into the Chinese chessboard calibration. The experimental results show that the method can resist the interference caused by image distortion and non orthographic projection, and overcome the disadvantages of the existing calibration techniques.

Machine vision positioning technology is widely used in machine vision[3] positioning technology in the industrial field, such as circuit board solder joint precision positioning[4], mobile phone glass cutting precision positioning[5].The main advantage of this method is that it can eliminate the influence of image distortion and non orthographic projection, therefore, if the idea and idea are introduced into the machine vision positioning in the industrial field, it will be a valuable research direction.

6. References

[1] Yuandong Tian 2016 A simple Analysis of AhplaGo *J. ActaAutomaticaSinica.* **5** 671-675
[2] Zhanbin Wang 2008 Bbarrel distortion correction of wide-angle lens based on spline function *J. Opto-Electronic Engineering.* **4** 140-144

IOP Conf. Series: Journal of Physics: Conf. Series **1026** (2018) 012023 doi:10.1088/1742-6596/1026/1/012023

[3] Malamas E. N., Petrakis E. G. M. and Zervakis, M. 2003. A survey on industrial vision systems, applications and tools *J. Image and Vision Computing.* **2** 171-188

[4] Jun Zhou, Shuo Xu and Dawei Tu 2016 Machine Vision System Based Precise Positioning for PCB Solder Joints *J. Machine Building & Automation.* **3** 172-175

[5] Bocai Huang, Shaojun Liu and Daoping Huang 2012 Design and realization of the mark point visual positioning system on the CNC glass carving machine *J. Light Industry Machinery.***4** 61-64

Application Study of Fuzzy ARTMAP Neural Network in Robot Servo System

Guoxuan Chi[1]

[1]Beijing University of Posts and Telecommunications, Beijing, China

Email: chiguoxuan@gmail.com

Abstract. Robot vision servo control is one of the hotspots in robot research field. However, it is difficult to establish an accurate mathematical model for robot vision servo system. In this paper, a robot visual servo system based on Fuzzy ARTMAP is proposed by using computer vision technology as well as fuzzy theory. After creating a mathmatic model structer, we present a simulated experiment to verify the performance of this system, which shows the superiority of Fuzzy ARTMAP by comparing it with other traditional visual servo system.

1. Introduction

Robot visual servo is a technique which can indirectly detect the robot's current pose or its relative pose on the target body by using visual sensors and related algorithms. On this basis, robot localization control or trajectory tracking can be realized[1]. It is a new interdisciplinary subject involving neural network, automatic control, real-time system analysis and so on[2]. In recent years, with the rapid development of image processing, pattern recognition and other related fields, the information contained in images has been mined and applied more. The accuracy and reliability of visual servo are improved, which enhances the learning ability of the robot to the surrounding environment and enables it to make intelligent decisions according to the understanding of the environment and accomplish the assigned tasks. But robot vision servo system is a very complicated nonlinear system. First, the robot system itself is a highly nonlinear, strong coupling, time-varying dynamic system, and its control is complex. Because of the inaccuracy of measurement and modeling, the change of load and the influence of external disturbances, the accurate and complete kinematics model of robot can hardly be obtained[3]. In addition, the vision is also a complex sensory. High-complexity algorithms are usually needed to extract the feature information from a large amount of visual data. The video capture equipment such as video camera will also be affected by many factors. Low accuracy, incompatible light intensity and transmission noise make the visual information contain noise inevitably, which increases the difficulty of image processing. At present, the average camera sampling rate is not high. Besides, a plenty of time will be needed to transmit large amounts of visual data. Furthermore, since most image processing algorithm is non- linear, how to make the complex visual servo system achieve the desired target efficiently depends largely on the selected control strategy.

2. Discussion on Fuzzy ARTMAP

Adaptive Resonance Theory. The adaptive resonance theory network was proposed by Professor Carpenter of Boston University in the United States. Subsequently, Carpenter and his student, Grasberg S., put forward the ART neural network. As a self-organizing clustering algorithm that mimics the cognitive process of human brains, ART has its unique advantages in solving the increasingly complex clustering and classification of large amounts of data in modern times. Its applications include classifying and judging the states of complex systems, implementing data mining and knowledge discovery[4]. ARTMAP can complete real-time learning and adapt to non-stationary environment. The object designed by ART has a stable and fast recognition ability, and it can quickly adapt to the new object which does not learn. The object also has the capability of self-normalization, sometimes as a key feature, sometimes as a noise, depending on the proportion of certain features in the whole. It does not need to know the sample results in advance since it is unsupervised learning[5]. ART's starting point is to draw on people's cognitive processes and the characteristics of brain work. People can categorize and recognize things in a complex, non-stationary, and disturbed environment. This learning is often self-organizing and the accumulation and storage of knowledge are both rigid and flexible, that is, on the one hand, they can firmly preserve what they have learned, and on the other hand, they can learn a lot of new knowledge. The input data and the top-down learning expectation simultaneously perform input to the competition collaboration network, while bottom-up learning is the output of the competition collaboration network. At the same time, bottom-up learning can also be the input of the competition collaboration network while the top-down learning expectation is the output. Real perception can be generated through the studying and matching of this competition collaborative network.

Fuzzy Theory. Fuzzy theory is a mathematical theory, being established by studying and summarizing all kinds of intelligent activities in which human beings use language as media. It is based on fuzzy sets, whose basic spirit is to accept the existence of the phenomenon of ambiguity of facts and to deal with concepts of fuzzy uncertainty. Based on this theory, converting natural language statements or body motions, which are mostly fuzzy, into rigorous quantification data can be easier. Thus, the computer can handle these messages. The membership function was first proposed by professor L.A. Zadeh to express the fuzziness in a complicated system. With the help of the membership function, a fuzzy concept can be expressed as a quantitative representation. The matching of fuzzy concepts refers to the similarity between two fuzzy concepts, and the similarity degree of the two fuzzy concepts is also called the matching degree. There are two basic methods for calculating the matching degree, semantic distance and proximity. When the matching ratio is greater than a given threshold, the matching of the two fuzzy concepts is considered. The semantic distance can also be used to determine whether the two fuzzy concepts match. Generally, a decision could be made by checking if the semantic distance is less than a given threshold. In most cases, shorter distance means more similarities.

Fuzzy ARTMAP Neural Network. Fuzzy ARTMAP neural network combines fuzzy reasoning and self-resonance theory. It has both strong on-line adaptive ability and good class resolution, and it has good operability. By integrating fuzzy theory into ARTMAP dynamical systems, we can realize arbitrary mapping between arbitrary input and output space, so it is suitable for mapping and classification in high dimensional space. It achieves a new minimum maximum learning rule by matching pursuit process. That is, minimizing prediction error and maximizing coding compression rate. The two modules can be divided into the attention subsystem and the orientation subsystem. The attention subsystem for dealing with the mode of learning, fine-tuning will establish accurate internal value to the already familiar pattern representation; orientation subsystem is used for processing the emergence of new mode when there are not familiar with the incident, it pays attention to callback system and set up internal encoding new inside to represent unfamiliar events. Fuzzy ARTMAP is based on the form similarity between the fuzzy subset calculation and the ART neural network class selection method, and combines the advantages of ART and fuzzy logic. The Fuzzy ARTMAP neural network is composed of two Fuzzy ART models. They are ARTa and ARTb, which are connected by Fab. The Fuzzy ARTMAP neural network logic is shown in Figure 1.

3. Robot Servo System Based on Fuzzy ARTMAP Neural Network

Camera Imaging. As the eye of a robot, the camera realizes the transformation from the scene to the image, and realizes the conversion from the three-dimensional scene space to the two-dimensional image space. The camera is fixed on the machine at the end of the arm, the camera coordinate system relative to the manipulator coordinate geometry is fixed, but relative to the working space of the base coordinate system is changed, and this model can accurately see the local scene. The axis coincides with the optical axis of the camera. The robot takes the image of the target ball through the camera and extracts its features to form a closed loop. Fuzzy vision controller is based on the error of image features. According to the rule, fuzzy control obtains the expected change rate of several key joint positions of the robot. Robot visual servo control system is a multi-input multi output nonlinear system dynamics, it has the characteristics of time-varying, strong coupling and nonlinear, it is difficult to find the exact mathematical model, using the traditional control method based on object model is difficult to carry out effective control. In addition, the robot vision servo control system also has serious uncertainty. This uncertainty can be divided into two categories. One is the uncertainty caused by the calibration error of camera parameters. The other is the uncertainty of the dynamic characteristics of the robot itself due to load, friction and other factors. The robot vision servo control method must have strong robustness to these uncertainties. Fuzzy control is mainly based on the operating experience of operators instead of the mathematical model of the object so it is not sensitive to the change of the parameters of the controlled object and can effectively overcome the nonlinear, time-varying, coupling and other factors.

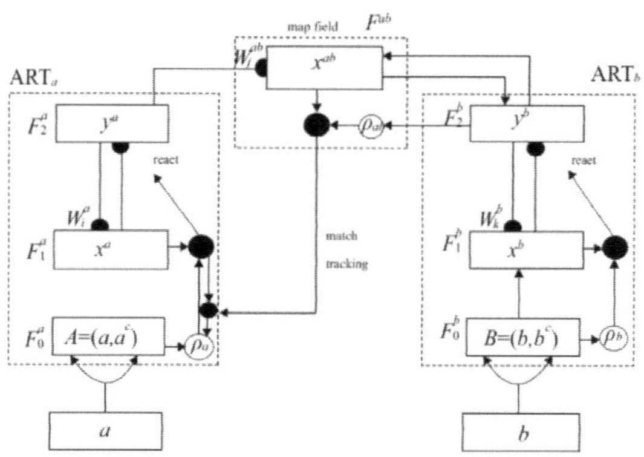

Figure 1. Fuzzy ARTMAP neural network structure.

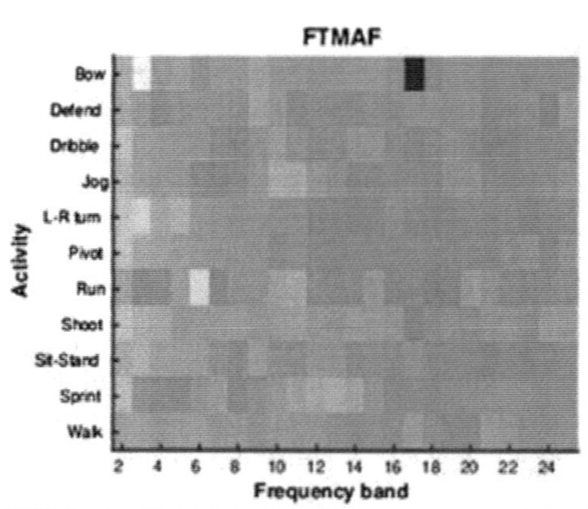

Figure 2. FTMAF method.

Up to now, many scholars have proposed different methods motion feature analysis and recognition which apply to different context. In this visual servo system, we use the FTMAF method of [1] to extract the frequency characteristics (i.e. Inter-frame features) of different motions by the correlation processing of the transform domain, which is showed in Fig 2. In addition, we also combine the position and Chroma of pixels with their neighborhood features (i.e. Intra-frame features) for the correlation analysis, and construct three-dimensional vector set I as the input data set for Fuzzy ARTMAP.

$$I = \{i_1, i_2, i_3, \ldots\}, \ \textit{in which} \ \ i_n = \{freq_n, pos_n, chroma_n\} \tag{1}$$

Design of Fuzzy Vision Controller. The fuzzy controller to the visual sphere in the image plane of the image features as the input language variable of robot visual servo system, the position of each joint manipulator is expected to change rate as the output signal corresponding to the output variables. When the desired value of the target image feature is expressed, the image characteristic value indicates that the robot hand reaches the target position. This paper sets the expected eigenvalue of the target. When the target center reaches the center of the imaging plane of the camera and the imaging radius of the target is small ball radius, it indicates that the robot end gripper reaches the desired position. The rule is based on the defined input and output language variables, and all possible state controllers should consider and indicate each of the best actions. The system is now analyzed, the first joint movement in the imaging plane will change the value of x and r, second and third joint movement will change the radius of R and Y values.

Table 1. Table of fuzzy control rule

	F-	QF-	N-	VN	N+	QN+	F+
F+	B+	VB+	VS+	Z	VS-	VB-	B-
QN+	B+	VB+	S+	Z	S-	VB-	B-
N+	VB+	VB+	S+	Z	S-	VS-	VB-
VN	VB+	VS+	S+	Z	S-	VS-	VB-
N-	VB+	VS+	S+	Z	S-	VS-	VB-
QN-	VB+	VS+	S+	Z	S-	VS-	VB-

Thus, the fuzzy vision controller determines J based on the values of x and r. Therefore, we can combine the values of variables x and r to construct a two-dimensional vector set O as the output data set of Fuzzy ARTMAP. After a series of simple linear operations based on fuzzy theory, the output vector can directly control the robots precisely.

$$O = \{o_1, o_2, o_3, \ldots\}, \ \ \textit{in which} \ \ o_n = \{x_n, r_n\} \tag{2}$$

In this paper, we focus on the design and processing flow in Fuzzy ARTMAP instead of mathematical analysis of fuzzy theory. To design a proper Fuzzy ARTMAP, we need to set some key parameters first.

$$\begin{cases} \alpha \in (0, +\infty) \\ \beta \in [0,1] \\ \rho \in [0,1] \end{cases} \quad \textbf{and} \quad \begin{cases} \sum_J \mathbf{w}_J = 1 \\ \mathbf{w}_J \in [0,1] \end{cases} \tag{3}$$

Note that the first parameter α is determined by the dynamics of the ARTMAP, β and ρ is the learning rate and threshold of the network respectively. Parameter wJ is the weight coefficient for the J-th pattern. In order to select the most matching pattern vector in the second layer of the ART network, we give the selection function as follows.

$$T_d = \frac{\sum_{d=1}^{3} min(i_d, w_d)}{(\alpha + \sum_{d=1}^{3} w_d)} \tag{4}$$

When the matching degree between input vector and every existing patterns are all less than the threshold, the system will add a new pattern vector. This mechanism is not only the basis for ensuring the incremental learning of neural networks, but also the foundation of system self-adaptability. If the system determines that the input vector matches a certain pattern vector, the weight of the pattern is

IOP Conf. Series: Journal of Physics: Conf. Series **1026** (2018) 012024 doi:10.1088/1742-6596/1026/1/012024

updated as follows, making the input vector more compatible with the pattern vector.

$$w_J^{new} = \beta \cdot \begin{pmatrix} min(i_J^1, w_J^1) \\ min(i_J^2, w_J^2) \\ min(i_J^3, w_J^3) \end{pmatrix} + (1-\beta) \cdot w_J^{old} \tag{5}$$

If there is a case that a classification is activated in ARTa or ARTb, the mapping domain Fab will be activated. The role of Fab is to contact ARTa and ARTb to perform match tracking. Let event A be "the J-th node F2a being activated", while the event B "the J-th node F2b being activated". Then, the output vector xab can be calculated by using the following formula.

$$x_{ab} = \begin{cases} min\left(y^b, w_J^{ab}\right) & if\ A \wedge B \\ w_J^{ab} & if\ A \wedge \neg B \\ y^b & if\ \neg A \wedge B \\ 0 & if\ \neg A \wedge \neg B \end{cases} \tag{6}$$

Simulation Experiments. The experimental simulation results in Fig 3. shows that the visual servo system based on Fuzzy ARTMAP has better performance than ordinary design when target modes are about 10 (set randomly as [10,20] uniform distribution).

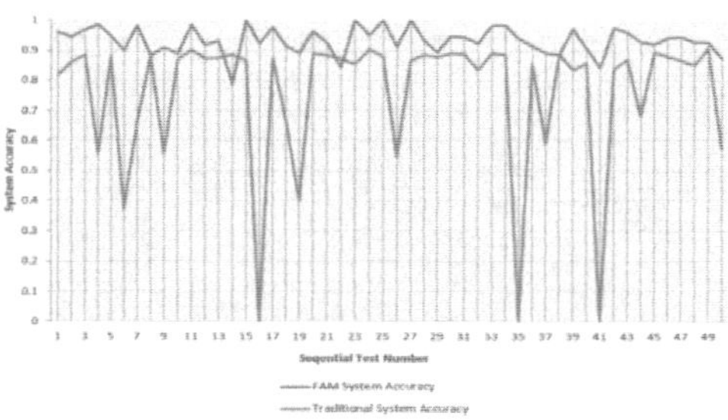

Figure 3. System Accuracy Test with Mode Number Around 15.

Compared with the previous visual servo system, the Fuzzy ARTMAP control system not only performs higher motion recognition accuracy, but also has strong flexibility and scalability. Although neural network is adopted in this system, the average system latency is still low enough, which could be clearly seen in Fig 4.

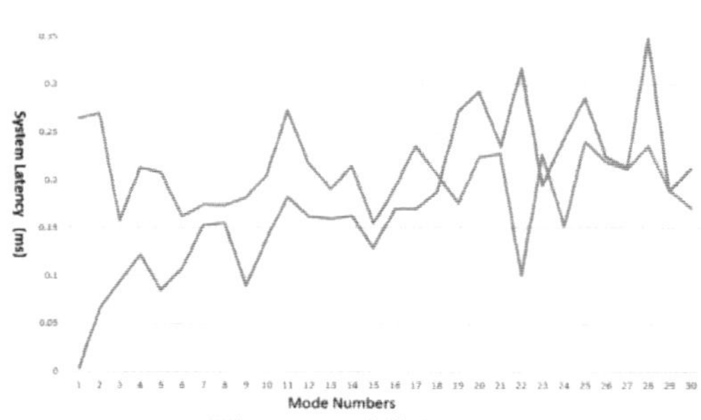

Figure 4. System Latency Test with Unfixed Mode Number.

In addition, the increased number of motions results in increased latency and reduced accuracy. However, as Fig 5. shows, even when the number of motion modes is more than 30, the accuracy of the system is still close to 85% which is contrasted with the severe decline of the traditional visual servo system accuracy.

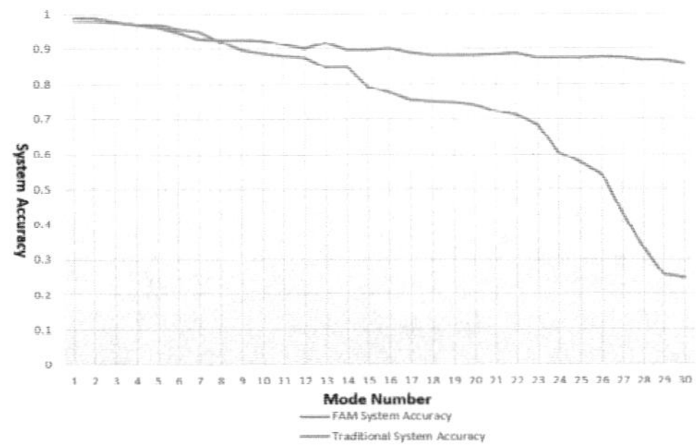

Figure 5. System Accuracy Test with Unfixed Mode Number.

We can conveniently study the performance of visual servo system based on image moments because some interface of the simulation program have been reserved for the system function expansion, which create favorable conditions for further research on visual servo system.

4. Conclusion

The Fuzzy ARTMAP system designed in this paper is applied to the robot visual servo control. It does not depend on the precise model of the object or the precise motion mode. The control rules come from the rules of users' daily life. It can figure out many problems caused by the nonlinearity, coupling and time variation of robot vision servo system. To design a robot which have more powerful recognition and processing ability, a Fuzzy ARTMAP program and some basic rules could be enough. System self-learning, scalable features not only ensure the stability of the system, but also provide a method to meet different needs of individual servo systems, which will be the development trend.

Now that the image feature extraction part of the system still costs so much computing resources. Our future work is to simplify the feature recognition model in order to reduce the time complexity as well as ensuring the accuracy.

5. References

[1] G Abebe, A Cavallaro and X Parra 2016 Robust multi-dimensional motion features for first-person vision activity recognition *J. Computer Vision & Image Understanding.* **149** 229-248

[2] Li Zhe 2015 AP Based Neurocomputational Spatial Uncertainty Measures *J. Hotogrammetric Engineering & Remote Sensing.* **74** 1573-584

[3] Wang Linglin, Liu Yongxin, Guo Peng and Zhang Hui 2016 Design of Ball Location Control System Using Industrial Robots Based on Computer Vision *J. Control Engineering of China.* **10** 1634-1638

[4] Wang Yongding and Nie Lina 2010 Design of Servo Steering of Intelligent Car Based on Fuzzy Control *J. Machinetool & Hydraulics.* **14** 75-78

[5] Yang Hui 2014 Design of Frequency Selective Surface Based on Evolving Fuzzy Neural *J. Modern Computer.* **7** 14-17

IOP Conf. Series: Journal of Physics: Conf. Series **1026** (2018) 012025 doi:10.1088/1742-6596/1026/1/012025

Design of cricket control system based on visual inspection

Liqian Wang[1]

[1]School of Mechanical & Electrical Engineering and Automation, Wuchang Shouyi University, Wuhan, 430064, China

Abstract. Aiming at nonlinear characteristic of the cricket system, the control system is designed based on MCU k60 and PD fuzzy controller. System test shows that the characteristic of dynamic response is good and the system steady state error is small. The integral control scheme of system is accurate, effective and stable, the path control and fixed-point control of the ball is accurate.

1. Introduction

Cricket system is a multivariable and strongly coupled nonlinear system and is a two-dimensional extension of the cue system. It provides a simply-constructed model to verify different kinds of control algorithms and process is Intuition. The cricket system is essentially a typical open-loop unstable complex nonlinear system, so that the control system of it is a good example of complex nonlinear systems control in practical engineering. Therefore, it is very significant to design a reasonable and effective model to control the cricket system.

2. The Control Principle

2.1. Hardware structure of cricket system

The hardware structure of the cricket system is shown in Figure 1. It is mainly composed of a camera, a ball and plate, a controller, connecting rod mechanism and a high-precision digital servo.

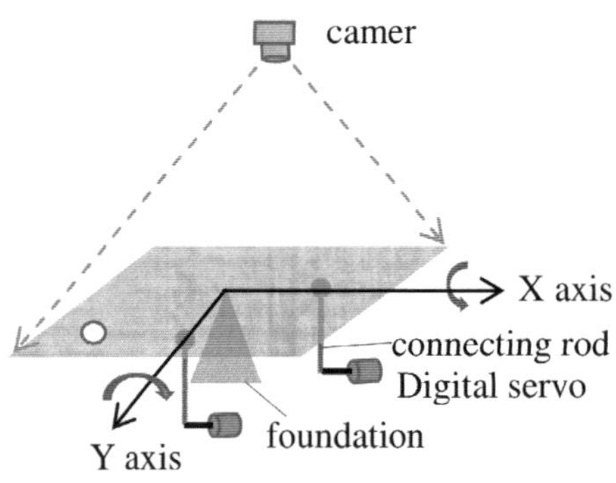

Figure 1. Mechanical structure.

In system, rectangular plate is supported by a triangular structure with a universal wheel at the top, and the universal wheel is located at the geometric center of the plate. The ball is placed over the plate and the moving trace of the ball need to be controlled. To achieve control requirements, the plate's air posture must be adjustable. Therefore coordinate system with the origin placed at the center of the plate is established in which the X axis and the Y axis along the plate's length and width, Regulating mechanism of the plate is provided by two connecting rods with universal joints, the top of the rods (universal joints)supporting at X axis and Y axis and the end of the rods connected to high-precision digital servo, so that the angle of the plate can be controlled by adjusting the rotation angle of the digital servo which is controlled by MCU K60. The ball position detection is realized by visual inspection, and the tilt angle of the plate is realized by the angle sensor.

2.2. Control principle of a cricket system

The block diagram of the system's control principle is shown in the following figure: after define the require moving trace or expected position of the ball, MCU driven digital servo rotating to change the plate attitude, at the same time by receiving the flat attitude data and ball position information data, MCU correcting output data after control algorithm calculating these data so that adjust.

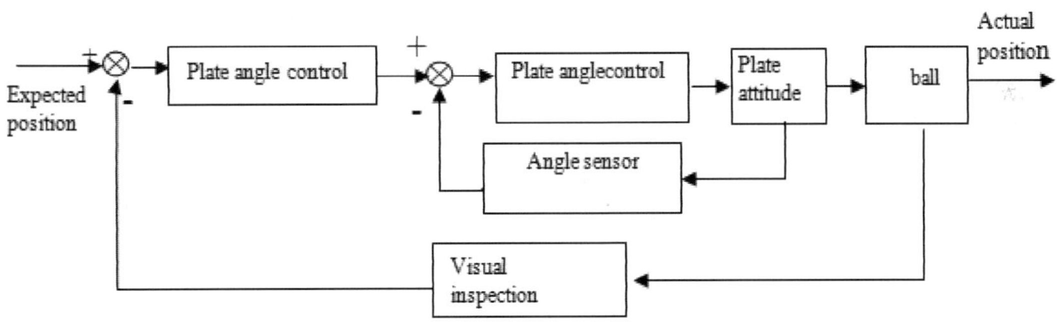

Figure 2. Mechanical structure.

(1)Visual inspection

The image data is obtained by camera OpenMV which provides interface library, then the actual position coordinate of ball is calculated by data-processing through sharpening, filtering and edge detection algorithms in image processing algorithms, through the data interface, the real-time information of the ball position coordinates will be obtain, and then the control signal will be obtain through the control algorithm.

(2) Plate attitude detection

According to the mechanical structure of the system, the plate is fixed on the Z axis and can rotate in space around the X axis and the Y axis. At this point, the ball has four degrees of freedom on the plate. For the space attitude of the plate, an angle sensor MPU6050 is used to detect the angle of rotation around the X axis and Y axis, then space attitude data of the plate is gained after data fusion calculating.

(3) Motion control

The control part of the cricket system is a digital system. Micro-controller K60, with 32 bit Floating-point arithmetic, realizes the control model algorithm: according to the expected position of the ball, actual position, the current attitude of the plate and the control algorithm, the rotation angle of the high precision digital servo is calculated, and the attitude of the plate is controlled by controlling the rotation angle of the digital servo.

The hardware structure of the control system is shown in figure 3:

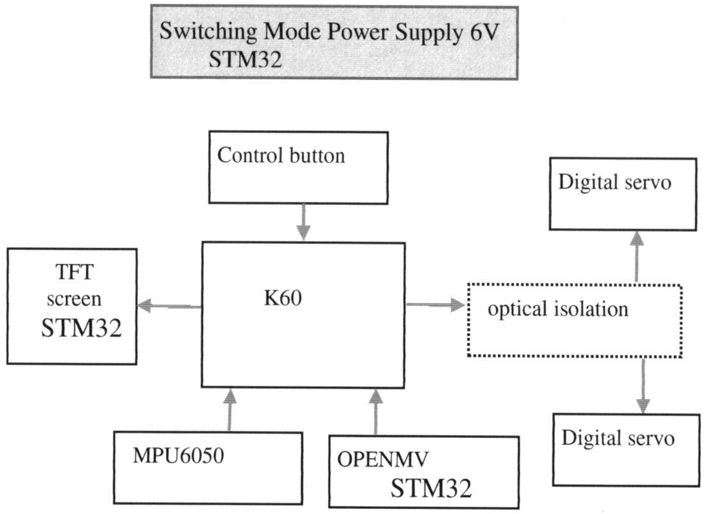

Figure 3. Hardware structure of control system.

3. Control Model Design

For nonlinear cricket system, good dynamic response needs accurate control of the moving trace of the ball by adjusting the attitude of the plate, small steady-state error of the control system lead to the accuracy position that the ball stopped. The PID classical control method can not obtain the desired control performance, so fuzzy control is adopted. Furthermore, the PD fuzzy controller has good dynamic characteristics, for the steady state error, motor vibration is used when the ball is near equilibrium point to eliminate the static error caused by pure PD control.

The controller structure is shown in figure 4:

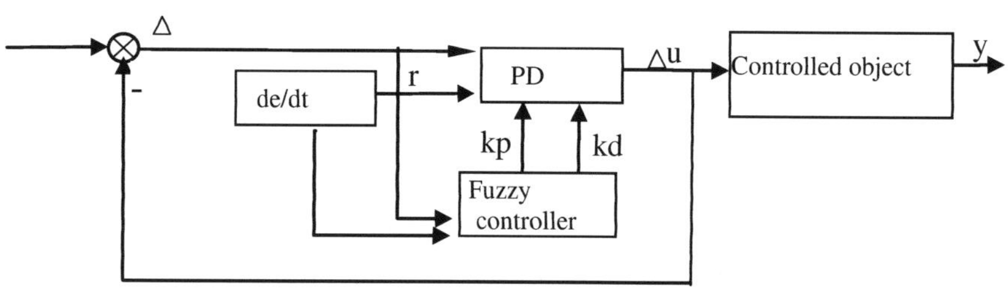

Figure 4. Controller structure.

In figure 4, Δe is position error, r is rate of change position error, Δu is the output of the controller. According to controller structure and system hardware structure, the ball moves in two-dimensional space, the control strategy is same when the ball moving along the X direction and moving along the Y direction. So the ball's moving in one dimensional direction can be controlled first, then get experiment data and then using it to modify the algorithm structure and parameters, so another ball's moving dimensional direction can also be controlled.

Determination of input variables: Centered on the origin of the coordinate system, the distance between small ball coordinate and center point is one dimension.

if E=NB and EC=PB then U =PS
(R1 =[(NB)E×(PB)EC]×(PS)U)
if E=NM and EC=PB then U= PS
(R2 =[(NM)E×(PB)EC]×(PS)U)
……
if E=PM and EC=NB then U= NS
(R48 =[(PM)E×(NB)EC]×(NS)U)
if E=PB and EC=NB then U= NS
(R49=[(PB)E×(NB)EC]×(NS)U)

After the control rules are completed, the variable, the change rate of small ball position is another one-dimensional variable, input variables are: ext=xt-x0, eyt=yt-y0; ecxt=ext-ex (t-1), ecyt=eyt-ey (t-1). According to the hardware structure and characteristics of the system and experimental data, the basic universe of positional error and its rate of change are[300, 300], [-50, 50]. According to the input, taking 7 language variables to describe position, change rate of position and control quantity change: A1: (PB), A2(PM), A3(PS), A4(0), B1(NS), B2(NM), B3(NB).

According to the movement characteristics of the ball in the device, that is the position of the ball, change rate of the position and the experimental data, the control status table is obtained:

query table is set up according to the rules, and the program is completed in keil5 environment and implemented by MCU K60 with ARM kernel.

4. System Verification

On the basis of the above work, the hardware andsoftware of the system were designed and tested: the expected position are origin of coordinate, coordinate position (100,100) and (-200,-100),the initial position of ball is random. The response performance of the system was shows in table 1.

Table 1. Experimental data

expected position (coordinate position)	error（mm）	take time	Test times
（0,0）	0.6,0.5	<15s	>10 times
（+100,+100）	0.7,0.7	<15s	>10 times
（-200,-100）	0.8,0.6	<15s	>3 times

The test data show that the control system is stable, the movement of the ball was controlled and the system error is acceptable.

5. Conclusions

The system test shows that the performance of dynamic response speed and steady-state error meet the control requirements, the moving trace of ball can be controlled, the average error of position is less than 0.6mm, the system takes less than 15 seconds to reach stability ,it shows that the control scheme is accurate and effective.

As a typical multivariable and nonlinear control object, this system can be used as a hardware carrier to verify various control algorithms and It is helpful to study control theory.

6. References

[1] Chenxi Zhai and Hongxing Li 2016 Direct adaptive fuzzy sliding mode control of cricket system *J. Computer Simulation.* **33** 383-388

[2] Hongrui Wang and Yantao Tian 2009 Parameter self tuning back stepping control of a cricketsystem *J. Control and Decision.* **5** 749-758

[3] Qian Wang and Zhenshun Sun 2004 Research on PD fuzzy control algorithm of cricket control system *J. Electric Drive.* **4** 23-25

Thinking About the Development of Offshore Wind Farm in Northeast China

Xia Lin[1,2], Peng Wang[1,2,*] and Yonghai Yu[1,2]

[1]National Marine Environmental Monitoring Center, Linghe Street, Dalian, China
[2]Key Laboratory of Sea-Area Management Technology. SOAr, Linghe Street, Dalian, China

Email: {PWang, Linxia, yhyu}@nmemc.org.cn

Abstract. The wind energy is the new developed energy on a large scale, and the offshore wind energy resource in in northeast China is abundant, which has great value of development. This paper not only states the meaning of offshore wind farm through introducing the planning location of offshore wind farm and the general information of wind energy, but also discuss the environmental problem brought by project from the view of birds and marine fishery resources. What is more, it offers a proposal about further strengthening study on its resource impact and provides a reference for offshore wind farm in northeast China.

1. Background

Among different types of renewable energy, offshore wind energy ranks among the top in terms of technology readiness, large-scale development possibility, and prospects for commercial development[1]. Featuring constantly evolving technologies, increasing unit capacity, and significantly decreased operating cost, offshore wind power generation needs no land resources, and thus has become a new field of wind resource development with great potential and promising development prospects[2][3]. At the national level, the National Development and Reform Commission has set the goal of building offshore wind farms providing 1000 ×104kW wind power, including 500×104kW installed wind power by 2020.

On the one hand, China boasts abundant offshore wind resources, and on the other hand, China's coastal areas, which are economically developed, have a high demand for electricity. In such a context, developing offshore wind resources will play an important role in satisfying power demand, improving energy structure, reducing environmental pollution, coping with global climate change, and promoting sustainable and healthy economic development. As of the end of 2013, China had installed 112 offshore wind turbines, with total installed capacity reaching 29×104kW. By the end of 2015, grid-connected wind power capacity had reached 12,900×104kW[4].

In northeastern China where wind resources are abundant, onshore wind farms have appeared at a rapid speed, but offshore wind farms are still in its infancy. Actually, the offshore area of Dalian, especially of Zhuanghe and Huayuankou, has rich wind resources which encourage the development of wind power resources. Given this, Dalian has been actively promoting the construction of wind farms, responding to the state's call of developing renewable energy resources, and promoting local social and economic development at the same time.

2. Planned Offshore Wind Farms in Northeastern China

The offshore area of Dalian City, Liaoning Province is the main target of wind farm construction in northeastern China. It's planned to build seven wind farms with a total installed capacity of 2.2 million kW at two sites. Among the seven planned wind farms, Zhuanghe Wind Farm III is now under construction upon approval, and other farms are in the process of feasibility assessment or environmental impact assessment.

Table 1. Statistical Table of Recommended Offshore Wind Farm Locations

Wind farm name			Water depth: 5m~30m (Year of construction: 2011~2015)				Water depth: 5m~30m (Year of construction: 2016~2020)				Remarks
			Distance from the shore (km)	Depth of water (m)	Area (km^2)	Installed capacity (kW)	Distance from the shore (km)	Depth of water (m)	Area (km^2)	Installed capacity (kW)	
Site I	Huayuankou Wind Farm I		12	5~14	65	300,000					Located at fish farms
	Huayuankou Wind Farm II		12	5~12	65	300,000					
Site II	Zhuanghe Wind Farm I		12	7~14	46	200,000					
	Zhuanghe Wind Farm II		18	10~20	64	300,000					
	Zhuanghe Wind Farm III		22	12~22	65	300,000					
	Zhuanghe Wind Farm IV	IV1	31	19~26	77	350,000					
	Zhuanghe Wind Farm IV	IV2					40	25~30	44	200,000	
	Zhuanghe Wind Farm V		32	18~27	56	250,000					
Total						2,000,000				200,000	

3. Overview of Offshore Wind Resources in Northeastern China

The above-mentioned wind farms are located in the East Asian monsoon region where SSE, S, SSW, NNW, N and NNE contribute to over 60% of the wind energy in most of the project area, and the figure even stands above 80% in some areas.

Data collected by No. 1 anemometer tower near the sites show that the annual average wind speeds at 85m, 90m, and 100m are 6.86m/s, 6.88m/s and 6.92m/s respectively which indicate annual average wind power density values of 351.3W/m2, 354.6W/m2 and 360.7W/m2. Data collected by No. 2 anemometer tower show that the annual average wind speeds at 85m, 90m, and 100m are 6.61m/s, 6.64m/s and 6.67m/s respectively, indicating annual average wind power density values of 310.6W/m2, 313.5W/m2 and 318.9W/m2. Data collected by No. 3 anemometer tower show that the annual average wind speeds at 85m, 90m, and 100m are 7.20m/s, 7.22m/s and 7.26m/s respectively, and the associated annual average wind power density values are 422.0W/m2, 425.9W/m2 and 433.3W/m2.

Generally speaking, offshore wind resources in northeastern China are abundant enough for the construction of wind farms.

4. Main Resource and Environmental Problems Related to Wind Farm Construction

4.1. Effects on birds

A number of islands (e.g. Shicheng Island and Haiwang Nine Islands in Zhuanghe) 20 km west of Zhuanghe Wind Farm III act as the breeding grounds of endangered species black-faced Spoonbill, and Class II protected species yellow-billed Egret and pelagic cormorants.

Wind farms will mainly affect the aforesaid birds in two ways. Firstly, wind farms will drive away cormorants. Secondly, the construction of wind farms will increase the number of species in the project area which indicates more food for birds. Besides, static facilities like booster stations will provide perching places for abundant bird species like gulls, changing the original avian community structure which mainly manifests as increased activities of gulls.

1) Construction of wind farms will generate suspended solids, waste water, waste oil, solid wastes, as well as noise pollution and light pollution which will affect the eyesight of birds diving to prey, such as cormorants. However, thanks to the deep water and low population of birds in the project area which is relative large, birds can easily find alternative areas to prey. Therefore, the project's effects on birds diving to prey are minor. It's worth noting that solid wastes generated during construction are likely to be eaten by birds, leading to their death due to poisoning or indigestion. Noise will also affect birds. However, since the effects of mechanical noise are limited within a 100m radius in the daytime and an 800m radius at night, and the offshore area around the project sites is much larger, noise will actually have a minor effect on birds flying by (which will keep away from noise) and birds on the breeding grounds.

2) After the wind farms are constructed, the arrays of large turbines may pose a threat to birds flying low at night during migration. Measured data show that the sound power level of a single turbine falls in the range of 85～100dB, and since most water birds are highly sensitive to moving things and noise which acts as warnings, they will keep away from noise when flying by.

4.2. Effects on fishing

The two construction sites, which are mostly 16~26m in depth, have been acting as fishing areas for a long time, especially for the fishing of spawns during the spring flood period, and Spanish mackerel during the winter flood period.

Nekton like fish is very mobile, and thus will swim away from pollutants. Therefore, increased suspended solids generated during the construction will not seriously affect marketable fish living in and around the sites, except that their distribution will be affected to some extent. Besides, construction activities may divide or even destroy the habitats of fish, posing threats to the ecosystem and commercial fish. During the construction, shipping activities will become much more frequent, and water quality will decrease. And even after the construction, the damaged ecological environment will take time to recover. As a result, fishing activities cannot be carried out during this period of time. According to administrative regulations concerning ocean utilization, offshore wind farms do not exclude non-related activities, which means the utilization and protection of fishery resources, such as fishing and reproduction, are allowed in offshore wind farms. However, to ensure the safe operation of wind turbines, ships, regardless of their types, are forbidden to anchor or fish around the foundations and near cables. As a result, the fishing and fish farming areas will be greatly reduced, causing loss to involved parties to some extent.

In accordance with the Provisions on the Protection of Submarine Cables and Pipelines (Decree No. 24 of the Ministry of Land and Resources of the People's Republic of China), activities such as sand excavation, drilling, piling, anchoring, clubbing, bottom trawling, netting, fish farming and other activities that may pose threats to the safety of submarine cables are forbidden within a 50m distance away from cables.

A large number of submarine cables are required by the project, and according to the 50m limit requirement, only some small areas in the two sites can be used for fish farming and fishing. This means decreased available fishery resources, fishing time, and fishing areas. Therefore, the construction of the wind farms will have a big impact on local fishermen.

However, from another point of view, prohibition of fishing and fish farming will actually help the conservation of fishery resources in the project are. Special reports on the effects of underwater noise and electromagnetic radiation indicate that noise and electromagnetic radiation generated during the construction will have a minor effect on fishery resources. Therefore, the project will, to some extent, promote the rational utilization of fishery resources.

5. Suggestions on Wind Power Development

5.1. Feasibility studies on the construction of offshore wind farms shall be carried out actively.

In northeastern China where wind resources are abundant, onshore wind farms have appeared at a rapid speed, but offshore wind farms are still in its infancy. Actually, the development of offshore wind resources plays an important role in satisfying electricity demand, improving energy structure, reducing environment pollution, coping with global climate change, and promoting the sustainable and healthy economic development of northeastern China. In the current context where renewable energy resource development is encouraged by the government, relevant parties should seize this opportunity, actively promoting the construction of wind farms in northeastern China. To do so, feasibility studies shall be carried out to find out how to avoid areas subject to frequently occurring sea-ice hazards and areas home to intensive resources such as migrant birds, so as to minimize economic loss caused by sea-ice hazards, and the damage of project construction to fishery and birds, thus promoting the development of offshore wind resources in northeastern China in a sustainable way.

5.2. Fishery resource conservation in the project area shall be studied.

Underwater test results indicate that noise and electromagnetic radiation generated during construction will have a minor impact on fishery resources. Although the project will reduce fish farming and fishing area; on the other hand, a fishery resource conservation area will come into being naturally, playing a positive role in promoting the rational utilization of fishery resources. In addition, fishery resource conservation studies may be carried out in the project area.

5.3. Effects of project construction on birds shall be studied.

Since there are important bird habitats around the planned sites, it's suggested to further study the possible effects of the project on birds of different species and birds having different migration routes during the construction period and operating period, so as to avoid threats to birds in the course of developing clean energy resources.

6. References

[1] Shihao Jiang, Kai Liu and Jichao Bian 2014 How Blue Economic Zone Promotes the Development of the Low Carbon Economy *J. Marine Information*. **3** 32-35

[2] Li Xu, Feng Li and Hongbing Peng 2015 Relationship between Offshore Wind Power Development and the Environment in China: A Study *J.China Population, Resources and Environment.* **1** 135-38

[3] Yanrong Zhou, Wei Zhang and Qiang Song 2015 An Analysis of the Status Quo of Global Offshore Wind Resource Development and Problems Concerning Sea Area Utilization *J. Ocean Development and Management.* **1** 135-38
[4] Baiqiong Liu, Min Xu and Qing Liu 2015 Main Problems Concerning China's Offshore Wind Resource Development and Suggested Solutions *J. Ocean Development and Management.* **3** 7-12

IOP Conf. Series: Journal of Physics: Conf. Series **1026** (2018) 012027 doi:10.1088/1742-6596/1026/1/012027

Study on the Application of Electronic Equipment in Voice Disease Prevention

Yingying Zhang[1]

[1]Musical Department, Heihe College, Heilongjiang, China

Abstract. Electronic auxiliary equipment in many fields have been used in the prevention of disease, the vocal voice is more and more widely used, in the daily vocal training and voice disorders are mostly because the pronunciation method caused by improper, therefore, prevention and treatment of voice diseases is the scientific sound, let the human body the vocal organs reasonable movement in the vocal motor; followed by active prevention, effective treatment and daily life care. "Sound" is just like his life for the singer. How to protect it, keep it, and make it shine forever is the dream of every singer. In this paper, the application of electronic equipment, the use of voice disease prevention, through the scientific method of voice, physical health, diet and other aspects of the use of electronic auxiliary equipment, discusses the protection of artistic voice, singing can find the correct way of voice health.

1. The Important Role of Voice Disease Prevention

1.1. Voice Disease Prevention is Good for Maintaining Good Breathing

The physical vibration wave. Vocal production is no exception, but also because of the breath of vibration, vibration throat vocal cords and sound. So far, the vibration of the vocal sound Public opinions are divergent. theory, unable to agree on which is right. One point of view is that the subglottic pressure reaches a certain level. When the glottis is closed, the air is blown out and the air sounds out. This view is called the theory of muscular elasticity by voice medicine, also known as aerodynamic theory. There is also a view that vocal cord vibration is controlled by the brain, controlled by the nerve, is the brain recurrent laryngeal nerve, rhythmic rhythmic impulses, so that the vocal cord vibration. Voice medicine says this theory is the theory of neural duration. Most scholars believe that the neutralization of the two principles produces the human voice, that is, under the control of the innervation of the human brain, the sound source of the vocal cord is affected by the force of the breath of the lungs. Based on this understanding, then scientific sound is under the control of positive thinking, good breathing support the human voice action. As early as the Tang Dynasty, our ancestors realized that breath played an important role in the process of human voice. An Duan Tang festival in "Yuefu miscellany" said: "good singer, must adjust the gas." In addition to the essence of ancient wise judgment, today there are a lot of fresh experience. China's famous performing artist Mr. Cheng Yanqiu said: "the air feel void by the pubic region, head, waist, shoulders relaxed two." A comprehensive overview of the parts, status, sensations, and methods of singing breathing movements is given. China's vocal music educator Tang Xuegeng said: "the so-called Dantian gas, is the use of isolation (i.e. diaphragmatic breathing) manipulation." These claims are similar to the breathing requirements in western traditional singing. In 1893, the French Academy of music published "notes on singing methods"): "when singing, the belly should contract when breathing. This is the opposite of the general breathing and breathing, and the upper abdomen is slightly bulging."." The famous German singer Nelly Lehmann, summed up his decades of experience singing said: "when I put the inspiratory abdomen and diaphragmatic indentation, chest with ribs to keep breath." Overview of excellent art of singing, all good pronunciation, the breathing problems are through abdominal pressure (abdominal muscle contraction force), on intra-abdominal organ thrust force will be the role of bottom-up conduction to the diaphragm, making it to the bottom of lung, the power storage formed to support voice in the air the lungs a gas pipe. Due to the movement of the diaphragm and tension, outward expansion of the support under the wings rib, such breathing state, breathing air in vocal organs time, easy to produce "bernuolli acoustic effect", the coordination between the appropriate sub vocal cords, throat, trachea and vocal organs of muscle and cartilage action, to breath smooth, smooth, mellow sound rich, such as freely flowing style of writing, like connecting beads, move among force and soft has flexibility, expressive[1].

1.2. Prevention of Voice Disorders Can Produce Beautiful Sounds

To make a good sound, you must have a good voice, and in the daily life, to prevent sound illness, you can guarantee the singer can make beautiful sounds. Each person's voice conditions are not the same, strength, width, tone sound intensity are different, should be based on sound characteristics, personal abilities, the person. The start of training time should not be too long, not too wide range, sound intensity is not too strong, don't sing too much, too difficult to work, to sing soft and coherent melody, in order to achieve flexible laryngeal muscle action and steady breath control, and the coordination of action, to do flexible vocal exercises in natural sound in the region, according to the requirements of training, training to overcome the blindness in sound, emphasize the consciousness, don't expect the short term to improve the ability of sound, and we must focus on the gradual adjustment of vocal skill internal coordination. The vocal function of the human body is closely related to hearing. Beginners must pay attention to aural training and establish the correct sound concept in auditory memory, which is the basis of human voice training[2].

2. Commonly Used Voice Disease Prevention Methods

2.1. Exercise Actively

The glottis is a small physiological organ in the human body. The health condition of the whole body influences the voice directly [3]. Therefore, we should pay attention to exercise and improve the ability of preventing disease and disease. Choose proper sports according to your body condition, such as walking, playing ball, swimming, skating and so on. Sports and voice training should not be too close to prevent fatigue caused by congestion of the vocal cords. Exercise should be step-by-step, step by step, persevere, but also pay attention to protection at any time, can not be careless, in the case of maintaining a healthy voice, steadily and steadily improve the ability to sing pronunciation[4].

2.2. Perform Positive Recovery after Singing

There are three methods for the treatment of voice diseases, that is, local treatment, systemic regulation and vocal therapy. Specific practices are as follows:

(1) no sound. For acute inflammation, edema and other diseases should be banned sound rest treatment. In addition to the normal treatment of chronic inflammation, due to the arrangement of vocalization and silence to rest. Due to functional disease or psychogenic voice disorder disease with sound based correction. After the operation of the throat, the time should not be too long. According to the recovery after the operation, under the guidance of the doctor, the voice exercises are performed slowly, so that it is very beneficial to the recovery of the throat structure and function. According to voice, can be taken: A. absolute silence; a small amount of B. C (note softly; speech time control); D. natural sound area vocal exercises; E singing (note that the control time, sound intensity) of different degrees of gag practice. In the process of voice rest, some people think that "whisper" is a good

method, which is a wrong understanding. In the whisper, the larynx muscles do not get a rest, and the vocal cords are closed, which hinders the restoration of the normal vocal function. (2) drug therapy. It is divided into two forms: internal and local points. Oral administration: for acute inflammation such as congestion, edema, secretion increase, internal or external injection of antibiotics, hormones and other anti-inflammatory and analgesic drugs. Local point: through the suction device, the drug directly suction throat lesions, to achieve anti-inflammatory, analgesic and moisturizing role, in addition, can also be applied to the affected area with intravenous drip and clothing. (3) surgical treatment. This is the throat with proliferative lesions of the treatment methods, such as: the treatment of vocal nodules, polyps and other diseases, can take this method. (4) Chinese medicine treatment. Including drugs, acupuncture, ear acupuncture, massage, qigong, etc.. Drug treatment: according to different disease symptoms, follow the doctor's advice to take the pill, powder, decoction. Acupuncture: strengthens the meridian function by acupuncture, achieves the adjustment nerve, completes the treatment goal. Treatment of voice diseases common points are: sudden, Tianding, skylight, Renying, Lian Quan, gas house, water inrush etc. With acupuncture Shaoshang, Shao Chong, little Ze, Hegu etc.. Deficiency of Qi and blood, hypofunction of function and period of disease [5].

Chronic disease long with weak stimulation, acute disease of evil outside attack, hyperthyroidism, stage of disease process together with strong stimulation. Auricular acupuncture: Acupuncture ear acupuncture points to treat systemic diseases. Voice fatigue and inflammation take throat area, rhinitis take nasal region, insomnia take occipital area, subcortical region. Massage: using two hands on the body surface of the meridian points to push, take, rub, buckle, point, cut, quiver, knock, etc. massage action, the principle of acupuncture and acupuncture the same. Massage plays an important role in strengthening the function of organs and eliminating the fatigue of singing. Before singing, you can gently massage the springs, the sky, the people, the Zusanli, and so on, so as to achieve the purpose of stimulating excitement. Sing after recovery by heavy Hegu, Shaoshang, little Ze hole, to suppress the excitement, sedative. Qigong therapy in China: some Qigong boxing, in the aerodynamic condition and singing breathing requirements have many similarities, the vocal music work to learn Qigong, be of great advantage, have a positive effect on the recovery of fatigue and treatment of voice diseases. (5) physical therapy. Through medical devices, the use of physical factors, such as electricity, light, wax, water, sand, mud and other functions on the body surface, or diseased parts, to promote humoral response, the mobilization of nerve function, to achieve the purpose of disease prevention and treatment.

2.3. Pay Attention to Health Care in Daily Life

Life care is an important measure for prevention of voice diseases, it is to extend the life of art, enhanced voice ability are closely related, China's famous Peking Opera artist Mr. Mei Lan fang has accumulated rich experience in the protection of the throat, throat using hand. He said: "the spirit of fun, ataractics, diet, living in cold and warm care, and uniform practice voice Paul voice, thank you all, from low rise, their fifth full, sing the story." Absorbing the experience of our forefathers, we sum up the following problems: to protect the voice in life, we need to pay attention to the following aspects. First, To have a healthy physique and physique .Physical health is the material basis of singing, have strong body will have a loud voice, so vocal workers usually have to pay attention to physical exercise, in order to ensure healthy people. In addition, the function of the body play a role by psychological factors about people, the person's mood affects voice through sympathetic mood, sympathetic nerve, has a positive effect on the fatigue elimination of the throat instead, depression can cause voice fatigue. Therefore, vocal music workers should pay more attention to the exercise of personal psychological awareness, pay attention to the cultivation of morality and conduct, and constantly improve the level of cultural knowledge, create a perfect spiritual world, and develop a healthy spiritual temperament. Second. Prevent colds Colds and flu are common diseases, and the incidence is very high among vocal music workers. In addition to regular physical exercise, we should also pay attention to environmental sanitation. Eliminate the source of infection and prevent colds. Changes in temperature, pay attention to change clothes, sweat on the body, be careful not to catch cold by the wind. Third. Life must be regular The work, the rest, the diet, the sleep in life must be orderly. Form regular habits and habits. Sings every time, not be interrupted, according to voice the ability to gradually increase the singing time, and quite, singing is not excessive, the rest should be reasonable. The principle is not only to ensure the sound of the voice, but also to get the necessary exercise. Diet should be quantitative timing, before singing can not eat too full, so as not to affect the movement of the diaphragm. Sleep restores the tired body and makes the working song rest, so it's important for the singer to get plenty of sleep. Fourth. Pay attention to dietetic hygiene. Don't be greedy for alcohol or smoking, and should pay attention to the choice of stimulating food. Cold hot drinks, hot pepper, mustard, raw onions, raw garlic, greasy food great stimulation to the throat mucosa, promote the secretion, sound is not smooth, the sound is not clean, as there is always the feeling of throat phlegm. These foods are selected according to the individual's adaptation, less or no need. Five. Voice protection. The sound is a sound change in human development reflects changing during the throat hyperemia, edema, exudation, glottal incompetence, dysphonia. It should reduce the use of the voice activity, pay attention to rest, since the late, according to individual circumstances can vocalize, but the breath intensity, sound intensity, width range should be strictly controlled.

3. Electronic Auxiliary Equipment in Voice Disease Prevention

3.1. Correct the Wrong Sounding Method Through the Electronic Auxiliary Equipment

Voice diseases caused by the wrong sounding method are the most common and the main cause of the disease. The error caused by vocal voice disorders, often by medical treatment does not work, only to complete the treatment by the doctor and vocal teacher can cooperate, which by the vocal music teacher correct vocal method is particularly important. The occurrence and development of things have objective laws, and so do singing. When singing, because the respiratory control and the sympathetic method are not correct, causes the vocal organ to be uncoordinated, causes the throat burden to be overweight, the muscle group function loses the balance, the sound organ can lose the normal work ability. Double burden of laryngeal muscle, can cause local congestion, laryngeal muscle fatigue, chronic throat disease can occur for a long time, which is the root cause of voice disorders. Therefore, through the electronic auxiliary equipment, for correcting the wrong voice, prevention of voice diseases have a better effect.

3.2. Use Electronic Equipment to Reduce Voice Abuse

A person's energy is limited, the movement ability of the vocal organs is limited, all overload of work, will make the brain and muscle fatigue, caused by the accumulation of lactic acid metabolites, so vocal exercise time should not be too long. Unrealistic pursuit of treble, voice, or arbitrary expansion of the vocal range is more difficult, their songs will make beyond strength and power, and vocal fatigue by excessive shock, which if not corrected, a long time can play mucosal hyperemia and edema, hypertrophy, sound with the summary of laryngeal diseases. Some beginners, as eager to learn, often in suffering from throat diseases still vocalize, voice abuse will aggravate the vocal cord edema, difficult to treatment, delay the disease, serious person can make the vocal submucosal hemorrhage, which caused The consequence is that electronic equipment is used to monitor the abuse of voice, so as to prevent voice diseases [6].

To learn vocal music, we must adhere to the principle of gradual progress, obey the guidance of teachers, cooperate closely with teachers and students, and play a sound "middle voice" area, then expand the vocal range. Sing songs that suit your abilities. The length of exercise should also be determined according to the specific conditions of the individual's physique and technical level. If you feel tired, you should rest immediately.

3.3. Through Electronic Assistive Devices for Scientific Sound Practice

All walks of life have their own basic skills, vocal music is no exception, the basic skills of vocal music is pronunciation exercises, teachers teach the correct pronunciation method, students should adhere to practice every day. Scientific vocal methods can make the singing organs healthy and enhance their vocal ability, and play a positive role in the prevention of professional diseases in singing. If you can not adhere to scientific practice, singing level can not be improved. Some students feel that their sound conditions are good, mouth sound, do not have to practice every day, this idea is extremely wrong. When young, good physique, voice is very loud, but the conditions of talent can not take the place of scientific knowledge, superior gift bar.

4. Conclusions

Is it difficult to afford long-term, arduous task of singing, there is no scientific method, singing in the heavy pressure, it is easy to suffer from voice disorders, voice is also easy to prematurely senile, electronic auxiliary equipment for auxiliary singing voice, voice to prevent disease.

5. Acknowledgements

This thesis is a research project of philosophy and Social Science in Heilongjiang Province “Scientific training and health care of voice”, Project number is 16YSC03, Project category is Youth Project.

6. References

[1] Li J C 2006 Current status of voice medicine research in China *J. Chinese Medical Digest (otorhinolaryngology).* **8** 21-22
[2] Du J Q 2008 The role of calculation in the prevention of voice disorders *J. Chinese Journal of Integrative Medicine otorhinolary ngology.* **5** 106-110
[3] Li J 2006 CO2 laser laryngeal microsurgery for laryngeal on benign and malignant lesions *J. Audiology and Speech Pathology Journal.* **6** 55-58
[4] Yu P 2004 Comparison of different gender objective parameters of the voice analysis *J. Journal of Audiology and speech pathology.* **3** 19-20
[5] Hu T 2007 Electronic auxiliary equipment in diagnosis of voice diseases by *J. Journal of Audiology and speech pathology.* **19** 98-102
[6] Wu X H 2006 Pay preventive application on electronic equipment for computer voice diseases Chinese *J. Journal of Otolaryngology skull base surgery.***5** 33-34

Parameter Reconstruction of a Circuit System from Partial Eigeninformation

Wenting Wan[1]

[1]School of Mathematics and Physics, Jingchu University of Technology, Jingmen, Hubei, 448000, China

Email: wentingwan@163.com

Abstract. In this paper, the inverse eigenvalue problem of reconstructing a LC circuit system from its some parameters and incomplete eigenpairs is considered. The necessary and sufficient condition for the solvability of the problem is derived by solving a kind of general inverse eigenvalue problems. The computational expressions of the solution, the related numerical algorithm and a numercical example are given.

1. Introduction

The inverse eigenvalue problem is an important method for designing the circuit or the spring-mass system. Roughly speaking, an inverse eigenvalue problem is to reconstruct realizable system from prescribed natural frequencies or modalities or some parameters. The problem arises in a remarkable variety of applications, such as circuit analysis, electrical oscillation, vibroacoustics, structural design, parameters identification, vibration control, applied mechanic, etc. The problems are transferred into inverse eigenvalue problems for Jacobi matrices in mathematics. In recent years, some new results have been obtained on the construction of a Jacobi matrix, see [1][2][3][4]. Using two sets of eigenvalues or two incomplete eigenpairs, the inverse eigenvalue problems have been studied[5][6][7]. Problems [8][9][10] have been solved by one eigenpair or two eigenpairs to determine corresponding systems, and numerical algorithms for solving the problems were proposed. Referring to the actual application problems, this paper discusses a class of inverse eigenvalue problem, and constructs the LC circuit system from its some parameters and two defective modes. The necessary and sufficient condition for the existence and uniqueness of the solution is obtained, moreover, the expressions of the solution and the related numerical algorithms are presented.

Consider a LC circuit system whose inductors and capacitors are connected in Figure1. Generalized eigenvalue equation for the system is $(-\omega^2\boldsymbol{M}+\boldsymbol{K})\boldsymbol{X}=\boldsymbol{O}$, where $(\omega i,\boldsymbol{X})$ is an eigenpair of the system, $i=\sqrt{-1}$, $\omega\in\boldsymbol{R}$, $\omega\neq 0$,

$\boldsymbol{X}=(x_1,x_2,\cdots,x_n)^T\in\boldsymbol{R}^n$,

capacitance $c_i>0(i=1,2,\cdots,n)$,

inductance $l_i>0(i=1,2,\cdots,n)$,

capacitance matrix is: $K = diag(\frac{1}{c_1}, \frac{1}{c_2}, \cdots, \frac{1}{c_n})$, inductance matrix is:

$$M = \begin{bmatrix} l_1 & -l_1 & 0 & \cdots & 0 & 0 \\ -l_1 & l_1+l_2 & -l_2 & \cdots & 0 & 0 \\ \vdots & \vdots & \vdots & \ddots & \vdots & \vdots \\ 0 & 0 & 0 & \cdots & -l_{n-2}+l_{n-1} & -l_{n-1} \\ 0 & 0 & 0 & \cdots & -l_{n-1} & l_{n-1}+l_n \end{bmatrix}.$$

The problem we considered is as follows.

Problem Consider a LC circuit system whose inductors and capacitors are connected in Figure1, capacitance $c_i \in \boldsymbol{R}^+ (i = 1, 2, \cdots, p)$, inductance $l_i \in \boldsymbol{R}^+ (i = 1, 2, \cdots, p)$ are known. Given $\alpha, \beta \in \boldsymbol{R}$ ($\alpha \neq 0, \beta \neq 0$, $|\alpha| \neq |\beta|$) and vectors

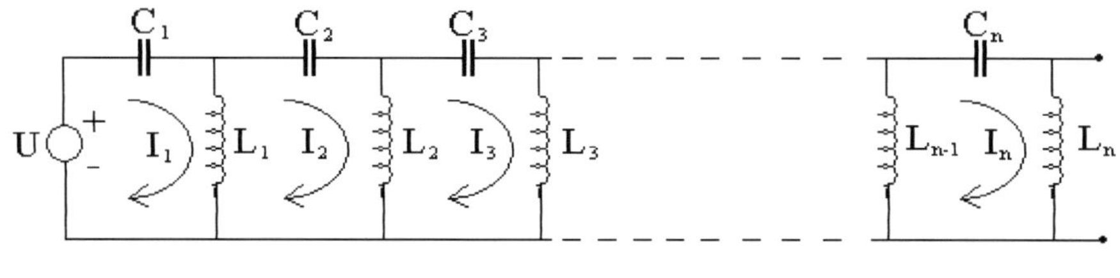

Figure 1. A LC circuit system.

$X_2 = (x_{p+1}, x_{p+2}, \cdots, x_n)^T$,

$Y_2 = (y_{p+1}, y_{p+2}, \cdots, y_n)^T \in \boldsymbol{R}^{n-p}$,

find the remaining parameters of the system, that is, solve capacitance c_i, inductance $l_i > 0 (i = p+1, p+2, \cdots, n)$

and vectors $X_1 = (x_1, x_2, \cdots, x_p)^T$,

$Y_1 = (y_1, y_2, \cdots, y_p)^T \in \boldsymbol{R}^p$, such that

$$(-\alpha^2 M + K)X = O,$$
$$(-\beta^2 M + K)Y = O, \tag{1}$$

where vectors X and Y are $X = (X_1^T, X_2^T)^T$, $Y = (Y_1^T, Y_2^T)^T \in \boldsymbol{R}^n$, K and M are respectively capacitance matrix and inductance matrix, $(\alpha i, X)$ and $(\beta i, Y)$ ($i = \sqrt{-1}$) are two eigenpairs of the system.

2. Solvability of the Problem

If $\omega \neq 0$, $c_i > 0 (i = 1, 2, \cdots, n)$, then the equation $(-\omega^2 M + K)X = O$ is equivalent to the equation $K^{-1}MX = \frac{1}{\omega^2}X$, where KM^{-1} is $n \times n$ Jacobi matrix.

In this section, we will discuss the condition for the solvability of Problem. We first introduce the following lemmas for later discussion.

Lemma 1[11] Let $(\lambda, \boldsymbol{X})$ be eigenpair of $n\times n$ Jacobi matrix $\boldsymbol{J}$, $\boldsymbol{X}=(x_1, x_2, \cdots, x_n)^T \in \boldsymbol{R}^n$, then,

(1) $x_1 x_n \neq 0$;

(2) If some $x_i = 0$, then, $x_{i-1}x_{i+1} < 0$, $i = 2,3,\cdots,n-1$.

Assume that an $n\times n$ matrix B is invertible. Since the equation $\boldsymbol{AX}=\lambda \boldsymbol{BX}$ is equivalent to the equation $\boldsymbol{B}^{-1}\boldsymbol{AX}=\lambda \boldsymbol{X}$, combined with the properties of Jacobi matrix and invertible matrix, we have the following lemma.

Lemma 2. Assume that an $n\times n$ Jacobi matrix $\boldsymbol{A}$ is symmetric, an $n\times n$ diagonal matrix B is invertible. Suppose that $(\lambda, \boldsymbol{X})$, $(\mu, \boldsymbol{Y})$ ($\lambda \neq \mu$) are two eigenpairs of generalized eigenvalue equation $\boldsymbol{AX}=\lambda \boldsymbol{BX}$, then, $\boldsymbol{X}^T\boldsymbol{BY}=0$.

Matrices $\boldsymbol{M}$, $\boldsymbol{K}$, $\boldsymbol{X}$, $\boldsymbol{Y}$ can be expressed the following partitioned matrices.

$$\boldsymbol{M}=\begin{bmatrix} \boldsymbol{M}_1 & -l_p\boldsymbol{e}_p\boldsymbol{\varepsilon}_1^T \\ -l_p\boldsymbol{\varepsilon}_1\boldsymbol{e}_p^T & \boldsymbol{M}_2 \end{bmatrix},$$

$$\boldsymbol{K}=\begin{bmatrix} \boldsymbol{K}_1 & \boldsymbol{O} \\ \boldsymbol{O} & \boldsymbol{K}_2 \end{bmatrix}, \quad \boldsymbol{X}=\begin{bmatrix} \boldsymbol{X}_1 \\ \boldsymbol{X}_2 \end{bmatrix}, \quad \boldsymbol{Y}=\begin{bmatrix} \boldsymbol{Y}_1 \\ \boldsymbol{Y}_2 \end{bmatrix},$$

where $p\times p$ Jacobi matrix $\boldsymbol{M}_1$ is the top left corner matrix of $\boldsymbol{M}$, $(n-p)\times(n-p)$ Jacobi matrix $\boldsymbol{M}_2$ is the bottom right corner matrix of $\boldsymbol{M}$, $p\times p$ diagonal matrix $\boldsymbol{K}_1$ is the top left corner matrix of $\boldsymbol{K}$, $(n-p)\times(n-p)$ diagonal matrix $\boldsymbol{K}_2$ is the bottom right corner matrix of $\boldsymbol{K}$, $p\times 1$ unit vector $\boldsymbol{e}_p=(0,0,\cdots,1)^T$, $(n-p)\times 1$ unit vector $\boldsymbol{\varepsilon}_1=(1,0,\cdots,0)^T$, $\boldsymbol{X}_1$, $\boldsymbol{Y}_1 \in \boldsymbol{R}^p$, $\boldsymbol{X}_2$, $\boldsymbol{Y}_2 \in \boldsymbol{R}^{n-p}$.

Formula (1) can be rewritten as follows.

$$\begin{cases} -\alpha^2\boldsymbol{M}_1\boldsymbol{X}_1+\alpha^2 l_p x_{p+1}\boldsymbol{e}_p+\boldsymbol{K}_1\boldsymbol{X}_1=\boldsymbol{O}, \\ \alpha^2 l_p x_p\boldsymbol{\varepsilon}_1-\alpha^2\boldsymbol{M}_2\boldsymbol{X}_2+\boldsymbol{K}_2\boldsymbol{X}_2=\boldsymbol{O}. \end{cases}$$

$$\begin{cases} -\beta^2\boldsymbol{M}_1\boldsymbol{Y}_1+\beta^2 l_p y_{p+1}\boldsymbol{e}_p+\boldsymbol{K}_1\boldsymbol{Y}_1=\boldsymbol{O}, \\ \beta^2 l_p y_p\boldsymbol{\varepsilon}_1-\beta^2\boldsymbol{M}_2\boldsymbol{Y}_2+\boldsymbol{K}_2\boldsymbol{Y}_2=\boldsymbol{O}. \end{cases} \tag{2}$$

From (2), combined with lemma 2, the problem solution can be transformed into solving the following problem I and II:

Problem I Given $\boldsymbol{K}_1$, $\boldsymbol{M}_1$ and positive real number α, β ($|\alpha| \neq |\beta|$), $\boldsymbol{X}_2$, $\boldsymbol{Y}_2 \in \boldsymbol{R}^{n-p}$ and l_p are known, solve vector $\boldsymbol{X}_1$, $\boldsymbol{Y}_1 \in \boldsymbol{R}^p$ which satisfied $x_1 \neq 0$, $y_1 \neq 0$ and

$$\begin{cases} (\boldsymbol{M}_1-\dfrac{1}{\alpha^2}\boldsymbol{K}_1)\boldsymbol{X}_1=l_p x_{p+1}\boldsymbol{e}_p, \\ (\boldsymbol{M}_1-\dfrac{1}{\beta^2}\boldsymbol{K}_1)\boldsymbol{Y}_1=l_p y_{p+1}\boldsymbol{e}_p. \end{cases} \tag{3}$$

Problem II Given positive real number α, β ($|\alpha| \neq |\beta|$), $\boldsymbol{X}_2$, $\boldsymbol{Y}_2 \in \boldsymbol{R}^{n-p}$ and l_p are known, solve $\boldsymbol{K}_2$ and $\boldsymbol{M}_2$, such that

$$\begin{cases}(\boldsymbol{M}_2-\frac{1}{\alpha^2}\boldsymbol{K}_2)\boldsymbol{X}_2=l_p x_p \boldsymbol{\varepsilon}_1,\\ (\boldsymbol{M}_2-\frac{1}{\beta^2}\boldsymbol{K}_2)\boldsymbol{Y}_2=l_p y_p \boldsymbol{\varepsilon}_1,\\ \boldsymbol{X}_1^T\boldsymbol{K}_1\boldsymbol{Y}_1+\boldsymbol{X}_2^T\boldsymbol{K}_2\boldsymbol{Y}_2=\boldsymbol{O},\end{cases} \tag{4}$$

where vectors $\boldsymbol{X}_1=(x_1,x_2,\ldots,x_p)^T$, $\boldsymbol{Y}_1=(y_1,y_2,\ldots,y_p)^T\in\boldsymbol{R}^p$ are the solution of Problem I.

Let $\phi_p(\omega)=\det(\boldsymbol{M}_1-\frac{1}{\omega^2}\boldsymbol{K}_1)$, $\phi_0(\omega)=1$, $\phi_1(\omega)=l_1-\frac{1}{\omega^2 c_1}$, $i\times i$ determinant $\phi_i(\omega)$ is the ith order principal minor determinant of $\phi_p(\omega)$.

Obviously, based on the result of $\phi_{i-1}(\omega)$, we can expand determinant $\phi_i(\omega)$ by the ith row and get

$$\phi_i(\omega)=(l_{i-1}+l_i-\frac{1}{\omega^2 c_i})\phi_{i-1}(\omega)-l_{i-1}^2\phi_{i-2}(\omega),\quad i=2,3,\ldots,n-p. \tag{5}$$

Let $u_i=x_i-x_{i-1}$, $v_i=y_i-y_{i-1}$,

$h_i=\alpha^2 y_i u_{i+1}-\beta^2 x_i v_{i+1}$,

$f_i=\alpha^2 y_i u_i-\beta^2 x_i v_i$,

$g_i=\alpha^2\beta^2(u_{i+1}v_i-v_{i+1}u_i)$,

$i=p+1,p+2,\ldots,n$,

$s=\boldsymbol{X}_1^T\boldsymbol{K}_1\boldsymbol{Y}_1$,

$t=\frac{g_{p+1}}{h_{p+1}}x_{p+1}y_{p+1}+\sum_{i=p+2}^{n}\frac{g_i}{h_i}x_i y_i(\prod_{j=p+1}^{i-1}\frac{f_j}{h_j})$,

where $u_{n+1}=-x_n$, $v_{n+1}=-y_n$.

Theorem 1. Systems of linear equations $(\boldsymbol{M}_1-\frac{1}{\omega^2}\boldsymbol{K}_1)\boldsymbol{X}_1=l_p x_{p+1}\boldsymbol{e}_p$ have a unique solution if and only if $\phi_p(\omega)\neq 0$, and the solution is $x_i=x_{p+1}\phi_p^{-1}(\omega)\phi_{i-1}(\omega)\prod_{j=i}^{p}l_j$, $i=1,2,\ldots,p$.

Proof. Systems of linear equations $(\boldsymbol{M}_1-\frac{1}{\omega^2}\boldsymbol{K}_1)\boldsymbol{X}_1=l_p x_{p+1}\boldsymbol{e}_p$ have a unique solution if and only if $\det(\boldsymbol{M}_1-\frac{1}{\omega^2}\boldsymbol{K}_1)\neq 0$, that is $\phi_p(\omega)\neq 0$.

If $\phi_p(\omega)\neq 0$, then,

$$\boldsymbol{X}_1=(x_1,x_2,\ldots,x_p)^T$$
$$=l_p x_{p+1}(\boldsymbol{M}_1-\frac{1}{\omega^2}\boldsymbol{K}_1)^{-1}\boldsymbol{e}_p=l_p x_{p+1}[\det(\boldsymbol{M}_1-\frac{1}{\omega^2}\boldsymbol{K}_1)]^{-1}\cdot(\boldsymbol{M}_1-\frac{1}{\omega^2}\boldsymbol{K}_1)^*\boldsymbol{e}_p.$$

Because $(\boldsymbol{M}_1 - \frac{1}{\omega^2}\boldsymbol{K}_1)^* \boldsymbol{e}_p$ is algebraic cofactor of the ith row elements of the matrix $\boldsymbol{M}_1 - \frac{1}{\omega^2}\boldsymbol{K}_1$, and then from (5), we have

$$\boldsymbol{X}_1 = (x_1, x_2, \mathrm{L}\ , x_p)^T = l_p x_{p+1}\phi_p^{-1}(\omega)[\phi_0(\omega)\prod_{j=1}^{p-1} l_j, \phi_1(\omega)\prod_{j=2}^{p-1} l_j, \phi_2(\omega)\prod_{j=3}^{p-1} l_j, \mathrm{L}\ , \phi_{p-1}(\omega)]^T .$$

Therefore, $x_i = x_{p+1}\phi_p^{-1}(\omega)\phi_{i-1}(\omega)\prod_{j=i}^{p} l_j$, $i = 1, 2, \mathrm{L}\ , p$.

From Theorem 1, combined with lemma 1, the conditions for ProblemIhas unique solution is easy to derive.

Theorem 2. ProblemIhas a unique solution if and only if $\phi_p(\alpha) \neq 0$, $\phi_p(\beta) \neq 0$ and $x_{p+1} y_{p+1} \neq 0$, and the solution is $x_i = x_{p+1}\phi_p^{-1}(\alpha)\phi_{i-1}(\alpha)\prod_{j=i}^{p} l_j,$

$$y_i = y_{p+1}\phi_p^{-1}(\beta)\phi_{i-1}(\beta)\prod_{j=i}^{p} l_j, \quad i = 1, 2, \mathrm{L}\ , p . \tag{6}$$

Theorem 3. ProblemII has a unique solution if and only if

(1) $h_i \neq 0$, and h_i, f_i, g_i have the same sign, $i = p+1, p+2, \mathrm{L}\ , n$;

(2) $s + t l_p = 0$;

and the solution is

$$l_i = l_p \prod_{j=p+1}^{i} \frac{f_j}{h_j}, \quad i = p+1, p+2, \mathrm{L}\ , n,$$

$$c_{p+1} = \frac{h_{p+1}}{l_p g_{p+1}}, \quad c_i = \frac{h_i}{l_p g_i} \prod_{j=p+1}^{i-1} \frac{h_j}{f_j}, \quad i = p+2, p+3, \mathrm{L}\ , n . \tag{7}$$

Proof. The first two equations of systems of linear equations (4) can be rewritten as follows.

$$\begin{cases} u_{i+1}\alpha^2 l_i + x_i \dfrac{1}{c_i} = u_i \alpha^2 l_{i-1}, \\ v_{i+1}\beta^2 l_i + y_i \dfrac{1}{c_i} = v_i \beta^2 l_{i-1}, \end{cases} \quad i = p+1, p+2, \mathrm{L}\ , n , \tag{8}$$

where $u_{n+1} = -x_n$, $v_{n+1} = -y_n$.

Systems of linear equations (8) have a unique solution if and only if coefficient determinant

$$\begin{vmatrix} u_{i+1}\alpha^2 & x_i \\ v_{i+1}\beta^2 & y_i \end{vmatrix} \neq 0 ,$$

that is $h_i \neq 0$,

and the solution is $l_i = \frac{f_i l_{i-1}}{h_i}$, $\frac{1}{c_i} = \frac{g_i l_{i-1}}{h_i}$, $i = p+2, p+3, \cdots, n$.

And for $l_i > 0$, $c_i > 0$, h_i, f_i, g_i have the same sign, $i = p+1, p+2, \cdots, n$.

In addition, the expressions of l_i and $\frac{1}{c_i}$ can be expanded as follows.

$$l_i = l_{i-1}\frac{f_i}{h_i} = l_{i-2}\frac{f_{i-1}}{h_{i-1}}\cdot\frac{f_i}{h_i} = \cdots = l_p \prod_{j=p+1}^{i} \frac{f_j}{h_j},$$

$i = p+1, p+2, \cdots, n$,

$$\frac{1}{c_{p+1}} = l_p \frac{g_{p+1}}{h_{p+1}}, \text{ that is } c_{p+1} = \frac{h_{p+1}}{l_p g_{p+1}},$$

$$\frac{1}{c_i} = l_{i-1}\frac{g_i}{h_i} = l_{i-2}\frac{f_{i-1}}{h_{i-1}}\cdot\frac{g_i}{h_i} = \cdots = l_p \frac{g_i}{h_i}\prod_{j=p+1}^{i-1} \frac{f_j}{h_j},$$

that is

$$c_i = \frac{h_i}{l_p g_i}\prod_{j=p+1}^{i-1}\frac{h_j}{f_j}, \quad i = p+2, p+3, \cdots, n.$$

Therefore,

$$\boldsymbol{X}_2^T \boldsymbol{K}_2 \boldsymbol{Y}_2 = \sum_{i=p+1}^{n} \frac{1}{c_i} x_i y_i$$

$$= l_p \frac{g_{p+1}}{h_{p+1}} x_{p+1} y_{p+1} + l_p \sum_{i=p+2}^{n} \frac{g_i}{h_i}\left(\prod_{j=p+1}^{i-1}\frac{f_i}{h_i}\right) x_i y_i$$

$$= l_p\left[\frac{g_{p+1}}{h_{p+1}} x_{p+1} y_{p+1} + \sum_{i=p+2}^{n} \frac{g_i}{h_i} x_i y_i \left(\prod_{j=p+1}^{i-1}\frac{f_i}{h_i}\right)\right] = t l_p.$$

By the third equation of systems of linear equations (4), we have $s + t l_p = 0$.

From theorem 2 and theorem 3, we have the necessary and sufficient condition for unique solution of the problem.

Theorem 4. Problem has a unique solution if and only if

(1) $\phi_p(\alpha) \neq 0$, $\phi_p(\beta) \neq 0$ and $x_{p+1} y_{p+1} \neq 0$;

(2) $h_i \neq 0$, and h_i, f_i, g_i have the same sign, $i = p+1, p+2, \cdots, n$;

(3) $s + t l_p = 0$.

3. Numerical Algorithm and Example

From above discussion, the algorithm for solving Problem is presented as follows.

Algorithm. Given diverse real number α, β, $\{c_i\}_{i=1}^{p}$, $\{l_i\}_{i=1}^{p}$, and vectors $\boldsymbol{X}_2$, $\boldsymbol{Y}_2 \in \boldsymbol{R}^{n-p}$, construct $\{c_i\}_{i=p+1}^{n}$, $\{l_i\}_{i=p+1}^{n}$, and vectors $\boldsymbol{X}_1$, $\boldsymbol{Y}_1 \in \boldsymbol{R}^{p}$.

Step 1. From recurrence formula (5), compute $\phi_i(\alpha)$, $\phi_i(\beta)$, $i = 1, 2, \cdots, p$,

If $\phi_p(\alpha) = 0$ or $\phi_p(\beta) = 0$ or $x_{p+1} y_{p+1} = 0$, go to step 6.

Step 2. From formula (6), compute x_i and y_i , $i = 1, 2, \ldots, p$.
Step 3. Compute h_i, $i = p+1, p+2, \ldots, n$.
If some $h_i = 0$, $i = p+1, p+2, \ldots, n$, go to step 6.
Step 4. Compute f_i, g_i, $i = p+1, p+2, \ldots, n$.
If h_i, f_i, g_i ($i = p+1, p+2, \ldots, n$) have different sign, go to step 6.
Step 5. Compute $s + tl_p$. If $s + tl_p \neq 0$, go to step 6.
Step 6. The solution can not be determined uniquely, end the algorithm.
Step 7. From formula (7), compute l_i, c_i, $i = p+1, p+2, \ldots, n$.

Example. Given $\alpha = 1.0457$, $\beta = -0.7384$, $n = 9$, $\{c_i\}_{i=1}^4 = \{2, 1, 0.5, 1\}$, $\{l_i\}_{i=1}^4 = \{1.5, 1, 0.7, 0.3\}$, and

$$\boldsymbol{X}_2 = (x_5, x_6, x_7, x_8, x_9)^T = (-0.2378, -0.2683, -0.2067, 0.5261, 0.4847)^T,$$

$$\boldsymbol{Y}_2 = (y_5, y_6, y_7, y_8, y_9)^T = (0.6729, 0.5216, -0.2594, 0.0249, 0.1498)^T.$$

We need to construct $\boldsymbol{M}$, $\boldsymbol{K}$ and design a LC circuit system.
By Algorithm, we get

$$\{\phi_i(\alpha)\}_{i=1}^4 = \{1.0428, -0.5967, -0.9658, 0.2098\},$$

$$\{\phi_i(\beta)\}_{i=1}^4 = \{0.5829, -1.8619, 3.0820, -1.6587\},$$

$\phi_4(\alpha) \neq 0$ and $\phi_4(\beta) \neq 0$, $x_5 y_5 \neq 0$,

$$\boldsymbol{X}_1 = (x_1, x_2, x_3, x_4)^T = (-0.3570, -0.2482, 0.1420, 0.3284)^T,$$

$$\boldsymbol{Y}_1 = (y_1, y_2, y_3, y_4)^T = (-0.1278, -0.0497, 0.1586, -0.3751)^T.$$

$$\{h_i\}_{i=5}^9 = \{-0.0421, -0.0791, -0.1758, -0.0370, -0.0398\},$$

$$\{f_i\}_{i=5}^9 = \{-0.2807, -0.0395, -0.1055, -0.0616, -0.0398\},$$

$$\{g_i\}_{i=5}^9 = \{-0.0701, -0.0198, -0.3516, -0.0616, -0.0398\},$$

$h_i \neq 0$, and h_i, f_i, g_i have the same sign, $i = 5, 6, 7, 8, 9$,

$$s + tl_p = -4.8572\text{e}-017 \approx 0,$$

Then, $\{l_i\}_{i=5}^9 = \{2.0025, 1.0006, 0.6003, 1.0006, 1.0000\}$,

$$\{c_i\}_{i=5}^9 = \{1.9991, 1.9994, 0.4997, 0.9995, 0.9999\}.$$

Using Matlab, it is not difficult to get all generalized eigenvalues of $(-\omega^2 \boldsymbol{M} + \boldsymbol{K})\boldsymbol{X} = \boldsymbol{O}$.
The eigenvector which corresponds to eigenvalue $\alpha i = 1.0457i$ is

$$\boldsymbol{X} = (-0.3571, -0.2482, 0.1420, 0.3284, -0.2377, -0.2683, -0.2067, 0.5261, 0.4847)^T,$$

and the eigenvector which corresponds to eigenvalue $\beta i = -0.7384i$ is

$$\boldsymbol{Y} = (-0.1278, -0.0497, 0.1586, -0.3751, 0.6730, 0.5216, -0.2594, 0.0249, 0.1498)^T.$$

These data indicate that the algorithm is very effective.

4. Conclusions

The problem is converted to an inverse eigenvalue problem for Jacobi matrix. The solvability condition and numerical method of this inverse eigenvalue problem are given. Numerical results show that the algorithm work well.

5. Acknowledgements

This work is supported by the Scientific Research Fund Project of Jingchu University of Technology No. QN201607.

6. References

[1] Xiaoqian Wu and Erxiong Jiang 2012 A new algorithm on the inverse eigenvalue problem for double dimensional Jacobi matrice *J.Linear Algebra Appl.* **7** 1760-1770
[2] Gusein Sh. G. 2012 On an inverse problem for two spectra of finite Jacobi matrices *J. Appl Math Comput.* **14** 7573-789
[3] Ying Wei 2015 Inverse eigenvalue problem of Jacobi matrix with mixed data *J. Linear Algebra Appl.* **466** 102-116
[4] Yu Bai and Guangsheng Wei 2016 A new inverse three spectra theorem for Jacobi matrices J. *Linear Algebra Appl.* **493** 301-312
[5] Graham M.L. and Gladwell Gbadeyan J.A. 1985 On the inverse problem of the vibrating string or rod *J. Mech Appl Math.* **38** 169-174
[6] Peter N. and Frank U. 1997 Inverse eigenvalue problems associated with spring-mass systems *J. Linear Algebra Appl.* **254** 409-425
[7] Xiantong Huang, Xiyan Hua and Lei Zhang 2007 Physical parameters reconstruction of a fixed-fixed mass-spring system from its characteristic data *J. Comput Appl Math.* **2** 645-655
[8] Ram Y.M. Graham M.L. and Gladwell 1994 Constructing a finite element model of a vibrating rod from eigendata *J. Sound Vib.* **2** 229-237
[9] Zhengjian Bai 2008 Constructing the physical parameters of a damped vibrating system from eigendata *J. Linear Algebra Appl.* **2-3** 625-656
[10] Xia Tian and Hua Dai 2009 Inverse mode problems for the finite element model of a vibrating rod *J. Appl Math Comput.* **2** 479-86
[11] Hua Dai 1994 Inverse eigenvalue problem for Jacobi matrices *J. Comput. Phys.* **4** 451-456

The Identification Method of Transformer Excitation Inrush Current Based on the Decay Characteristics of Aperiodic Component

Jianqin Feng[1], Pei Zhang[1,*] and Zhifei Chen[2]

[1]School of Electrical and Information Engineering, Zhengzhou University of Light Industry, Dongfeng Rd 5, Zhengzhou, China
[2]School of Management and Engineering, Zhengzhou University, Science Avenue 100, Zhengzhou, China

Email: zhangpei_zzuli@163.com

Abstract. Focusing on the optical current transformer secondary side current aperiodic component decay characteristics, a novel identification method of excitation inrush current based on aperiodic component dynamic decay characteristics was proposed, and the traditional transformer excitation inrush current identification method based on the various static characteristics were break through. First, considering the transformer losses were exist and the truth of characteristics of excitation inrush current was not a periodic function, the aperiodic component was calculated by taking different cycles of the excitation inrush current derivation. Second, taking the (k+1)th cycle of excitation inrush current integration as an example, the five different forms excitation inrush current integration were found, and the five forms aperiodic component function expression were written. Then, taking derivation of five-stage aperiodic component and the aperiodic component decay speed expression was got. Compared with the normal operation and short circuit faults occurred, using the aperiodic component decay dynamic time constant to identity the excitation inrush current. Combined with the related simulation, not only verified this novel identification method was effective, but also provides a new way for the application of dynamic characteristics in excitation inrush current identification and relay protection.

1. Introduction

The inrush current produced when transformer no-load switching, its value could up to the 4~8 times of rated current. The inrush current could cause the winding deformation of the transformer, harm the insulation of the transformer, reduce the reliability of the power supply and cause the unwanted operation of transformer differential protection[1], which has been widely concerned for a long time.

At present, many recognition methods of transformer inrush current have been proposed [1-7]. Reference [2] proposed a method of multi feature about identifying inrush current in the feature space based on the summary characteristics of each feature value of differential current. Reference[1] obtained the imaginary part of the fundamental wave according to the full wave Fourier algorithm and half wave Fourier algorithm, recognized inrush current by using the correlation coefficient. Reference

IOP Conf. Series: Journal of Physics: Conf. Series **1026** (2018) 012029 doi:10.1088/1742-6596/1026/1/012029

[3] proposed a classification algorithm based on virtual equivalent inductance to identify transformer inrush current.

These identification methods focused on the use of many static characteristics, such as the existence of discontinuous angle, the contents of second harmonic or third harmonic, the symmetrical of waveform.

Any method which takes the static characters as the distinguishing feature is easily disturbed by noise factors. When considering the various static characters, the inrush current was assumed as constant in the sampling process, which is contradictory to the actual attenuation process of inrush current. The static characters existed both in fault process and normal operation. The only different is the contents [4]. In a word, the present methods could not solve the problem of identifying inrush current because the complexity of inrush current.

A novel method to identify the inrush current is proposed based on the dynamic characteristics of the aperiodic component of the inrush current in this paper.

2. The Aperiodic Component of Inrush Current

2.1. The Definition of Aperiodic Component of Inrush Current Considering Attenuation

In the case of ideal transformer (ignoring the transformer loss), the inrush current is a complete periodic function and does not decay. So, the DC component Iμ0 could be obtained by integrating the inrush current in any period, such as formula 1[5].

$$I_{\mu 0} = \frac{1}{2\pi}\int_0^{2\pi} i(\xi)d\xi \tag{1}$$

Where i is the inrush current of the transformer, ξ is the integral variable.

The aperiodic component of the inrush current of the ideal transformer is constant in this time.

Considering the actual attenuation of inrush current, the inrush current is no longer a simple periodic function. Therefore, different aperiodic components could be calculated according to the integral of the inrush current in different periods[6].

According to the traditional integral method, this research proposed a novel method to calculate the aperiodic component considering the decay situation as shown in formula (2).

$$i_{DC}(t) = \frac{1}{T}\int_t^{t+T} i(\xi)d\xi \tag{2}$$

Where t is the time variable, T is the power frequency period.

2.2. The Expression of Excitation Inrush Current

The inrush current is calculated by the relation between the flux and the current of the transformer. Assuming t_j is the alternating time between the duration period and the intermittent period. t_0 is the moment of the transformer no-load closing(the initial moment of the first intermittent period). t_1 is the initial moment of the first duration period. t_2 is the initial moment of the second intermittent period. t_3 is the initial moment of the second duration period. tm is the final moment of the last duration period(the final moment of the whole inrush current).

Therefore, $t_{2k} \le t < t_{2k+1}$ is the intermittent period, $t_{2k+1} \le t < t_{2(k+1)}$ is the duration period. (*k=0, 1,* ..., $\frac{m}{2}-1$).

According to the relationship between the flux and the current of the transformer, the method Fourier decomposition method is used to decompose the inrush current into the periodic component and the free component in the reference.

The inrush current is due to the transformer core saturation, the relation between the magnetic flux and voltage could be described as formula(3) when the amplitude of rated voltage and the amplitude of rated flux as the base value to represent the voltage of u and Φ.

$$u = \frac{d\phi}{dt} \tag{3}$$

Assuming the transformer no-load closing in the $t=0$, and the voltage in the transformer equals $u = u_m \sin(\omega t + \alpha)$. Where u_m is the peak voltage.

The mathematical description of inrush current $i(t)$ is formula (4).

$$i(t)=i_p(t)+i_{np}(t) \tag{4}$$

Where i_p and i_{np} are the forced component and free component of inrush current respectively, as shown in formula (5) and (6).

$$i_p(t)=\begin{cases} 0, & t_{2k} \le t < t_{2k+1} \\ I_{m\mu}\sin(\omega t+\alpha-\beta_\mu), & t_{2k+1} \le t < t_{2(k+1)} \end{cases} \tag{5}$$

Where $I_{m\mu}$ is the amplitude of the forced component, α-$\beta\mu$ is the phase angle of the inrush current during the duration period.

$$i_{np}(t)=\begin{cases} 0, & t_{2k} \le t < t_{2k+1} \\ I_{np0}(t_{2k+1})e^{-(t-t_{2k+1})/\tau_\mu}, & t_{2k+1} \le t < t_{2(k+1)} \end{cases} \tag{6}$$

Where I_{np0} is the initial value of the free component, τ_μ is the decay time constant of aperiodic component.

2.3. The Aperiodic Component Expression

Taking the integral of the inrush current in the k+1 period as an example, the aperiodic component of the inrush current obtained.

(1) When t was in $[kT,\ t_{2k+1}]$, because $t_{2k+1}-kT < t_{2k+3}-(k+1)T$, the range of $t+T$ was $(k+1)T \le t+T < t_{2k+1}+T$, the integral value of aperiodic component in one cycle was the integral value of aperiodic component in $k+1$ cycle. The integral region is shown in Figure1(a).

(2) When t was in $[t_{2k+1},\ t_{2k+3}-T]$, the range of $t+T$ was $t_{2k+1}+T \le t+T < t_{2k+3}$, the integral region is shown in Figure1(b).

(3) When t was in $[t_{2k+3}-T,\ t_{2k+4}-T]$, the range of $t+T$ was $t_{2k+3} \le t+T < t_{2k+4}$, the integral region was shown in Figure1(C).

(4) When t was in $[t_{2k+4}-T,\ t_{2k+2}]$, the range of $t+T$ was $t_{2k+4} \le t+T < t_{2k+2}+T < (k+2)T$, the integral region is shown in Figure1(d).

(5) When t was in $[t_{2k+2},\ (k+1)T)$, the integral region is all of the inrush current of next period. The integral region is shown in Figure1(e).

IOP Conf. Series: Journal of Physics: Conf. Series **1026** (2018) 012029 doi:10.1088/1742-6596/1026/1/012029

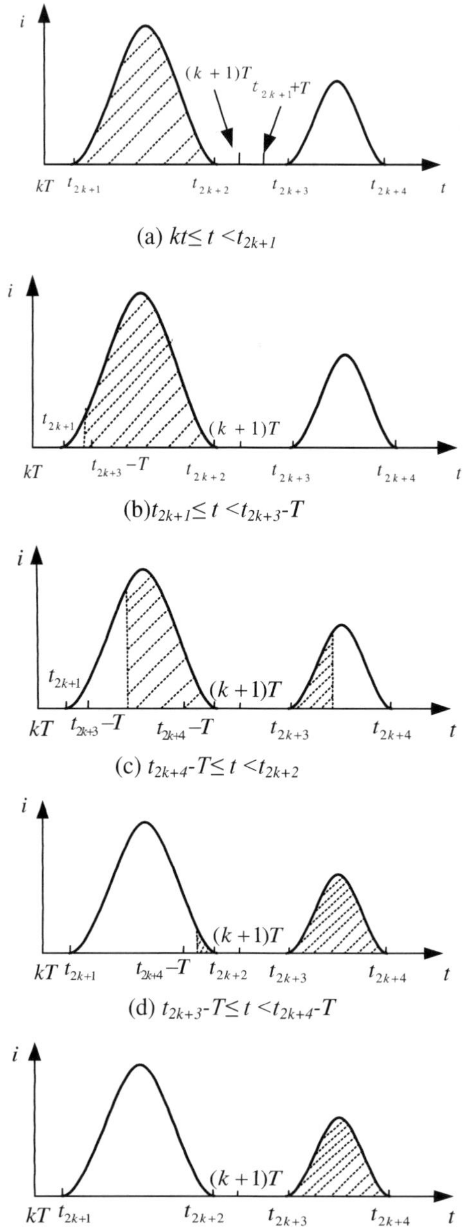

(a) $kt \leq t < t_{2k+1}$

(b) $t_{2k+1} \leq t < t_{2k+3}-T$

(c) $t_{2k+4}-T \leq t < t_{2k+2}$

(d) $t_{2k+3}-T \leq t < t_{2k+4}-T$

(e) $t_{2k+2} \leq t < (k+1)T$

Figure 1. The integral cases of (k+1)th period of inrush current.

Based on the 5 cases above, the aperiodic component i_{DC} was obtained according to formula (2)~(6), as shown in formula (7).

IOP Conf. Series: Journal of Physics: Conf. Series **1026** (2018) 012029 doi:10.1088/1742-6596/1026/1/012029

$$i_{DC}(t)=\begin{cases} 1/T\{-(I_{m\mu}/\omega)[\cos(\omega t_{2k+2}+\alpha-\beta_\mu)-\cos(\omega t_{2k+1}+\alpha-\beta_\mu)]- \\ \quad \tau_\mu I_{np0}(t_{2k+1})[e^{-(t_{2k+2}-t_{2k+1})/\tau_\mu}-1]\}, \quad \boxed{kT\le t<t_{2k+1}}; \\ 1/T\{-(I_{m\mu}/\omega)[\cos(\omega t_{2k+2}+\alpha-\beta_\mu)-\cos(\omega t_{2k+3}+\alpha-\beta_\mu)]- \\ \quad \tau_\mu I_{np0}(t_{2k+1})[e^{-(t_{2k+2}-t_{2k+1})/\tau_\mu}-e^{-(t-t_{2k+1})/\tau_\mu}]- \\ \quad \tau_\mu I_{np0}(t_{2k+2})[e^{-(t+T-t_{2k+3})/\tau_\mu}-1]\}, \quad \boxed{t_{2k+1}\le t<t_{2k+3}-T} \\ 1/T\{-(I_{m\mu}/\omega)[\cos(\omega t_{2k+2}+\alpha-\beta_\mu)-\cos(\omega t+\alpha-\beta_\mu)]- \\ \quad \tau_\mu I_{np0}(t_{2k+1})[e^{-(t_{2k+2}-t_{2k+1})/\tau_\mu}-e^{-(t-t_{2k+1})/\tau_\mu}]\}, \\ \quad \boxed{t_{2k+3}-T\le t<t_{2k+4}-T}; \\ 1/T\{-(I_{m\mu}/\omega)[\cos(\omega t_{2k+2}+\alpha-\beta_\mu)-\cos(\omega t+\alpha-\beta_\mu)]- \\ \quad (I_{m\mu}/\omega)[\cos(\omega t_{2k+4}+\alpha-\beta_\mu)-\cos(\omega t_{2k+3}+\alpha-\beta_\mu)]- \\ \quad \tau_\mu I_{np0}(t_{2k+1})[e^{-(t_{2k+2}-t_{2k+1})/\tau_\mu}-e^{-(t-t_{2k+1})/\tau_\mu}]-\tau_\mu I_{np0}(t_{2k+2}) \\ \quad [e^{-(t_{2k+4}-t_{2k+3})/\tau_\mu}-1]\}, \quad \boxed{t_{2k+4}-T\le t<t_{2k+2}}; \\ 1/T\{-(I_{m\mu}/\omega)[\cos(\omega t_{2k+4}+\alpha-\beta_\mu)-\cos(\omega t_{2k+3}+\alpha-\beta_\mu)]- \\ \quad \tau_\mu I_{np0}(t_{2k+2})[e^{-(t_{2k+4}-t_{2k+3})/\tau_\mu}-1]\}, \quad \boxed{t_{2k+2}\le t<(k+1)T}; \end{cases} \quad (7)$$

The characters of aperiodic component variation of inrush current

In order to have an intuitive description of aperiodic components, the aperiodic component change curve was shown in Figure2 according to the formula (6). Two periods was elected for detail analysis at initial decay period and the end of decay period.

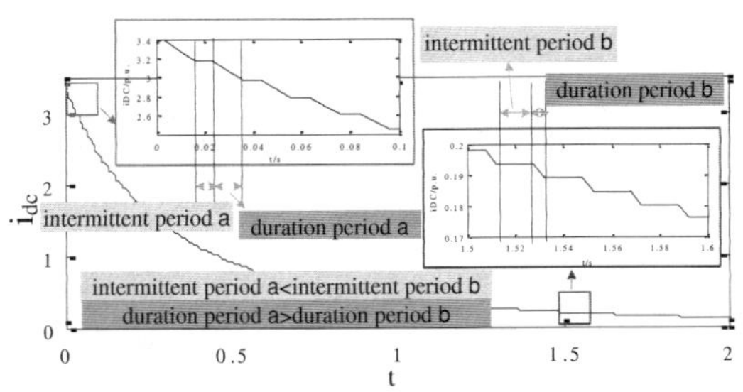

Figure2. The change curve of inrush current aperiodic component.

According to Figure2, the aperiodic component decay rule is close to an exponential function and the decay time is more than 2 seconds. Based on the statistical data, the inrush current of transformer aperiodic components could last tens of seconds or even a few minutes. The aperiodic component of inrush current decays very slow.

According to the local curve of Figure2, the aperiodic component does not decay in the intermittent period. The aperiodic component decays in the duration period.

According to the two sampling curves in the beginning of period and the end of period, the duration period is longer than the intermittent period at the beginning, but become shorter and shorter. Therefore, the aperiodic component attenuation velocity decreased. Namely, the decay rate of the aperiodic component is the fastest at the beginning of the inrush current.

In summary, the inrush current aperiodic components decay speed is slower than the aperiodic component of short-circuit current decay time, which means the dynamic time constant of inrush current aperiodic component is greater than the dynamic time constant of short-circuit current aperiodic component.

3. The Attenuation Velocity of Aperiodic Component of Inrush Current

Taking derivation of aperiodic components (7) analysis, the attenuation velocity of aperiodic component expressions was obtained as shown in formula (8).

$$i'_{DC}(t)=\begin{cases} 0; & \boxed{kT \le t < t_{2k+1}}; \\ 1/T[-I_{np0}(t_{2k+1})e^{-(t-t_{2k+1})/\tau_\mu}+I_{np0}(t_{2k+2})e^{-(t+T-t_{2k+3})/\tau_\mu}]; & \boxed{t_{2k+1} \le t < t_{2k+3}-T}; \\ \{-T(\mathrm{I}_{m\mu}/\omega)[\cos(\omega t_{2k+2}+\alpha-\beta_\mu)-\cos(\omega t+\alpha-\beta_\mu)]- \\ \quad \tau_\mu T I_{np0}(t_{2k+1})[e^{-(t_{2k+2}-t_{2k+1})/\tau_\mu}-e^{-(t-t_{2k+1})/\tau_\mu}]\}^{-2}\times \\ \quad [T\,\mathrm{I}_{m\mu}\sin(\omega t+\alpha-\beta_\mu)] & \boxed{t_{2k+3}-T \le t < t_{2k+4}-T}; \\ 1/T[-I_{np0}(t_{2k+1})e^{-(t-t_{2k+1})/\tau_\mu}+I_{np0}(t_{2k+2})e^{-(t+T-t_{2k+3})/\tau_\mu}]; & \boxed{t_{2k+4}-T \le t < t_{2k+2}}; \\ 0; & \boxed{t_{2k+2} \le t < (k+1)T}; \end{cases} \tag{8}$$

The attenuation velocity of aperiodic component of inrush current obtained according to formula (8), as shown in Figure3.

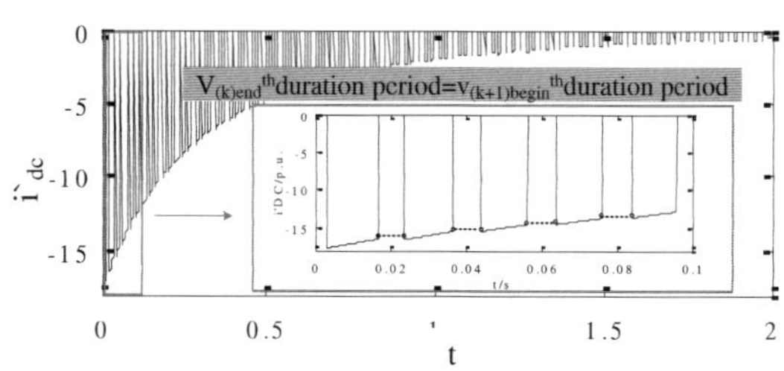

Figure3. The attenuation velocity of inrush current aperiodic component.

According to Figure3, the change rate of aperiodic component is negative. The negative sign represents the aperiodic component is decay.

The whole attenuation velocity decreased, but in the intermittent period, the aperiodic component attenuation velocity is zero, which means the aperiodic component does not decay. The aperiodic component decay rate continued to decrease in the duration period. In the two adjacent cycles, the initial decay rate of the next cycle is equal to the end decay rate of the previous cycle[7].

4. Simulation Analysis

MATLAB was used to simulate the electromagnetic calculation of power system in this research. The parameters of the system were shown in Figure4. The power supply was represented by a lumped parameter, and the lines were represented by Bergeron models.

IOP Conf. Series: Journal of Physics: Conf. Series **1026** (2018) 012029 doi:10.1088/1742-6596/1026/1/012029

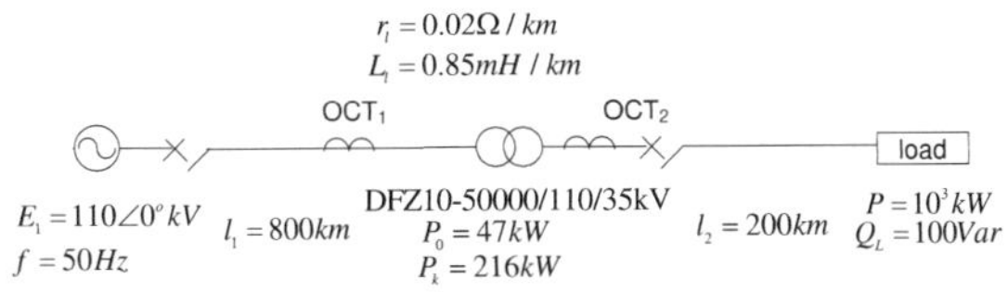

Figure 4. Model and parameters of simulation system.

(1) The decay rate of aperiodic component of short circuit current

The peak current of primary side was 89.2A when the transformer was operation normally. Because the circuit impedance tends to zero when the secondary side of the transformer occurred short-circuit fault, while the transformer primary side waveform decreased, the ultimately ended of the attenuation peak was 798.5A, which was the 8.95 times than the normal operation of the current peak value.

The component of the short circuit current aperiodic component was obtained by using the Fourier decomposition method, as shown in Figure5.

The short-circuit fault occurred in 0.5s in Figure5. According to the flux conservation principle, the transformer primary side current increased rapidly in 5 cycles because the existence of inductance. After reaching the peak value, the waveform shows attenuation trend because the existence of the loop impedance. The component of aperiodic component in the current of the primary side is exponentially decaying.

In this research, the data of the aperiodic component of short circuit current is used to exponential fitting. The fitting curve is shown in formula (9).

$$i_{np} = 9.794e^{-89.37t} \tag{9}$$

Where the time constant τ=0.011

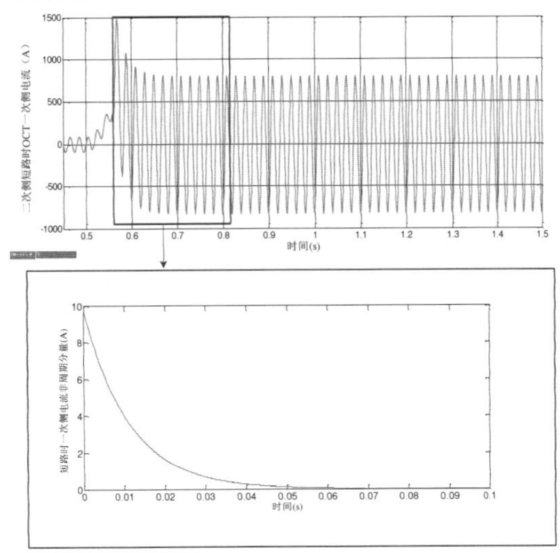

Figure 5. Short circuit current and its aperiodic components of primary side.

(2) The decay rate of aperiodic component of inrush current

In this research, the least square method is used to fit the aperiodic components in each period and the quantitative changes of aperiodic components of dynamic time constant obtained[8]. The dynamic time constant fitting curve of aperiodic component is shown in Table 1.

IOP Conf. Series: Journal of Physics: Conf. Series **1026** (2018) 012029 doi:10.1088/1742-6596/1026/1/012029

Table 1. The inrush current dynamic time constant of the aperiodic component fitting curve

T	1	2	3	4	5	6
τ	0.228	0.232	0.237	0.241	0.247	0.253
T	7	8	…	99	100	…
τ	0.257	0.261	…	0.635	0.698	…
τ	0.257	0.261	…	0.635	0.698	…

According to Table1, the time constant of the aperiodic component of inrush current value increased gradually from the first cycle and the minimum time constant of the initial period is 0.228.

Because the aperiodic component does not decay in the discontinuous period and only decay in the duration period, the average decay rate of aperiodic component in one cycle is slower and the time constant is also greater.

According to the time constant in Table1 and formula (9), it is found that the minimum time constant 0.228 of the aperiodic component of the inrush current is greater than the time constant 0.011 of the aperiodic component of the short-circuit current.

The aperiodic component of the inrush current of the transformer has the characteristics of slow attenuation and long duration. It provides a contrast with the characteristics of fast attenuation and short duration of the short circuit current aperiodic component.

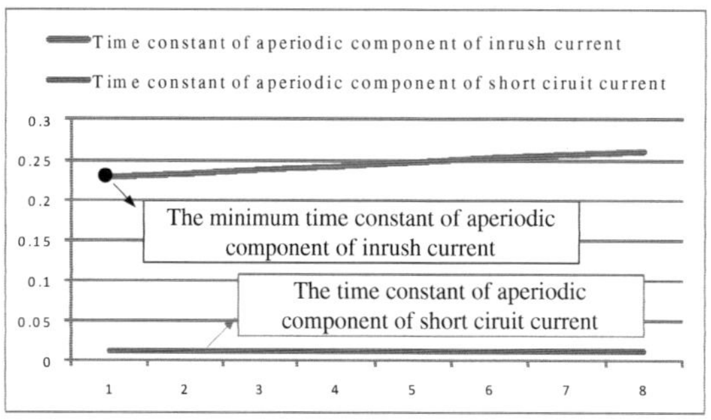

Figure 6. Variation rules of time constant of aperiodic component of inrush current and short circuit current.

5. The Identification Method of Inrush Current

Based on the above analysis, it is possible to distinguish the inrush current by using the different dynamic time constant of the aperiodic component of the inrush current and the time constant of the aperiodic component of the short circuit current. The flow chart is shown in Figure7.

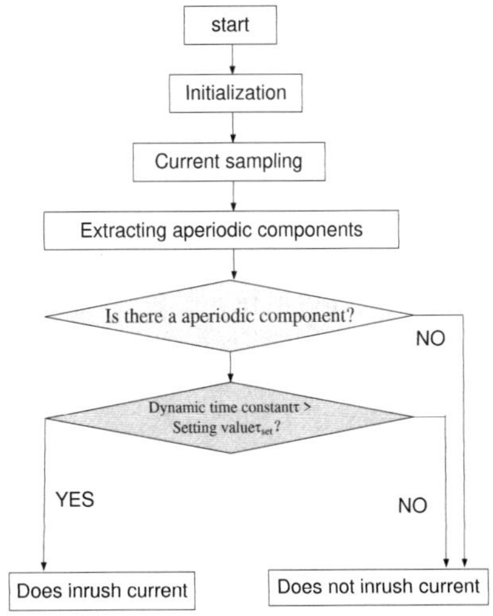

Figure 7. The discriminant flow chart of inrush current.

According to Figure7, the transformer relay protection device sampling current first, extracting the aperiodic components by using the least square method second, and judging whether there is an aperiodic sampling current. There is no inrush current if there is no discrimination of aperiodic components. If there is discrimination of aperiodic components, then go to further recognition according to the time constants. If the time constant is greater than the setting value, the case is inrush current. If the time constant is less than the setting value, the case is not inrush current.

The lasting time of aperiodic component of short-circuit current was small, usually lasting tens of milliseconds[9] [10]. But the lasting time of aperiodic component of inrush current was long, usually lasting several seconds or even minutes [11]. Therefore, the range of could adopt 0.15~0.5 according to the different transformer capacities.

6. Conclusions

This research proposed a novel method to identify inrush current based on attenuation characteristics of aperiodic components according to the attenuation characteristics of aperiodic component. Some conclusions are as follow.

(1) The aperiodic component does not decay in the intermittent period of inrush current. The aperiodic component does decay in the duration period of inrush current. The velocity attenuation of kth cycle at the end of the period is equal to the decay rate of $(k+1)^{th}$ cycle at the beginning of the period.

(2) The decay rate of aperiodic component of the inrush current is slower than that of the short circuit current, that is, the time constant of the dynamic component of the aperiodic component of the inrush current is greater than that of the short circuit current.

(3) The inrush current occurred according to judging the dynamic attenuation time constant of the extracted aperiodic component is greater than the setting value.

7. Acknowledgements

The work here is supported by the National Natural Science Foundation of China(51507157), and the work here is supported by the National Natural Science Foundation of China(51607158). Supported by the Graduate's Scientific Research Foundation of Zhengzhou University of Light Industry.

8. References

[1] Yuan An, Lulin Tian and Jiajun Liu 2008 Transformer excitation inrush identification based on morphology-wavelet theory *J. Electric Power Automation Equipment.* **7** 41-45
[2] Xue and Wang 2012 Synthetic transformer inrush identification based on characteristic space *J. Electric Power Automation Equipment.* **11** 83-86
[3] Tao Zheng, Geye Lu and Yanjie Zhao 2016 A discriminating algorithm for identifying inrush of UHV voltage-regulating transformer based on virtual equivalent inductance *J. Transactions of China Electrotechnical Society.* **7** 118-125
[4] Zhengqing Han and Shuping Liu 2007 Magnetizing inrush identification based on differential current characteristics *J. Electric Power Automation Equipment.* **9** 51-4
[5] Jianping Jiang and Yiguang Jiang 2013 Detection method of weak signals via stochastic resonance *J. Ship Electronic Engineering.* **5** 143-145
[6] Hongquan Ji, Yihan Yang and Yansong Li 2007 Study on line aperiodic component current comparison pilot protection based on optical current transducers *J. Proceedings of the CSEE.* **19** 45-49
[7] Lin Teng and Qingquan Jia 2002 Optical current transducer and its application in protective relaying *J. Power System Technology.* **1** 31-33
[8] Lihong Xue and Dongping Li 2015 Research on piecewise linear fitting method based on least square method in 3D space points *J. The Open Automation and Control Systems Journal.* **7** 1575-1579
[9] Jianqin Feng, Sifang Huang and Hailong Song 2014 Short-Circuit current aperiodic components and their application in relay protection *J. Electrotechnics Electric.* **12** 35-8
[10] Yong Feng, Shiming Liu and Wei Chen, etc 1999 Study on the new algorithm of decaying DC components *J. RELAY.* **3** 14-15
[11] Xidong and Xu 2005 Research on power transformer longitudinal differential protection *J. Doctoral Dissertation of Zhejiang University.* **2** 54-55

Research on Inrush Current Mechanisms of Single Phase Transformers

Jianqin Feng[1], Pei Zhang[1,*] and Zhifei Chen[2]

[1]School of Electrical and Information Engineering, Zhengzhou University of Light Industry, Dongfeng Rd 5, Zhengzhou, China
[2]School of Management and Engineering, Zhengzhou University, Science Avenue 100, Zhengzhou, China

Email: zhangpei_zzuli@163.com, 1161679542@qq.com

Abstract. Based on a transformer equivalent circuit which is nonlinear but approximately piecewise linear, a set of differential equations are given to constrain the change of flux in the transient process after a no-load transformer is switched on. By solving the differential equations, a group of mathematical expressions are obtained to depict the change law of flux during the transient process. Thereafter, the mathematical expressions are changed into an alternative form of iterative formulae describing change of flux at different time periods, for digital simulation of the inrush current. According to the simulation result, the change curves of flux and current and their components are given, and their typical local change characteristics and corresponding reasons are analyzed. Finally, the reason resulting in decaying of inrush current, the change law of decaying rate of inrush current, and the factors affecting decaying rate of inrush current are respectively analyzed by means of mathematical derivation, Some of the conclusions of the existing researches are confirmed in theory, some incorrect or inaccurate viewpoints are corrected and some new conclusions are obtained in this paper.

1. Introduction

A large excitation current can be generated when the transformer is empty load, and the numerical value can be 4 ~ 8 times that of the rated current. Excitation inrush current leads to transformer winding deformation, the damage of transformer insulation, the lower the quality of power supply, and cause of transformer differential protection misoperation [1][2] [3], thus widely attention for a long time. At present, the research on transformer field flow is mainly focused on the identification method of excitation current flow, which produces a series of research results 4] [5] [6] [7] [8]. However, due to the non-linearity, randomness, confusion and diversity of the excitation current, the existing solutions are not perfect. Therefore, it is necessary to thoroughly study the mechanism of excitation flow and fully grasp its change law, so as to facilitate the correct recognition of excitation current flow.

At present, the research on the mechanism of excitation current flow is relatively small, and it is not thorough enough, either qualitative or oversimplified. Literature analyzed the excitation flow of nondestructive transformers without taking into account the attenuation of flow. Literature only qualitatively analyses the effect of transformer winding resistance and system resistance on the attenuation of excitation current, but lacks strict theoretical derivation. Obtained by mathematical deduction [9][10] and literature to the same conclusion, analysis on the excitation inrush current decay is still qualitative, and in this paper, through calculating the nonlinear inductance of transformer excitation loop inductance average linearization method is too simplified, not suitable for descriptions of excitation inrush current attenuation law. In addition, the literature [11] [12] indicates that the transformer winding resistance and system resistance result in the decrease of the core deflator, which is the reason for the attenuation of excitation current, and the calculation method of the flow attenuation is given. This paper argues that bias only within the duration of the inrush current by the index attenuation law, its attenuation speed depends on the excitation circuit time constant, bias during flow pulsation for basically remain unchanged, as a result the attenuation of bias flow duration is the only reason why excitation inrush current decay to zero, these views worthy of further discussion.

In this paper, starting from the nonlinear equivalent circuit of single-phase transformer excitation circuit, through the study of the piecewise linearization of transformer magnetizing curve, through solving the differential equation, the mathematical description about the change rule of excitation inrush current is obtained, and further analyzes the change mechanism of excitation inrush current, the different conclusions.

2. Mathematical Model

The equivalent circuit of single-phase transformer airload is shown in Figure 1.

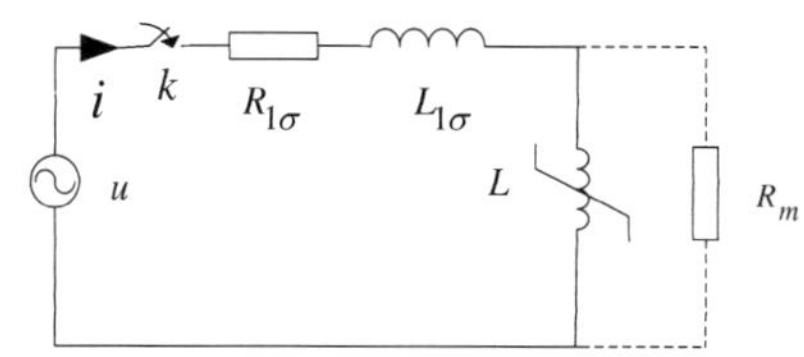

Figure 1. Equivalent circuit for single phase transformers when switching on without load.

In Figure 1, u as the power supply voltage, i as the excitation current, R1σ for transformer winding resistance, L1σ as the leakage inductance of the transformer primary winding, the Rm to reflect transformer core eddy current loss and magnetic hysteresis loss equivalent resistance, L for excitation inductance of the transformer.

Since the eddy current loss and hysteresis loss are very small, the resistance Rm is very large, so it is ignored. Excitation inductance is non - linear inductance. In order to simplify the problem, the magnetization curve is divided into piecewise linearization, as shown in Figure 2.

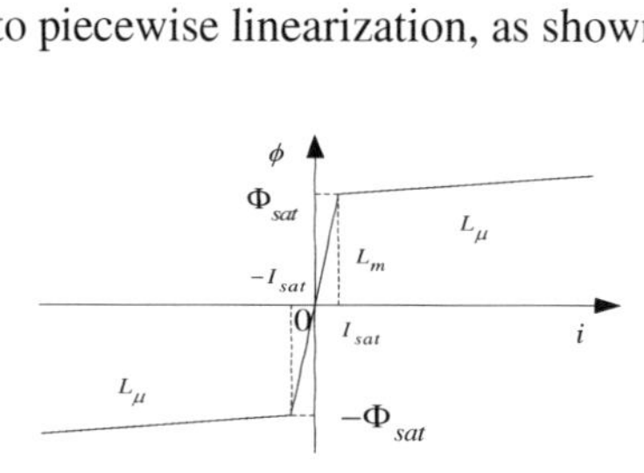

Figure 2. Approximate magnetization curve for transformers.

Figure 2, the Lm for transformer core, the excitation inductance is not satisfied when Lμ is iron core saturation excitation inductance, φ as the core of main magnetic flux, Φsat for saturation magnetic flux, Isat for corresponding saturation current saturation magnetic flux. The nonlinear excitation inductance can be approximated using equation (1).

$$L=\begin{cases} L_m, & |\phi| \le \Phi_{sat} \\ L_\mu, & |\phi| \ge \Phi_{sat} \end{cases} \tag{1}$$

As can be seen from Figure 2, the relationship between magnetic flux and current is

$$i=\begin{cases} \phi / L_m, & \phi \le \Phi_{sat} \\ (\phi-\Phi_{sat}) / L_\mu + I_{sat}, & \phi \ge \Phi_{sat} \\ (\phi+\Phi_{sat}) / L_\mu + I_{sat}, & \phi \le -\Phi_{sat} \end{cases} \tag{2}$$

Given that Isat is small, it is ignored. Make

$$\Phi_s' = \begin{cases} \Phi_{sat}, & \phi \ge \Phi_{sat} \\ -\Phi_{sat}, & \phi < -\Phi_{sat} \end{cases} \tag{3}$$

There are

$$i=\begin{cases} \phi / L_m, & |\phi| \le \Phi_{sat} \\ (\phi-\Phi_s') / L_\mu, & |\phi| > \Phi_{sat} \end{cases} \tag{4}$$

Set the transformer at the time of t=0, the initial phase Angle of the power supply voltage is α, and the power supply voltage u=Umsin(ωt+α), and the parameters of each parameter are indicated by the scalar value, and the voltage equation of the excitation circuit is

$$\frac{d\phi}{dt}+L_{1\sigma}\frac{di}{dt}+R_{1\sigma}i=U_m\sin(\omega t+\alpha) \tag{5}$$

In equation (4) and equation (5), the magnetic flux equation in the transient process of the transformer no-load is obtained, as shown in equation (6).

$$\begin{cases} (1+\frac{L_{1\sigma}}{L_m})\frac{d\phi}{dt}+\frac{R_{1\sigma}}{L_m}\phi=U_m\sin(\omega t+\alpha), |\phi| \le \Phi_{sat} \\ (1+\frac{L_{1\sigma}}{L_\mu})\frac{d\phi}{dt}+\frac{R_{1\sigma}}{L_\mu}\phi= \\ \qquad U_m\sin(\omega t+\alpha)+\frac{R_{1\sigma}}{L_\mu}\Phi_s', \quad |\phi| \ge \Phi_{sat} \end{cases}$$

(6)

3. The Change Law of Excitation Current Flow

3.1. Mathematical Description

The differential equation (6) can be used to describe the mathematical expression of the change of ferromagnetic flux. See type (7).

$$\phi = \begin{cases} \dfrac{U_m}{\omega\sqrt{\left(\dfrac{R_{1\sigma}}{\omega L_m}\right)^2 + \left(1+\dfrac{L_{1\sigma}}{L_m}\right)^2}} \sin(\omega t + \alpha - \arctan\dfrac{\omega(L_m + L_{1\sigma})}{R_{1\sigma}}) + \Phi_0 e^{-R_{1\sigma}t/(L_m+L_{1\sigma})}, & |\phi| \le \Phi_{sat} \\ \dfrac{U_m}{\omega\sqrt{\left(\dfrac{R_{1\sigma}}{\omega L_\mu}\right)^2 + \left(1+\dfrac{L_{1\sigma}}{L_\mu}\right)^2}} \sin(\omega t + \alpha - \arctan\dfrac{\omega(L_\mu + L_{1\sigma})}{R_{1\sigma}}) + \Phi_s^{'} + \Phi_1 e^{-R_{1\sigma}t/(L_\mu+L_{1\sigma})}, & |\phi| > \Phi_{sat} \end{cases} \tag{7}$$

Type, Φ0 and Φ1 is the integration constant Φ, decided by the boundary conditions of differential equation.

Make

$$\xi_m = \sqrt{\left(\frac{R_{1\sigma}}{\omega L_m}\right)^2 + \left(1+\frac{L_{1\sigma}}{L_m}\right)^2} \tag{8}$$

$$\xi_\mu = \sqrt{\left(\frac{R_{1\sigma}}{\omega L_\mu}\right)^2 + \left(1+\frac{L_{1\sigma}}{L_\mu}\right)^2} \tag{9}$$

$$\Phi_m = U_m / \omega\xi_m \tag{10}$$

$$\Phi_{m\mu} = U_m / \omega\xi_\mu \tag{11}$$

$$\beta_m = \arctan[\omega(L_m + L_{1\sigma}) / R_{1\sigma}] \tag{12}$$

$$\beta_\mu = \arctan[\omega(L_\mu + L_{1\sigma}) / R_{1\sigma}] \tag{13}$$

$$\tau_m = (L_m + L_{1\sigma}) / R_{1\sigma} \tag{14}$$

$$\tau_\mu = (L_\mu + L_{1\sigma}) / R_{1\sigma} \tag{15}$$

The equation (7) can be expressed as

$$\phi=\begin{cases}\Phi_m \sin(\omega t+\alpha-\beta_m)+\Phi_0 e^{-t/\tau_m}, \\ \qquad |\phi|\le\Phi_{sat} \\ \Phi_{m\mu}\sin(\omega t+\alpha-\beta_\mu)+\Phi_s^{'}+\Phi_1 e^{-t/\tau_\mu}, \\ \qquad |\phi|>\Phi_{sat}\end{cases} \tag{16}$$

Considering the ωL»R1σ, Lm»L1σ, Lm+L1σ»R1σ, so ξm≈1, βm≈0.5π, τm can be thought of infinity, e-t/τm≈1, (16) can be simplified to

$$\phi=\begin{cases}\Phi_m \sin(\omega t+\alpha-\pi/2)+\Phi_0, & |\phi|\le\Phi_{sat} \\ \Phi_{m\mu}\sin(\omega t+\alpha-\beta_\mu)+\Phi_s^{'}+\Phi_1 e^{-t/\tau_\mu}, \\ \qquad |\phi|>\Phi_{sat}\end{cases} \tag{17}$$

Make

$$\phi_p=\begin{cases}\Phi_m \sin(\omega t+\alpha-\pi/2), & |\phi|\le\Phi_{sat} \\ \Phi_{m\mu}\sin(\omega t+\alpha-\beta_\mu), & |\phi|>\Phi_{sat}\end{cases} \tag{18}$$

$$\phi_{np}=\begin{cases}\Phi_0, & |\phi|\le\Phi_{sat} \\ \Phi_s^{'}+\Phi_1 e^{-t/\tau_\mu}, & |\phi|>\Phi_{sat}\end{cases} \tag{19}$$

The transformer main flux can be expressed as

$$\phi=\phi_p+\phi_{np} \tag{20}$$

According to equation (20), ferroflux is composed of two parts, φp and φnp, during the transient process after the empty load of the transformer.

φp is the sinusoidal quantity of the frequency with the power supply voltage, which is generated by the power supply excitation, which is called the mandatory flux component, and its variation rule is shown in equation (18). φnp is the free flux component, which makes the transformer magnetic waveform biased to the time axis side, so it is also called partial magnetic field, and its variation rule is shown in equation (19). It can be seen from equation (18) that the forced magnetic flux component changes according to two different sinusoidal rules during the duration of flow and the duration of flow. During the flow pulsation amplitude and phase Angle of Φm and α-βm respectively to the duration respectively Φmμ and α-βμ. It can be seen from equation (19) that the free flux component is not periodic component, and the variation law of the duration of flow and flow duration is different. During the flow pulsation, free flux component is a dc component, its value is Φ0. In flow duration, free flux component includes a dc component and an exponentially decaying aperiodic component law, the value of the dc component for Φs, , aperiodic component of the initial value of Φ1, decay time constant for τμ. Φ0 and Φ1 according to enter the flow pulsation periods and Φ into the initial flow duration time of the initial conditions determine, represented as a unity

$$\phi(\mathrm{t}-0)=\phi(\mathrm{t}+0) \tag{21}$$

The change law of excitation flow is obtained by equation (2) and equation (17)

$$i=\begin{cases}\dfrac{\Phi_m}{L_m}\sin(\omega t+\alpha-\pi/2)+\dfrac{\Phi_0}{L_m}, & |\phi|\le\Phi_{sat}\\ \dfrac{\Phi_{m\mu}}{L_\mu}\sin(\omega t+\alpha-\beta_\mu)+\dfrac{\Phi_1}{L_\mu}e^{-t/\tau_\mu}, & |\phi|>\Phi_{sat}\end{cases} \tag{22}$$

Because Lm is large, the excitation current i≈∞ during the flow. Make

$$I_{m\mu}=\Phi_{m\mu}/L_\mu \tag{23}$$

$$I_{\mathrm{np0}}=\Phi_1/L_\mu \tag{24}$$

The equation (22) can be simplified

$$i=\begin{cases}0, & |\phi|\le\Phi_{sat}\\ I_{m\mu}\sin(\omega t+\alpha-\beta_\mu)+I_{\mathrm{np0}}e^{-t/\tau_\mu}, & |\phi|>\Phi_{sat}\end{cases}$$

(25)

Make

$$i_p=\begin{cases}0, & |\phi|\le\Phi_{sat}\\ I_{m\mu}\sin(\omega t+\alpha-\beta_\mu), & |\phi|>\Phi_{sat}\end{cases} \tag{26}$$

$$i_{\mathrm{np}}=\begin{cases}0, & |\phi|\le\Phi_{sat}\\ I_{\mathrm{np0}}e^{-t/\tau_\mu}, & |\phi|>\Phi_{sat}\end{cases} \tag{27}$$

There are

$$i=i_p+i_{\mathrm{np}} \tag{28}$$

According to the equation (28), the excitation current flow is composed of the force component ip and the free component inp. The force current component corresponds to the forced magnetic flux component, and its variation rule is shown in equation (26). The free current component corresponds to the free flux component, and its variation rule is shown in equation (27). During the flow interval, the force component, the free component and the flow of the excitation current are 0. In the duration of flow, the force component is changed by sine, the amplitude is Imμ, and the phase Angle is α-βμ. The free component is the aperiodic component of the exponential decay, whose initial value is Inp0 and the time constant is τμ. The full current of the current is superimposed by the two. The initial value of the aperiodic component Inp0 can be determined by equation (24) or by equation (29).

$$i(\mathrm{t}-0)=i(\mathrm{t}+0) \tag{29}$$

3.2. Change Curve

Sets the remanence of transformer core before switching to Φr, by type (21), the closing moment core flux is still Φr, namely φ=Φr. Because Φr<Φsat, |φ|<Φsat, so the transformer no-load closing the first core for a period of time will not be saturated, the transformer in phase flow pulsation, the core flux according to the type (17) of the first law of function change. As the change of magnetic flux, when $|\varphi|\geq\Phi sat$, transformer iron core saturation, the transformer into the flow duration, the core flux according to the type (17) of the second kind of function change. When magnetic flux changes to$|\varphi|<\Phi sat$, transformer core out of saturation state, and enter the flow pulsation period, until once again into the flow duration. Since then, the flow interval and the duration of continuous flow are repeated alternately, and the main magnetic flux of iron core alternates with two different functions.

Alternating moment for tj(j=0,1,···,m), t0 for transformer no-load closing moments when the first phase flow pulsation of the initial moment, t1 as the flow for the first time duration of the initial point, t2 for the second phase flow pulsation of the initial time, t3 for a second flow duration of initial point, and the tm for the last time the end of the flow duration time is end of the entire excitation inrush current, the t2k acuities were t2k≤t<t2k+1 is the flow pulsation stage, t2k+1≤t<t2(k+1) as the continuous phase flow. Here, the indices are going to be k=0, 1, ···, m/2-1.

According to different time periods of transformer flux variation, a set of iterative formula can be obtained, as shown in formula (30) ~ (40).

$$\phi(t) = \phi_p(t) + \phi_{np}(t) \tag{30}$$

$$\phi_p(t) = \begin{cases} \Phi_m \sin(\omega t + \alpha - \pi/2), \\ \qquad t_{2k} \leq t < t_{2k+1} \\ \Phi_{mu} \sin(\omega t + \alpha - \beta_u), \\ \qquad t_{2k+1} \leq t < t_{2(k+1)} \end{cases} \tag{31}$$

$$\phi_{np}(t) = \begin{cases} \Phi_0(t_{2k}), \qquad t_{2k} \leq t < t_{2k+1} \\ \Phi_s^{'} + \Phi_1(t_{2k+1}) e^{-(t-t_{2k+1})/\tau_\mu}, \\ \qquad t_{2k+1} \leq t < t_{2(k+1)} \end{cases} \tag{32}$$

$$\Phi_0(t_{2k}) = \begin{cases} \Phi_m \cos\alpha + \Phi_r, \qquad k = 0 \\ \Phi_{mu} \sin(\omega t_{2k} + \alpha - \beta_\mu) - \\ \quad \Phi_m \sin(\omega t_{2k} + \alpha - \pi/2) + \\ \quad \Phi_s^{'} + \Phi_1(t_{2k-1}) e^{-(t_{2k}-t_{2k-1})/\tau_u}, \\ \qquad k = 1, 2, \mathrm{L}\ , m/2 - 1 \end{cases} \tag{33}$$

$$\begin{aligned} \Phi_1(t_{2k+1}) = {} & \Phi_m \sin(\omega t_{2k+1} + \alpha - \pi/2) - \\ & \Phi_{mu} \sin(\omega t_{2k+1} + \alpha - \beta_\mu) + \\ & \Phi_0(t_{2k}) - \Phi_s^{'}, \quad k = 0, 1, \mathrm{L}\ , m/2 - 1 \end{aligned} \tag{34}$$

IOP Conf. Series: Journal of Physics: Conf. Series **1026** (2018) 012030 doi:10.1088/1742-6596/1026/1/012030

$$\begin{cases} t_0 = 0, \\ t_{2k+1} = \left[\arcsin(\Phi_s' - \Phi_0(t_{2k}))/\Phi_m - \alpha + \pi/2 \right] / \omega \\ \qquad\qquad\qquad\qquad k = 0,1,\mathrm{L}\ ,m/2-1 \\ \Phi_{m\mu}\sin(\omega t_{2k} + \alpha - \beta_\mu) + \Phi_1(t_{2k-1})e^{-(t_{2k}-t_{2k-1})/\tau_\mu} = 0 \\ \qquad\qquad\qquad\qquad k = 1,2,\mathrm{L}\ ,m/2 \end{cases} \tag{35}$$

$$i(t) = i_p(t) + i_{\mathrm{np}}(t) \tag{36}$$

$$i_p(t) = \begin{cases} 0, & t_{2k} \le t < t_{2k+1} \\ I_{m\mu}\sin(\omega t + \alpha - \beta_\mu), & t_{2k+1} \le t < t_{2(k+1)} \end{cases} \tag{37}$$

$$i_{\mathrm{np}}(t) = \begin{cases} 0, & t_{2k} \le t < t_{2k+1} \\ I_{\mathrm{np0}}(t_{2k+1})e^{-(t-t_{2k+1})/\tau_\mu}, & t_{2k+1} \le t < t_{2(k+1)} \end{cases} \tag{38}$$

$$I_{\mathrm{np0}}(t_{2k+1}) = \Phi_1(t_{2k+1}) / L_\mu \tag{39}$$

$I_{\mathrm{np0}}(t_{2k+1})$ can also be determined by formula (40).

$$I_{\mathrm{np0}}(t_{2k+1}) = -I_{m\mu}\sin(\omega t_{2k+1} + \alpha - \beta_\mu) \tag{40}$$

According to the type (30) ~ (40) flow flux forced component can be obtained, free weight, total flux and excitation inrush current force components, freedom, all current numerical solution, and then draw the transformer magnetic flux and the change of excitation inrush current curve. The change curve of magnetic flux and current is shown in FIGURE 3.

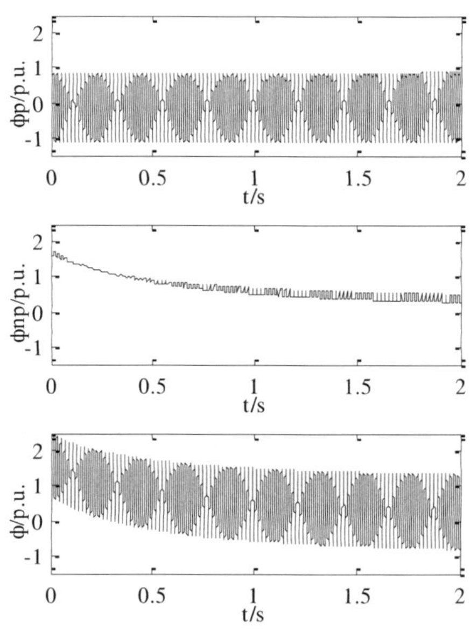

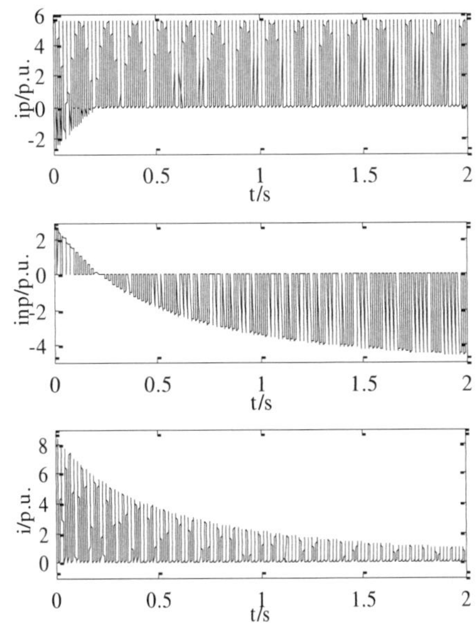

Φ_m=1.1(p.u.), $\Phi_{m\mu}$=0.821(p.u.), Φ_r=0.7(p.u.),
Φ_{sat}=1.2(p.u.),α=0^0,β_μ=0.4947π, τ_μ=0.192(s), L_μ=0.15(H)

Figure 3. Flux and current curves during inrush current.

Can be seen from Figure 3, mandatory stable flux component changes, positive amplitude for Φmμ, slightly less than the opposite amplitude Φm, its the width of the upper and lower envelope for Φm+Φmμ. The bias magnetic field shows the volatility and attenuation. The fluctuation range of the bias is smaller and smaller, until it goes to zero, then it will become bigger and bigger and eventually stabilize. Although the bias is in the fluctuation, the decrease and the increase alternate, but from the envelope, the trend of continuous attenuation, and gradually to zero. The whole flux envelope also shows a tendency of attenuation, and the width of the upper and upper envelope gradually becomes larger, until finally reaching stability. Envelope of the initial width is greater than 2Φmμ less than Φm+Φmμ, stable when the width of the 2Φm.

Can be seen from Figure 3, mandatory current component varies at different stages of change rule: in supersaturated phase, partly to timeline side, positive amplitude remains unchanged, peak reverse gradually decay to zero; In the undersaturated phase, it is completely biased to the time axis, and the amplitude is always stable. The free current component is positive in the supersaturation phase and gradually attenuates to zero. In the unsaturated phase, the value is negative and gradually increases, and eventually stabilizes. The full current is completely partial to the time axis in the whole current flow, and the peak shows the continuous attenuation trend and gradually goes to zero.

Points out, by the way, transformer no-load closing moments of the initial bias for φnp(0)=Φmcosα+Φr, are sufficient and necessary conditions of excitation inrush current

$$\left|\phi_{np}(0)\right| > \Phi_{sat} - \Phi_m \tag{41}$$

The excitation current is generated only when the initial flux satisfies formula (41). If |φnp(0)|<Φsat, excitation inrush current process includes undersaturated phase only, the greater the absolute value ofφnp(0), the longer the duration of undersaturated stage. If |φnp(0)|>Φsat, excitation inrush current process will include supersaturated phases and undersaturated stage, the greater the absolute value ofφnp(0), the longer the duration of supersaturated phases, then undersaturated phase duration is certain. Because of the initial bias by residual magnetism and no-load switching-in moment power transformer phase Angle to determine, there is a great deal of randomness, so that the

IOP Conf. Series: Journal of Physics: Conf. Series **1026** (2018) 012030 doi:10.1088/1742-6596/1026/1/012030

generation of excitation inrush current, size, and duration have great randomness.

The typical changes of magnetic flux and current and its components in the excitation current flow are shown in Figure 4.

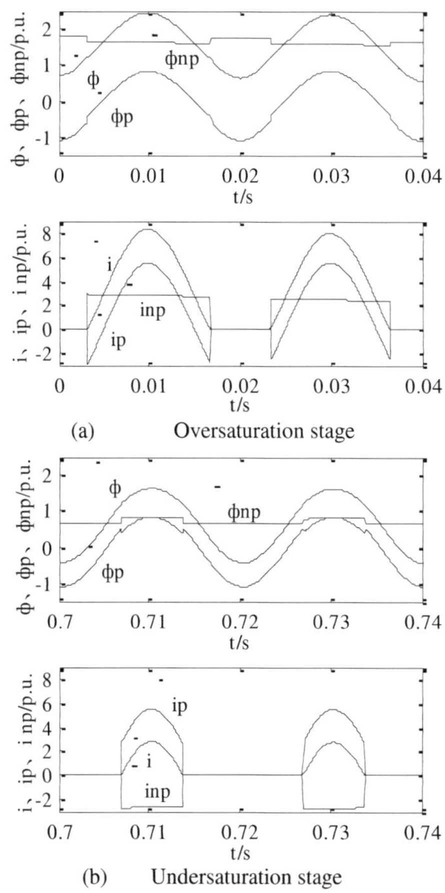

Figure 4. Local change characteristics of flux and current and their components.

4. Attenuation of Excitation Current

Because the bias magnetism determines the generation, size and attenuation of the excitation flow, the attenuation of the excitation current can be studied by the attenuation of bias magnetic.

4.1. The Reason for the Bias Attenuation

The change of magnetic bias can be divided into three situations in each workweek. In the duration of flow, the bias is changed by exponential rule. During the alternation of the current and the duration of flow, the bias changes. It can be seen that the factors that lead to partial magnetic attenuation are not external to two: the exponential rule change of the flow duration and the variation of the time bias of the current period.

(1) the influence of the change of flow in the current period

By type (32), the flow duration, bias by the constant term and index of two parts, the change of the bias depends on the index of the quantity, the first cycle flow continued during the period bias of the power frequency variation is

$$\Delta\phi_{np1.l} = -\Phi_{1(t_{2l-1})}[1-e^{-(t_{2l}-t_{2l-1})/\tau_\mu}] \tag{42}$$

Considering formula (23), formula (39) and formula (40), there are

$$\Delta\phi_{np1.l} = \Phi_{m\mu}\sin(\omega t_{2l-1}+\alpha-\beta_\mu)[1-e^{-(t_{2l}-t_{2l-1})/\tau_\mu}] \tag{43}$$

Make θj=ωtj+α,(j=0,1,···,m),There are

$$\Delta\phi_{np1.l} = \Phi_{m\mu}\sin(\theta_{2l-1}-\beta_\mu)[1-e^{-(\theta_{2l}-\theta_{2l-1})/\tan\beta_\mu}] \tag{44}$$

In supersaturated phase, sin(θ2l-1 -βμ)<0, thenΔφnp1.l<0, so the bias. In undersaturated phase, sin(θ2l-1-βμ)>0, then Δφnp1.l>0, so the bias. Thus, the change of the bias in the supersaturated phase flow duration will cause the attenuation of bias, but instead cause bias in undersaturated phase increases, it is clearly not lead to the root of the excitation inrush current decay to zero. The reason for this phenomenon is that during the supersaturated phase, the free current component is consistent with the partial magnetic direction, so the attenuation of the free current component causes the attenuation of the bias magnetic field. In the unsaturated phase, the free current component is the opposite of the partial magnetic field, and the attenuation of the free current component leads to the increase of the bias magnetic field.

It can be seen from equation (19) that when time goes to infinity, the exponential term tends to zero and the partial magnetic flux tends to saturate the flux. Therefore, if we ignore the change of the bias magnetic variation and only consider the variation of the continuous period of flow, the trend of bias magnetic variation is shown in Figure5.

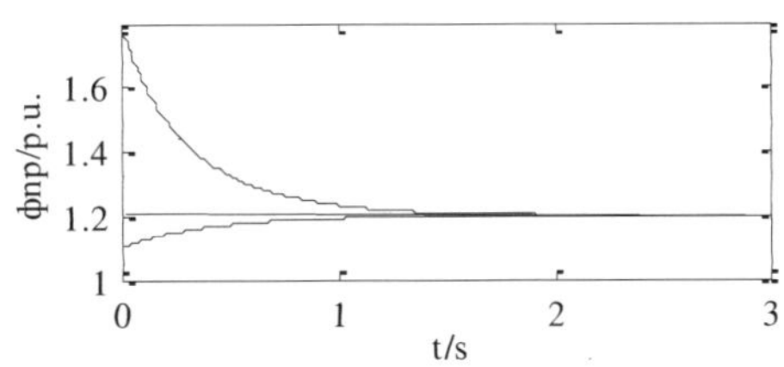

Figure 5. The change trend of bias flux if only considering the influence of exponential term.

Figure 5 shows that the bias in only considering flow duration changes, if the excitation inrush current process from the beginning of the supersaturated phases, the bias in the supersaturated phase continuous attenuation, at the end of the supersaturated phases, bias attenuation to the saturated magnetic flux, after constant; If starting from the undersaturated phase, the bias increases gradually until the saturation flux is reached and then remains constant. Thus it can be seen that the variation of the continuous phase bias is inclined to make the bias towards saturated magnetic flux instead of approaching zero.

(2) the influence of partial magnetic jump on the bias attenuation

The forced magnetic flux component changes continuously in the continuous period and the duration of the flow duration respectively. During the alternating current period and the duration of current flow, the forced magnetic flux component will change from one sinusoid to another sine curve, and then the magnetic flux component will be mutated. Since the total flux cannot be mutated, the bias will be changed accordingly, and its mutation is the same as that of the forced flux component, in the opposite direction.

It is not considered that the change of the magnetic flux component and the partial magnetic variation are shown in FIGURE 6.

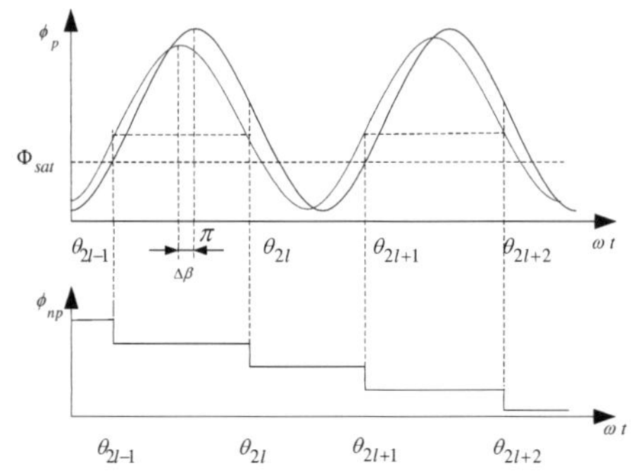

Figure 6. Mutations of forced flux and bias flux.

It can be seen from FIGURE 6 that the magnetic bias occurs twice in each phase of the cycle, and the total number of mutations in the l frequency cycle is

$$\begin{aligned}\Delta\phi_{np2.l} = [\Phi_m \sin(\theta_{2l-1} - \beta_m) - \Phi_{m\mu} \sin(\theta_{2l-1} - \beta_\mu)] + \\ [\Phi_{m\mu} \sin(\theta_{2l} - \beta_\mu) - \Phi_m \sin(\theta_{2l} - \beta_m)]\end{aligned} \tag{45}$$

It can be seen from FIGURE 6 that the waveform of the forced magnetic flux component of the duration of flow is symmetric about $\omega t = \pi - \Delta\beta$, so it is

$$\theta_{2l} = 2(\pi - \Delta\beta) - \theta_{2l-1} \tag{46}$$

The equation (42) and (43) can be obtained

$$\Delta\phi_{np2.l} = -2\Phi_m \sin\Delta\beta \sin(\theta_{2l-1} + \Delta\beta) \tag{47}$$

Because $\Delta\beta$ usually is very small, θ2l-1»Δβ, θ2l-1+Δβ≈θ2l-1, there are

$$\Delta\phi_{np2.l} = -2\Phi_m \sin\Delta\beta \sin(\theta_{2l-1}) \tag{48}$$

By type (48), when βμ =βm, Δβ=0, Δφnp2.l=0, a bias the total variation in the cycle is zero, bias is not decay; When βμ<βm , Δβ>0, Δφnp2.l<0, the change of the bias amount is negative, partial magnetic attenuation will happen.

Thus draw the conclusion: Δβ>0, bias amount of mutations in each cycle power frequency always negative, bias will be diminishing, until the end of the excitation inrush current.

It is shown in FIGURE 7 that the variation trend of bias magnetic variation is only considered when the deflection of the magnetic field is ignored.

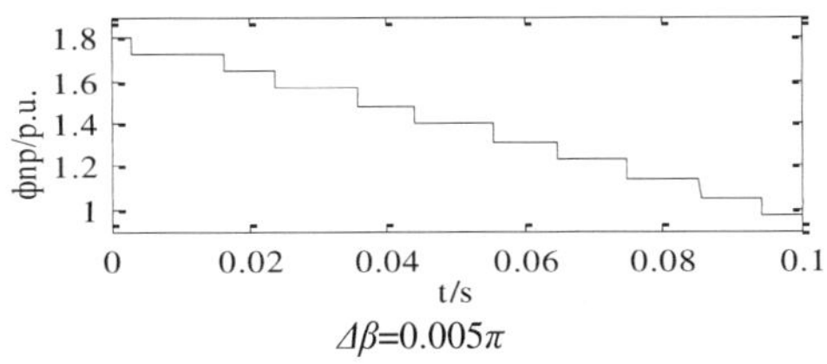

Figure 7. The change trend of bias flux if only considering its mutations.

(3) influence of transformer winding parameters

If you ignore the transformer primary winding resistance and leakage reactance, namely the R1σ=0, L1σ=0, βμ=βm=π/2, Δβ=0, τμ=∞, tanβμ=0, Φmμ=Φm, , by type (44) with type (45) Δφnp1.l=0, Δφnp2.l=0, so the bias will not decay. At this time, the bias of the duration of the current flow remains the same, and due to the same force flux waveform in the continuous period and the duration of flow, the deflection of the magnetic flux will not occur at the time of alternating current. Therefore, in the whole flow process, the partial magnetic flux is constant, and the whole magnetic flux is changed by the sinusoidal law with the bias axis, corresponding to the so-called non-destructive transformer.

If only to ignore the resistance of the transformer winding, namely the R1σ=0, tβμ=βm=π/2, Δβ=0, τμ=∞, tanβμ=∞, Φmμ<Φm, then there is stillΔφnp1.l=0, Δφnp2.l=0, bias is still will not decay. In this case, although the flow pulsation periods and flow flux component of the same phase duration of mandatory, but the amplitudes are not equal, when forced to magnetic flux component from a sine curve changes will still occur on a sine curve to another jump, is just a period of two jump are equal in size, in the opposite direction, the total amount of mutations still is zero. Accordingly, the bias magnetic field will fluctuate, but not attenuate, and its change curve is shown in Figure 8.

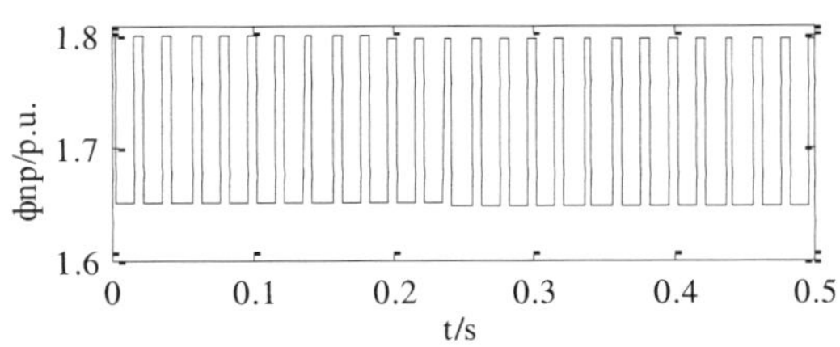

Figure 8. Leakage reactance's influence on bias flux.

In the case of R1σ≠0, βμ<βm, Δβ>0, thenΔφnp1.l≠0, Δφnp2.l<0 At this time, bias in flow duration according to the index law of change, change in on both sides of the boundary flow duration time, and under the joint action of both, gradually decay to zero, corresponds to the actual situation, the curve is shown in Figure 3.

To sum up, due to the existence of the transformer winding resistance (including resistance) system, the forced component in flow pulsation periods and the phase Angle difference of flow duration $\Delta\beta$ is not zero, the bias in the flow duration on both sides of the border moment of jump in transformer bias continues to decline, leading to continuous attenuation of excitation inrush current.

4.2. The Decay Rate of Bias

As mentioned above, the variation within a cycle consists of two parts: the variation of the duration of flow and the total number of mutations in the two boundary moments of flow duration.

The attenuation rate of partial magnetic field is denoted by the attenuation of the partial magnetic field in each period, and the attenuation rate of the l cycle is due to the exponential term

$$v_{np1.l} = (\Phi_m / \xi_\mu)\cos\theta_{2l-1}[1 - e^{-(\theta_{2l}-\theta_{2l-1})/\text{ctan}\,\Delta\beta}] \quad (49)$$

The attenuation velocity caused by a bias magnetic mutation is

$$v_{np2.l} = 2\Phi_m \sin\Delta\beta \sin\theta_{2l-1} \quad (50)$$

The total attenuation rate is zero

$$v_{np.l} = v_{np1.l} + v_{np2.l} \quad (51)$$

By type (49), The velocity component is related to vnp1.l, θ2l-1, Δβ and ξμ. θ2l-1 depends on the stage during which the current cycle is in the excitation flow, in the over-saturation stage, 0<θ2l-1<π/2, cosθ2l-1>0 and the velocity component vnp1.l. At the beginning of the flow, θ2l-1 is the smallest, and the cosθ2l-1 is the largest, so the velocity component vnp1.l is also the largest. As theθ2l-1 increases, the cosine cosθ2l-1 decreases and the velocity component vnp1.l becomes smaller and smaller. At the critical saturation stage, becauseθ2l-1=π/2, cosθ2l-1=0, the velocity component vnp1.l=0. After entering the undersaturated phase, π/2<θ2l-1<π, cosθ2l-1<0, so the speed component vnp1.l<0. As θ2l-1 goes from π over π/2, cosine theta of cosθ2l-1 goes from 0 to -1, so the absolute value of the velocity component vnp1.l. L is getting bigger and bigger. 2(π-θ2l-1) represents the width of the flow duration, its maximum value at the beginning of flow, reduce gradually, after therefore 2(π-θ2l-1) make the influence of velocity component vnp1. At the beginning of the freshness of the absolute value of l is the largest, with the passage of time gradually reduced after, more and more small. Parameter Δβ will affect the size of the velocity components vnp1.l the size of the time constant of the $c\tan\Delta\beta$, The smaller Δβ, the greater the time constant of $c\tan\Delta\beta$, velocity component vnp1.l. The greater the absolute value of l. The influence of parameter ξμ on vnp1.l is more intuitive. The larger the ξμ, the smaller the absolute value of vnp1.l. θ2l-1, Δβ and ξμ of vnp1.l. The influence is shown in Figure 9.

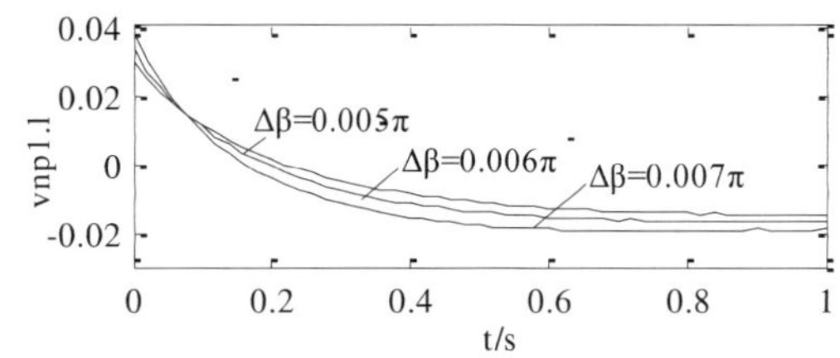

(a) ξ_μ=1.34

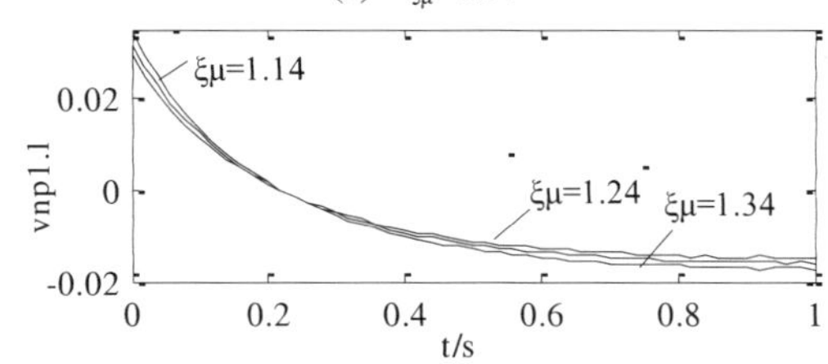

(b) $\Delta\beta$=0.005π

Figure 9. Influence of factor θ2l-1 and Δβ and ξμ on velocity component vnp1.l.

Figure 9 shows that the parameter is deduced ξμ of velocity component vnp1.l the influence of the relative to the Δβ effect is very small, can be ignored. Δβ is vnp1.l mainly.

By type (50), the velocity component vnp2.l and Δβ and θ2l-1 . In the whole field of excitation

IOP Conf. Series: Journal of Physics: Conf. Series **1026** (2018) 012030 doi:10.1088/1742-6596/1026/1/012030

flow, 0<θ21-1<π , sinθ21-1>0, so the velocity component vnp2.l is always positive. In the supersaturated phase, θ21-1 from small to large tends to π/2, sinθ21-1 monotonically increasing, gradually tending to 1, so the velocity component vnp2.l is getting bigger and bigger. At the critical saturation stage, because θ21-1=π/2 , sinθ21-1=1, the velocity component vnp2.l reaches its maximum value. In the undersaturated phase, θ21-1 tends to π over π/2, and the sinθ21-1 decreases monotonically, and the 1 tends to 0, so the velocity component vnp2.l is getting smaller and smaller. Since Δβ is very small, usually between 0 and π/2, sinΔβ monotone increasing, so the Δβ, the greater the velocity component vvnp2.l. θ21-1 and Δβ to vnp2.l. The influence is shown in Figure 10.

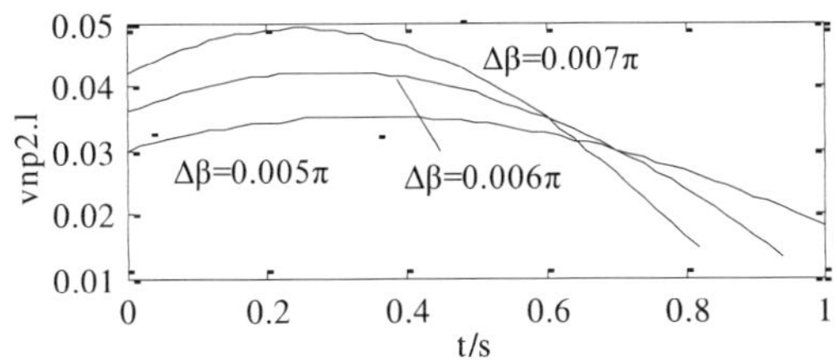

Figure 10. Influence of factor θ21-1and Δβ on velocity component vnp2.l.

Figure 10 shows that Δβ, the greater the velocity component vnp2.l. The faster the bias attenuation, the shorter the duration of excitation flow

According to the velocity component vnp1.l and vnp2.l, the total velocity vnp.l of partial magnetic decrement is obtained by equation (51). The relationship between the three is shown in Figure 11.

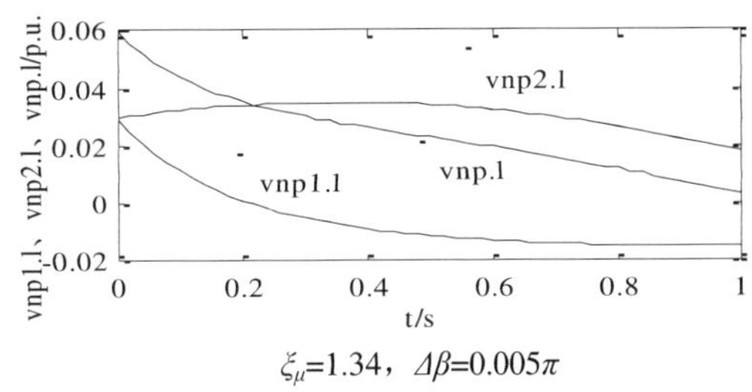

Figure 11. Comparison between velocity component vnp1.l and vnp2.l and velocity vnp.l.

As you can see from Figure 11, except that the value of the velocity component vnp1.l is roughly equivalent to vnp2.l at the beginning of the flow, In the flow process, it is shown that vnp1.l is less than vnp2.l . It can be seen that vnp2.l is the main factor determining the speed of partial magnetic decay.

Δβ as mentioned above, the parameters of vnp1.l and vnp2.l is larger, the influence of inevitable on total speed vnp.l has very big effect. The influence of ξμ on the secondary factor vnp1.l without affecting the main factor vnp2.l, and the effect on vnp1.l is very small, so the effect on the total speed vnp.l will be smaller and can be ignored. Δβ to vnp.l. The influence is shown in Figure 12.

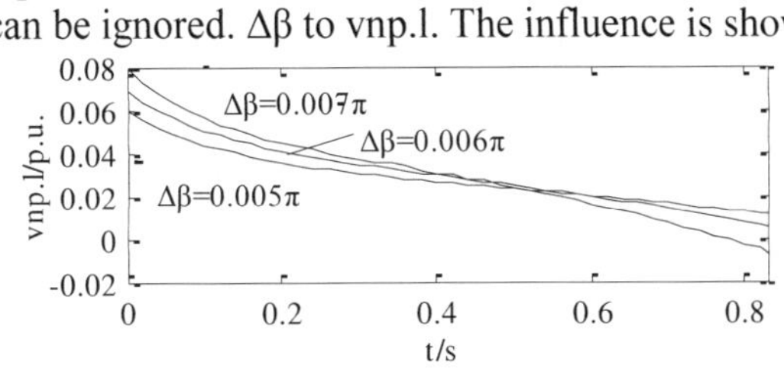

Figure 12. Influence of factor Δβ on velocity vnp.l.

It can be seen from FIGURE 12 that the velocity of partial magnetic decrement is the largest at the beginning of the current flow, and then the monotonousness decreases, and the velocity of the oversaturation phase decreases rapidly and the speed of the unsaturated phase decreases slowly. Δβ have bigger influence on the attenuation of bias, Δβ, the greater the attenuation of bias the greater speed, and vice versa.

Need to point out is, (49) type (50) is under the assumption that the forced flow duration waveform of magnetic component symmetrical about ωt=π-Δβ, under the condition of considering the influence of the bias flow duration attenuation, the condition is not established, so the need for them to make correction. The influence of bias magnetic attenuation on magnetic flux waveform is shown in Figure 13.

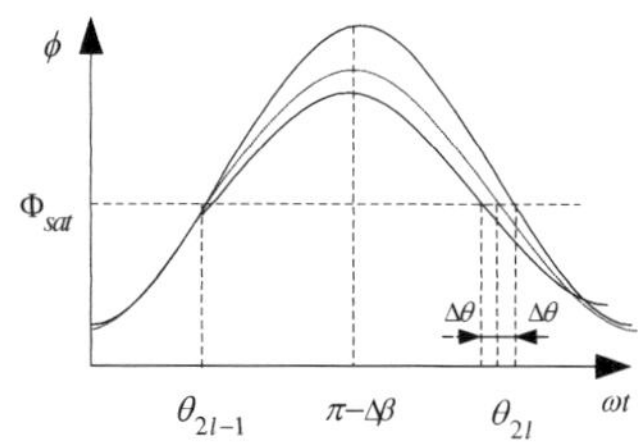

Figure 13. Influence of exponential term of bias flux on transformer flux waveform.

As can be seen from Figure 13 in supersaturated phase, caused by exponential term attenuation will lower bias flux waveforms, causing θ21 reach, in undersaturated phase, caused by exponential term bias increase will raise the flux waveforms, causing θ21.

Sets the offset of θ21 to Δθ, and moved forward positive, ward is negative, there are

$$\theta_{2l} = 2\pi - 2\Delta\beta - \theta_{2l-1} - \Delta\theta \tag{52}$$

According to the derivation, the velocity component vnp1.l still can be approximated by formula (49), and vnp2.l can be determined by equation (53).

$$v_{np2.l} = 2\Phi_m[\sin(\Delta\beta + \Delta\theta / 2) - \sin(\Delta\theta / 2) / \xi_\mu]\sin\theta_{2l-1} \tag{53}$$

In fact, due to the transformer saturation excitation circuit time constant is still large, in a flow of power frequency cycle duration bias caused by item index variation is very small, Δθ≈0, ignore mandatory flux waveforms of asymmetric flow duration is acceptable. Ignore Δθ bias attenuation speed before and after the contrast as shown in Figure 14.

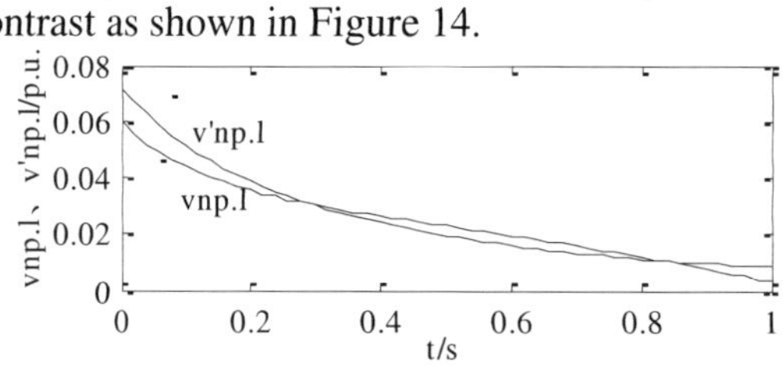

Figure 14. The difference of decaying rate of bias flux between considering factor Δθ and ignoring it.

Figure 14, v'np.l to ignore the bias before Δθ attenuation speed, by type (49), (51) and (53) to determine; vnp.l for ignoring Δθ after partial magnetic attenuation speed, by type (49), (50), and (51).

As can be seen from FIGURE 14, the difference between the two values is not significant, and the trend of change is also roughly the same. Therefore, it is not possible to ignore the asymmetry of the magnetic waveform in the duration of the current flow and not affect the correctness of the foregoing conclusion.

5. Conclusion

Based on the content of this paper, the following conclusions can be drawn:

(1) the sufficient and necessary condition for generating excitation current flow is the difference between the absolute value of initial partial magnetic flux and the amplitude of the forced flux component.

(2) the trend of partial magnetism is changed by the exponential rule in the duration of the current flow. The undersaturated stage increases exponentially.

(3) there is a tendency towards saturation flux in the continuous flow of the bias magnetic flux; During the flow interval, the bias is basically the same. Mutations occur at alternating periods of continuous flow and continuous flow.

(4) because of the existence of transformer winding resistance makes mandatory flux flow component of the flow pulsation period and duration of phase is not zero, cause a bias in the flow duration on both sides of the border moment of jump, lead to transformer bias continues to decline, which cause the continuous attenuation of excitation inrush current, and eventually lead to the end of the excitation inrush current.

(5) the change of partial magnetic flux in the duration of flow will only affect the attenuation velocity of the bias magnetic field, which will not cause the continuous attenuation of the bias magnetic, so it is not the decisive factor that causes the excitation current to end.

6. Acknowledgements

The work here is supported by the National Natural Science Foundation of China(51507157), and the work here is supported by the National Natural Science Foundation of China(51607158). Supported by the Graduate's Scientific Research Foundation of Zhengzhou University of Light Industry.

7. References

[1] Yuhuan Liu, Yuping Lu and Yubo Yuan 2007 A Novel Scheme Based on Flux Restraint Theory Used in Distinguishing Inrush Currents for UHV Transformers *J. Proceedings of the CSEE.* **34** 52-58

[2] Xuesong Zhang and Benteng He 2005 A New Method To Identify Inrush Current By Phasor Comparison *J. Proceedings of the CSEE.* **19** 43-47

[3] Hongliang Zong, Huafeng Jin, Zhengfei Zhu and Shaocun Zhang 2001 Transformer Inrush Detected By The Variation of Magnetizing Impedance *J. Proceedings of the CSEE.* **7** 91-94

[4] Xue Wang 2012 Synthetic transformer inrush identification based on characteristic space *J. Electric Power Automation Equipment.* **11** 83-86

[5] Jiangping Tan 2008 Identification of transformer magnetization inrush current based on magnetic flux track *J. Electric Power Automation Equipment.* **12** 77-80

[6] Zhengqing Han and Shuping Liu 2007 Magnetizing inrush identification based on differential current characteristics *J. Electric Power Automation Equipment.* **9** 51-54

[7] Zhijuan Lu, Wanshun Liu, Shiwu Xiao and Tao Zheng 2006 A Novel Scheme to Fast Discriminate Inrush Current from Internal Fault of Power Transformer *J. Proceedings of the CSEE.* **2** 47-51

[8] Haifeng Li, Gang Wang, Xiaohua Li and Shaopeng Hu 2005 Distinguish Between Inrush and Internal Fault of Transformer Based on Adaptive Wavelet neural network *J. Proceedings of the CSEE.* **7** 144-150

[9] Yubo Yuan, Dejia Li and Yuping Lu 2005 Physical Mechanism of Sympathetic Inrush of Transformer and Its Influence on Differential Protection *J. Automation of Electric Power Systems.* **6** 9-14
[10] Hongchun Shu, Xun He and Lixin Li 2006 Research on sympathetic inrush in operating transformer *J. Electric Power Automation Equipment.* **10** 7-12
[11] Jiansong Zhang, Benteng He and Xuesong Zhang 2005 Approach to Calculate Attenuated Magnetizing Inrush Current in Transformer *J. Automation of Electric Power Systems*. **12** 57-60
[12] Xuesong Zhang, Benteng He and Jiansong Zhang 2005 PrineiPle and Influeneing Faetors of the Transformer SymPathetic Inrush *J. Automation of Electric Power Systems.* **6** 15-19